Bioengineering: Tools, Methods and Applications

Bioengineering: Tools, Methods and Applications

Editor: Gretchen Kenney

www.callistoreference.com

Callisto Reference,
118-35 Queens Blvd., Suite 400,
Forest Hills, NY 11375, USA

Visit us on the World Wide Web at:
www.callistoreference.com

ISBN: 978-1-63239-890-1 (Hardback)

The publisher's policy is to use permanent paper from mills that operate a sustainable forestry policy. Furthermore, the publisher ensures that the text paper and cover boards used have met acceptable environmental accreditation standards.

Trademark Notice: Registered trademark of products or corporate names are used only for explanation and identification without intent to infringe.

Printed in the United States of America.

Cataloging-in-Publication Data

Bioengineering : tools, methods and applications / edited by Gretchen Kenney.
    p. cm.
Includes bibliographical references and index.
ISBN 978-1-63239-890-1
1. Bioengineering. 2. Synthetic biology. I. Kenney, Gretchen.
TA164 .B56 2017
660.6--dc23

Table of Contents

Preface

This book has been an outcome of determined endeavour from a group of educationists in the field. The primary objective was to involve a broad spectrum of professionals from diverse cultural background involved in the field for developing new researches. The book not only targets students but also scholars pursuing higher research for further enhancement of the theoretical and practical applications of the subject.

As bioengineering is emerging at a rapid pace, the contents of this book will help the readers understand the modern concepts and applications of this field. It also delves into the equipment used in bioengineering. Bioengineering is an inter-disciplinary field, with biology as the primary field and physics, chemistry, mathematics and computer sciences as its secondary contributors. This text attempts to understand the multiple branches that fall under this discipline and how much such concepts have practical applications. With its detailed analyses and data, this book will prove immensely beneficial to professionals and students involved in bioengineering at various levels.

It was an honour to edit such a profound book and also a challenging task to compile and examine all the relevant data for accuracy and originality. I wish to acknowledge the efforts of the contributors for submitting such brilliant and diverse chapters in the field and for endlessly working for the completion of the book. Last, but not the least; I thank my family for being a constant source of support in all my research endeavours.

Editor

Preface

Sustained Release of Antibacterial Agents from Doped Halloysite Nanotubes

Shraddha Patel [1,2], **Uday Jammalamadaka** [1], **Lin Sun** [1], **Karthik Tappa** [1] **and David K. Mills** [1,3,*]

Academic Editor: Ali Khademhosseini

[1]   Center for Biomedical Engineering and Rehabilitation Science, Louisiana Tech University, Ruston, LA 71272, USA; nenomole@gmail.com (S.P.); uja002@latech.edu (U.J.); lsu002@latech.edu (L.S.); kkt007@latech.edu (K.T.)

[2]   Wayne State University, St. John Hospital & Medical Center, 22101 Moross Rd, Detroit, MI 48236, USA

[3]   School of Biological Sciences, Louisiana Tech University, Ruston, LA 1272, USA

*   Correspondence: dkmills@latech.edu

Abstract: The use of nanomaterials for improving drug delivery methods has been shown to be advantageous technically and viable economically. This study employed the use of halloysite nanotubes (HNTs) as nanocontainers, as well as enhancers of structural integrity in electrospun poly-e-caprolactone (PCL) scaffolds. HNTs were loaded with amoxicillin, Brilliant Green, chlorhexidine, doxycycline, gentamicin sulfate, iodine, and potassium calvulanate and release profiles assessed. Selected doped halloysite nanotubes (containing either Brilliant Green, amoxicillin and potassium calvulanate) were then mixed with poly-e-caprolactone (PLC) using the electrospinning method and woven into random and oriented-fibered nanocomposite mats. The rate of drug release from HNTs, HNTs/PCL nanocomposites, and their effect on inhibiting bacterial growth was investigated. Release profiles from nanocomposite mats showed a pattern of sustained release for all bacterial agents. Nanocomposites were able to inhibit bacterial growth for up to one-month with only a slight decrease in bacterial growth inhibition. We propose that halloysite doped nanotubes have the potential for use in a variety of medical applications including sutures and surgical dressings, without compromising material properties.

Keywords: antiseptics; antibiotics; drug release; halloysite nanotubes; nanocontainers; poly-e-caprolactone; nanocomposite mats

1. Introduction

Among other natural materials studied for use in drug delivery, nanocoatings, and tissue-engineered scaffolds, halloysite has only recently been discovered. Halloysite nanotubes (HNTs) are naturally occurring clay nanoparticles found deposited in soils worldwide [1–3]. Halloysite is an economically viable raw material that can be mined from a deposit as a raw mineral and refined on location. Halloysite is structured as a two-layered aluminosilicate, chemically similar to kaolin and has a predominantly hollow nanotubular structure in the submicron range [2–5]. HNTs typically display an inner diameter ranging from 15 to 50 nm, an outer diameter from 30 to 50 nm, and a length between 100 to 2000 nm [1,2].

Halloysite nanotubes have high capillary forces; so they quickly adsorb numerous materials and within a wide range of pH. It has a negative electrical zeta-potential of *ca.* −50 mV, which imparts to HNTs suitable dispersibility in water-based polymers and other media. HNTs present a large outer surface area that may be functionalized and an inner lumen that can be loaded with different drugs or bioactive factors for sustained drug release [1,3,6]. This increases the drug effectiveness, without increasing concentration, as the drug is released slowly from the HNT lumen [1,6,7]. Drugs of smaller molecular size are typically vacuum-loaded within the inner lumen of the nanotube and drugs larger in molecular size can attach to the outer surface of halloysite. The inner lumen and outer surface are oppositely charged, and charge intensity is easily modified.

Halloysite nanotubes have been shown to be cytocompatible in studies using different cell types [2,4]. Human dermal fibroblasts cultured on HNT nanofilms showed no cytotoxic effects, proliferated and expressed tissue-specific proteins showing that they maintained their cellular phenotype on an HNT layered nanoparticle thin film [8]. Mesenchymal stem cells also thrived on HNT nanofilms [9] and a recent study has provided support for HNT biocompatibility [10]. These properties support the concept of the use of HNTs as a nanocontainer and nanocarrier able to entrap biologically active agents within the inner lumen, followed by their retention, and slow release [11]. Ruling *et al.* [12] and Patel [13] established the concept of using doped HNTs and the electrospinning technique to fashion drug doped polymer fiber composites (poly(lactic-co-glycolic acid) and polycaprolactone, respectively) for sustained release. Several recent studies have extended this work and fabricated electrospun HNTs/polymer composites for delivery of a diverse set of drugs. Xue *et al.* [14] doped metronidazole into HNT poly(caprolactone)/gelatin microfibers and showed that such a nanocomposite could extend drug release to 20 days *versus* only six days from plain microfibers [14]. Polylactic acid and halloysite composites have seen extensive study in conjunction with halloysite for applications ranging from drug delivery to tissue engineering [15,16]. Sun *et al.* [17] have taken this work further by showing the HNTs can be encapsulated in a drug-infused polyelectrolyte coating and doped into nylon-6. The composites were electrospun into fiber mats and sutures and reduced osteosarcoma cell proliferation *in vitro* [17]. Sharma *et al.* [14] offers an excellent review of nanofibers and their biomedical applications [18].

The current study demonstrates that a wide range of antibacterial agents can be loaded into the HNT lumen and slowly released. Furthermore, HNTs doped with chlorhexidine, povidone iodine, Brilliant Green were incorporated into PCL nanocomposite mats. These mats, when applied to confluent bacterial cultures, maintained an anti-bacterial growth inhibition field for up to one month. We propose that HNT doped-mats containing single or mixed sets of antibiotics/antifungals can be used for maintaining a sterile field in body cavities and on body surfaces or used to control or eliminate bacterial growth in contaminated wounds. Such scaffolds could also be loaded with growth factors (and other drugs) forming a mixed nanocomposite that would enhance healing leading to speedier wound repair and patient recovery.

2. Experimental Section

Materials

Halloysite nanotubes were purchased from Sigma Aldrich (St. Louis, MO, USA). The outside diameters of the halloysite vary from 50 to 70 nm, the average inner diameter is 15 nm and their lengths varies from 1 to 1.5 μm [6]. Brilliant Green is a topical antiseptic commonly used in Eastern Europe and Russia and is used to treat various skin and mucosal infections as well as sinus infections. Chlorhexidine is a chemical antiseptic that is effective on both Gram-positive and Gram-negative bacterial and is commonly used to treat skin infections, as a skin wash, and in mouthwash as a treatment for gingivitis. Iodine and povidone-iodine are commonly used antiseptics with povidone-iodine the universally preferred iodine antiseptic. Brilliant Green, chlorhexidine, iodine, and polycaprolactone were obtained from Sigma Aldrich (St. Louis, MO, USA) and povidone-iodine from Wal-Mart, Bentonville, AR. Amoxicillin and potassium calvulanate are both antibiotics grouped in a class of drugs called penicillins and were purchased from Macleods Pharmaceuticals, India. Doxycycline is a tetracycline antibiotic used to slow bacterial growth and was obtained from Shreya Life Science PVT. LTD. Roorkee, India. Nitrofurantoin is an antibiotic usually used in the treatment of urinary tract infections (Sigma Aldrich, St. Louis, MO, USA).

3. Methods

3.1. Halloysite Drug Loading

To study the release pattern of drugs from HNTs, chlorhexidine, povidone iodine, Brilliant Green, iodine, doxycyclin, amoxicillin and potassium calvulanate were loaded into halloysite nanotubes. For this set of experiments, 20 mg of each drug was dissolved in 1 mL of water or alcohol, and the mixture was sonicated until the drug was dissolved. Once the solution became transparent, 50 mg of halloysite powder was added. The mixture was sonicated again for 30 min. Then, the solution was placed in a vacuum chamber, and a vacuum applied for 20 min. When the vacuum is applied, air bubbles are removed from the halloysite and removal of the vacuum causes the drug solution to enter the halloysite lumen. After 20 min, the vacuum was stopped, and the tube containing doped halloysite was removed from the vacuum chamber and kept at room atmosphere for 20 min. This vacuum process was repeated three times, followed by washing with water to remove any unloaded drugs. As povidone-iodine comes prepared in solution form, HNTs (50 mg) were directly added into a 1 mL solution of povidone-iodine and loaded as described above. For all experiments, six samples per time or testing point were used. All experiments were repeated twice and results from both sets of experiments were in accord.

3.2. Fabrication of Drug Loaded Halloysite-PCL Scaffolds

3.2.1. Preparation of PCL Solution

A 9-wt% PCL-chloroform mixture was used for the producing electrospun PCL scaffolds. The PCL solution was prepared by sonicating PCL beads in chloroform. PCL beads were dropped one by one until a viscous homogenous PCL solution was formed. This took roughly 1.5 h and as chloroform evaporates quickly, additional chloroform was added to maintain volume.

3.2.2. Preparation of PCL and HNT Solution

For PCL mats, a 5 and 7-wt/% ratio of HNTs to PCL scaffold solution was used. It was prepared by the addition of 5 or 7-wt/% HNTs or antibacterial-doped HNTs (Brilliant Green, amoxicillin and potassium calvulanate) into the PCL-chloroform mixture. The PCL/chloroform/HNT solution was sonicated for 10 min, followed by electrospinning. Several variations on this method were also tested.

3.2.3. Electrospinning HNT/PCL Mats

The method of electrospinning was used to prepare antibacterial drug loaded HNT/PCL mats. The electrospinning set up consisted of a syringe pump, syringe, a collector plate, and a high voltage electricity source. The entire electrospinning set-up was placed in a Plexiglas housing to enclose and stabilize the path of the polymer jet.

Two different electrospinning configurations were used (Figure 1). For the fabrication of woven HNT/PCL fibers, 1 mL of the PCL/chloroform/HNT mixture was loaded into a 1 mL syringe. A syringe pump was used to apply constant pressure (1 mL/hr.) to the syringe and the PCL/chloroform/HNT solution released at 10 µL/minute using flat head needles to dispense the polymer solution. A high voltage (17–20 kV) was maintained between the tip of the syringe needle and the collector plate. A polymer jet is then formed, the solvent evaporates, and the loaded HNT/PCL mats were assembled on the surface of an aluminum collector plate (Figure 1A). The distance between the tip of the needle and the collector plate was maintained at 20 cm for assembly of the antibacterial drug loaded mats.

For scaffolds with an oriented HNT/PCL fiber arrangement, an electrospinning set-up similar to that for non-woven scaffolds was employed with the exception of the collecting apparatus (Figure 1B). For oriented fibrous scaffolds, a "U"-shaped collector, located 1 cm from the collector, was connected to the ground. The distance between the tip of the needle and the collector plate was maintained at 17 cm for assembly of the antibacterial drug loaded halloysite-PCL scaffolds. Briefly, 1 mL of the PCL/chloroform/HNT mixture was loaded into a 1 mL syringe and the pump was used to apply constant pressure (1 mL/hr.) to the syringe and the PCL/chloroform/HNT solution released at 10 μL/minute using flat head needles to dispense the polymer solution. Loaded PCL/HNT fiber nanocomposites were assembled as oriented fibers onto a rotating drum instead of the collector plate.

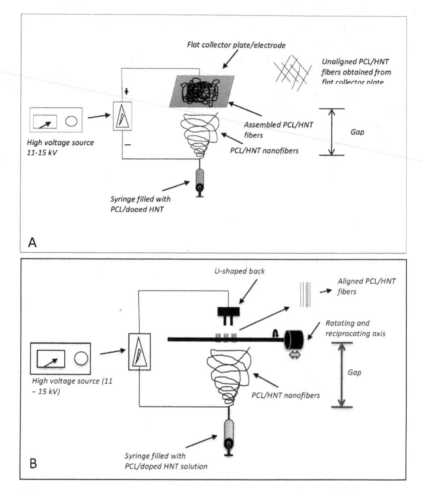

Figure 1. Electrospinning assembly set-up. (**A**) shows the assembly system that produced woven PCL fibers with no fiber orientation; (**B**) shows the assembly system that produced oriented (uniaxial) PCL fibers.

Amoxicillin, Brilliant Green and potassium calvulanate were successfully loaded in halloysite-Poly-lactic acid (PLA) and halloysite-polyethylene oxide (PEO) mats. However, these mats showed low tensile strength and were rejected for further testing (data not shown).

3.3. Drug Release from Drug Loaded Halloysite Nanotubes

Drug loaded HNTs were added to 1 mL of water and placed on a magnetic stirrer for 10 min. The tubes containing the solution were then centrifuged at 7000 rpm for 2 min. The supernatant was removed and 1 mL of water was added to the tube, and resuspended. The tube was again incubated on the magnetic stirrer for 10 min, followed by centrifugation at 7000 rpm for 2 min. This process was repeated at selected time intervals. Magnetic stirrer was removed and the amount of antibacterial drug released was measured with a UV Spectrometer (Cole Parmer, Vernon Hills, IL, USA). For all drug release studies, the assays performed were repeated twice.

3.4. Bacterial Studies

We used confluent cultures of *E. coli* and *S. aureus* to test the efficacy of HNT/PCL mats. In these bacterial experiments, 1 L of LB broth was prepared by mixing 10 grams of NaCl, 10 grams of tryptone, 5 grams of yeast extract, 15 grams of agar, and enough distilled water to make a final volume of 1 L. This LB broth was sterilized by autoclaving at 121 °C for 15 min. After autoclaving, LB broth was poured into Petri dishes, which were allowed to solidify over night. The next day, a loop of *E-coli* and *S. aureus* bacteria was spread on the LB broth plates. Bacteria were allowed to grow until confluent. At confluence, HNT/PCL nanocomposites loaded with *amoxicillin*, Brilliant Green, chlorhexidine, and provodine iodine, were applied to the surface of bacterial cultures. Serial cultures of bacteria were used to assess the duration of release from HNT doped scaffolds. HNT doped scaffolds were removed from spent bacterial cultures and applied to a fresh confluent bacterial culture, observed and photographed and the process repeated for over a one-month period.

4. Results

4.1. Morphology

Our results suggest that the speed of the target and the form of the collector plate dictates the degree of fiber anisotropy and diameter within the forming HNT/PCL scaffold. The addition of HNTs up to 15% did not significantly affect the fiber pattern or porosity (data not shown). With modest changes in target speed and collector design, scaffolds could be produced with a range of desired alignments, from random patterns with large pore size (Figure 2A,B) or at the slower speeds a more oriented fiber pattern (Figure 2C,D). Figure 2B,D shows clear differences in fiber diameter and porosity.

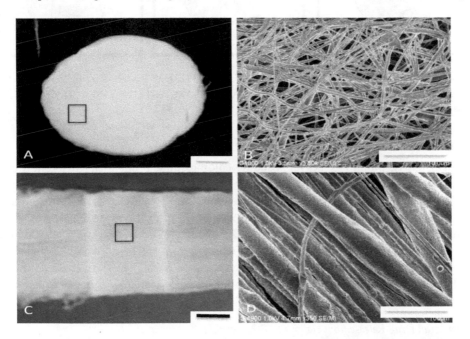

Figure 2. Fabricated PLC/Halloysite nanocomposites with 5% (**A,B**) and 7.5% HNTs (**C,D**). (**A**) Photograph of 5% HNT/PCL nanocomposite with a random fiber pattern. Area outlined by square shown under SEM microscopy in (**B**); (**B**) SEM image of 5% HNT/PCL nanocomposite, insert a single fiber at higher power; (**C**) 7.5% HNT/PCL nanocomposite with an oriented fiber pattern. Area outlined by square is shown at higher magnification in (**D**); (**D**) SEM image of 7.5% HNT/PCL nanocomposite, containing 5% HNTs in electrospun PCL. Note the PCL fibers have a defined orientation. Scale bars = 2 centimeters in (**A,C**), 10 microns in (**B**) and 100 microns in (**D**).

4.2. Contact Angle Measurements

The contact angle measurements of PCL and PCL/HNT nanocomposite scaffolds were performed. Pure PCL is hydrophobic in nature (Figure 3A). Results of contact angle measured showed that PCL/HNT nanocomposite scaffolds showed increased hydrophilicity with increased HNT content (*i.e.*, decreased the contact angle, Figure 3B,C). Decrease in contact angle implies more wetting of the surface. Wetting of the mats results in better water flux into the scaffold leading to better drug elution. Moreover, cell adhesion is favored on hydrophilic surfaces.

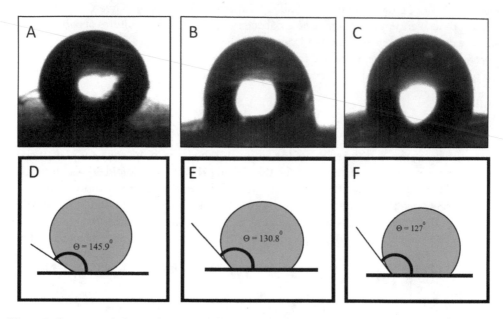

Figure 3. Contact angle for random-oriented PCL mats and PCL mats loaded with HNTs. (**A**) Pure PCL (145.9°); (**B**) PCL with 5% HNT (130.8°); and (**C**) PCL with 10% HNT (127°). *n* = 6. (**D–F**) graphically illustrated contact angles shown in (**A–C**).

4.3. Antibacterial Drug Release from Drug-Loaded Halloysite

The drug release profiles for povidone iodine, chlorhexidine, Brilliant Green, iodine, doxycyclin, amoxicillin and potassium calvulanate are shown in Figures 4–6. The concentration of the drug released from HNTs was measured by a UV spectrophotometer. As shown in Figure 4, 76% of povidone iodine from HNTs was released from HNTs in 6.5 h, 84.95% of chlorhexidine was released in 4 h, 96.52% of Brilliant Green was released in 5 h, and 92.68% of Iodine in 5 h. In Figure 5 release profiles of antibiotics are shown. 98.73% of doxycycline was released from HNTs in 4 h and HNTs loaded with both amoxicillin and potassium calvulanate showed a release rate of 94.83% over a 5-hour period. In almost all cases, despite extensive washing after loading, there was an initial burst of drug followed by a more gradual release pattern.

4.4. Antibacterial Drug Release from Halloysite-PCL Scaffold

Figure 6 shows drug release profiles from the Brilliant Green loaded halloysite-PCL scaffold and amoxicillin and potassium calvulanate loaded HNT/PCL scaffold, respectively. As shown in Figure 6, 99.95% of Brilliant Green was released from Brilliant Green loaded halloysite-PCL scaffold in 1.1 days. A sum total of 91.67% of *amoxicillin* and potassium calvulanate was released after 9.6 days from amoxicillin and potassium calvulanate loaded halloysite-PCL scaffold. Despite repeated rinses of doped scaffolds after loading, approximately, 15%–20% of the initial drug burst is probably associated with externally adsorbed dyes and antibiotics (Figure 6).

Figure 7 provides visual evidence of doped PCL/HNT scaffolds and drugs released (amoxicillin, Brilliant Green and amoxicillin) from loaded HNTs. Even after one month, antiseptics and antibiotics were still being released from PCL scaffolds (data not shown).

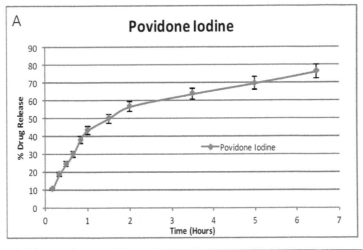

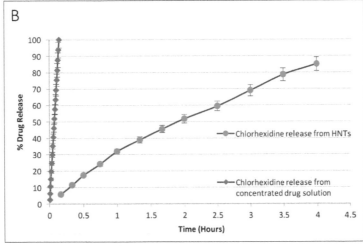

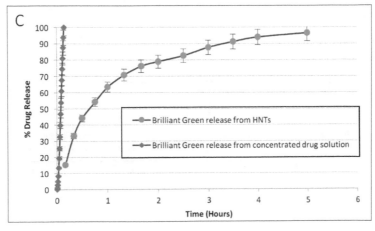

Figure 4. *Cont.*

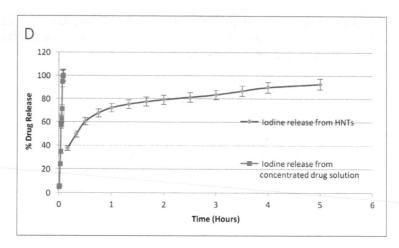

Figure 4. Release profile of antiseptics from HNTs. (**A**) Povidone iodine; (**B**) Chlorhexidine; (**C**) Brilliant green; (**D**) Iodine. $n = 4$ trials.

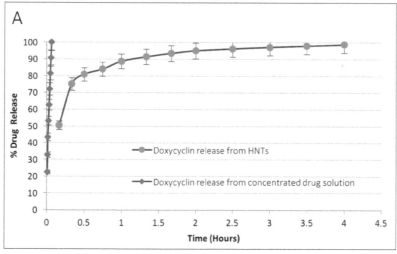

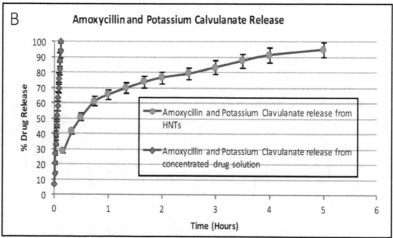

Figure 5. Antibiotic release from HNTs. (**A**) Doxycyclin; (**B**) Amoxycillin and Potassium Calvulanate. $n = 4$ trials.

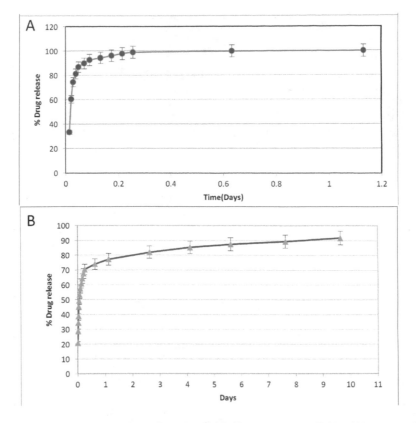

Figure 6. (**A**) Brilliant Green release from loaded halloysite-PCL scaffolds; (**B**) Amoxycillin and potassium calvulanate release from loaded halloysite-PCL scaffolds. $n = 3$ trials.

Figure 7. Fabricated PLC/Halloysite nanocomposite scaffolds loaded (and releasing) amoxicillin (**A**) Brilliant Green; (**B**) and Brilliant Green and amoxicillin (**C**). Photographed after one week. Brightfield microscopy, 4X.

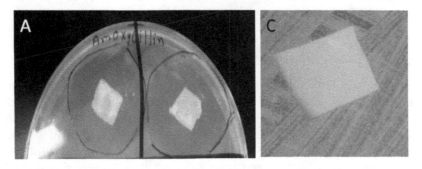

Figure 8. *Cont.*

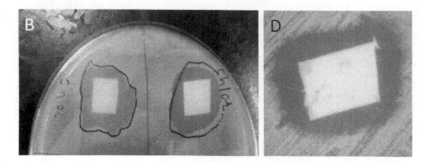

Figure 8. Doped HNT/PCL mats applied to confluent bacterial cultures. 5% HNT/PCL nanocomposites loaded with amoxicillin (**A**) and chlorhexidine (**B**) and applied to confluent cultures of *E. coli*. Black lines outline extent of anti-bacterial growth zone; (**C**) Zone of inhibition of *S. aureus* growth around empty halloysite-PCL scaffolds showing no antibacterial effect; (**D**) PCL mats with HNTs doped with Brilliant Green. (**A–D**) were photographed one week after application of doped mats.

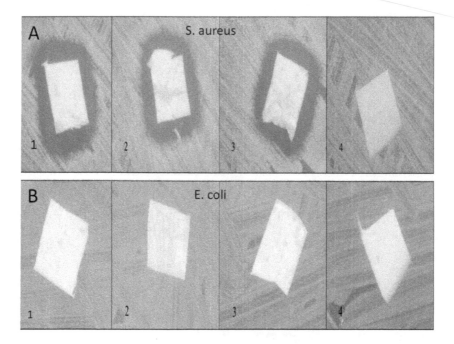

Figure 9. (**A**) *S. aureus* and (**B**) *E. coli* confluent cultures with scaffolds added and observed after 7 days. PCL mats prepared by combining the PCL solution and 7% HNTs before electrospinning, (**A1/B1**); PCL with 7% HNTs doped with Brilliant Green prepared by dispersing the HNTs in chloroform and then dissolving PCL within the solution using sonication, (**A2,3/B2,B3**); (**A4/B4**) PCL mats containing no Brilliant green or HNTs. Growth inhibition was observed in Brilliant Green doped HNT/PCL scaffolds and only on *S. aureus*. No observable difference between the two fabrication methods was observed.

4.5. Anti-Bacterial Effects of Doped HNT/PLC Nanocomposites

Confluent cultures of the bacteria, *E. coli* and *S. aureus*, were established and the anti-bacterial effects of doped HNT/PLC nanocomposites were assessed. Amoxicillin, Brilliant Green, chlorhexidine and povidone iodine were loaded into HNT/PCL scaffolds and studied for their ability to inhibit bacterial growth. Doped nanocomposites were effective or ineffective in creating zones of growth inhibition depending on the antibacterial applied and the bacterial strain. Zones of inhibition were apparent within hours after application and amoxicillin and chlorhexidine doped HNT/PCL mats remained effective on *S. aureus* up to one week (Figure 8). Variation in method of doped HNT/PCL solutions prior to electrospinning did not alter their effectiveness (inhibition zone size and duration of inhibition varied) (Figure 9).

5. Discussion

In this study, HNT-doped antimicrobials showed different release profiles but all displayed an overall pattern of slow release. Ruiling *et al.* [12] showed that tetracycline hydrochloride-doped HNTs could be electrospun with poly(lactic-co-glycolic acid) to form drug loaded mats that delivered sustained release of its contents, improved the tensile strength of the PLGA fibers, and did not provoke a cytotoxic response [12]. Nitya *et al.* [19] showed similar results using montmorillonite with impressive gains in material properties [19]. We have extended these observations to support the potential applicability of HNTs as a suitable drug-releasing vehicle in a polymer composite. HNTs provided sustained release for a wide array of substances including: anti-bacterial antibiotics, antiseptics, and disinfectants including chlorhexidine, povidone iodine, Brilliant Green, iodine, doxycyclin, and amoxicillin and potassium calvulanate. In addition, amoxicillin, Brilliant Green, chlorhexidine, povidone iodine, and potassium calvulanate were successfully loaded in halloysite-PCL, halloysite-PLA and halloysite-PEO mats. Doped HNT loaded PLA or PEO electrospun scaffolds were also fabricated. However, these mats showed low tensile strength and were rejected for further testing. However, both are resorbable materials and there may be an application using these materials as a drug-releasing platform where tensile strength is not a key consideration.

HNT/PCL mats could be spun in either a random or oriented fiber paper, and fibers also increased in coarseness with HNT content. Drugs and disinfectants could be easily doped into the PCL/HNT fiber mats and released in a sustained fashion. This suggests that halloysite could be used for loading different sets of antibacterials, disinfectants, drugs or active agents and delivered in novel combinations designed to treat specific affected tissues, wound, or diseased organs, or customized to meet a patient's needs. The amount of drug released from scaffolds is a function of its volume and concentration of HNTs. To have a balance between drug release and material properties, the concentration of HNTs must be limited to 7.5% by weight. Varying the total thickness of the scaffolds can provide with desired drug release volume. This can be achieved in two ways, electrospinning thicker scaffolds or having multiple layers of thinner scaffolds. Scaffolds made through latter method may have better drug release and material properties and also have ease of fabrication. Further investigation on this area is needed to have a definitive proof.

It can be observed that drug release was extended by hours upon loading them in HNTs. Incorporating these doped HNTs into PCL scaffolds has extended this release to days. It is to be noted that this release is from the doped HNTs on the surface of the scaffolds. As the scaffolds are metabolized, more drug release can be expected. Accounting the surface release and further release from matrix, it can be expected that drug release can be extended to weeks or months. Drug release and kinetics from PCL scaffolds are functions of solubility of drug, method of doping HNTs, concentration of doped HNTs in the scaffolds, thickness of scaffold, flux of fluids surrounding the scaffolds, *etc.*

Halloysite nanotubes are currently under active investigation as carrier and container materials for a diverse set of biomedical applications due to their morphology, size, and ability to deliver a diverse set of biologically active agents including antibiotics, disinfectants, anti-cancer agents, growth factors, *etc.* [12,20–26]. HNTs have been incorporated into various bone cements [7,23], hydrogels [24,25], and polymers [12–14,26]. In all cases, HNTs released antibiotics, growth factors or chemotherapeutic agents in a sustained fashion. The desirable cytocompatibility and tunable properties of clay halloysite nanotubes (HNTs) have been established [10,23–27], as well as its potential biocompatibility [10].

A strong driver supporting HNT incorporation into medical biomaterials (e.g., coatings, dressings, implants, *etc.*) is the growing incidence of hospital-acquired infections, the need for novel drug delivery systems, and nanoenhanced scaffolds. A potential HNT application is a flexible and tunable multi-layer wound dressing that possesses multiple functionalities including absorption, antibacterial/fungal protection, and tissue regeneration. The dressing could be used for both prophylactic and therapeutic interventions, as a dressing or wound packing material, or as a topical gauze or pad. The addition of doped HNTs provides enhanced material properties, potential to load multiple drug sets, and increased control over the release kinetics of loaded drugs (10–100+ hrs.) as compared to conventional compositions, properties ideal for the treatment of chronic un-healing wounds, multiple microbial infections, and or for multi-vector treatments.

Halloysite has several advantages over the existing competitor technology, carbon nanotubes. Carbon nanotubes (CNTs) have potential novel application in nanomedicine as biocompatible and supportive substrates, and as pharmaceutical excipients for creating versatile drug delivery systems [28]. Carbon nanostructures are one of the few classes of engineered nanomaterials that have received focused toxicological characterization [29,30]. At present, some emerging concepts of carbon-nanotoxicology can be identified with toxicity dependent on several different factors, such as their size, their shape, the surface characteristics and the amount of the substances present in the particles preparations [23]. In contrast to carbon nanotubes, halloysite nanotubes are significantly cheaper [2,6,7,11], does not provoke a cytotoxic cellular response [2,10,13,23–26], the inner and outer surfaces are modifiable, and the lumenal diameter fits many globular protein diameters. Furthermore, the large surface area, low cost, ease of loading, tunability and potential broad applicability for medical device development, and no limit on scalability for commercial applications, make HNTs superior to carbon nanotubes.

While we originally targeted HNT/PCL mats for anti-bacterial use, these nanocomposite scaffolds are also promising in several other biomedical application areas. El-Refaie *et al.*, were the first to propose the electrospinning technique as a vehicle as a drug delivery system using spun polylactic acid and poly(ethylene-co-vinylacetate) fibers for the delivery of tetracycline HCl [31]. Ruiling *et al.* [12] for electrospun PLGA/HNTs scaffolds [12], and Patel [13] for PCL/HNTs microfibrous mats, [13] demonstrated the potential of these scaffolds for sustained drug release. As a drug delivery system, drug release can be modified (potentially controlled) by varying production parameters, such as applied voltage, gap distance, flow rate, composition of polymer solution and collector plate design to obtain fibrous scaffolds with a desired diameter, porosity and fiber pattern. In tissue engineering, fibrous scaffolds have a large surface area, possess a rough surface topography, a malleable nature, and the ability to produce fiber architectures patterned after the character of a tissue's natural extracellular matrix (see Figures 3 and 5). These are design features that support cell adhesion, proliferation, and functionality and would promote neotissue formation [32]. The addition of HNTs to various polymers may enable development of a diverse set of HNT enhanced materials [12,14,19–27]. The capabilities of HNTs support their use in constructs that will localize the distribution of chemotherapeutics or biologically active agents at the targeted site and provide sustained drug or agent release, hence increasing drug efficacy and lowering toxicity of the released therapeutics.

6. Conclusions

This study employed the use of HNTs as nanocontainers, as well as enhancers of structural integrity, in electrospun PCL scaffolds. HNTs were loaded with amoxicillin, Brilliant Green, chlorhexidine, doxycycline, gentamicin sulfate, iodine, and potassium calvulanate. Selected doped halloysite nanotubes (containing either Brilliant Green, amoxicillin and potassium calvulanate) were mixed with PLC and electrospun into nanocomposite mats. Release profiles from doped HNTs and doped HNT/PCL nanocomposite mats showed a pattern of sustained release for all bacterial agents. Nanocomposites were able to inhibit bacterial growth for up to one-month with only a slight decrease in bacterial growth inhibition. We propose that anti-microbial drug doped HNT/PCL nanocomposites have the potential for use in a variety of medical applications including sutures and surgical dressings.

Acknowledgments: The authors wish to acknowledge the financial support of the Louisiana Governor's Biotechnology Initiative and the College of Applied and Natural Sciences, Louisiana Tech University.

Author Contributions: David K. Mills had the original idea for this project. Shraddha Patel conducted the majority of experimental analyses and drafted the manuscript. Uday Jammalamadaka, Lin Sun, Karthik Tappa assisted with the experimental analysis and all authors read, revised and approved the final manuscript. All authors read and approved the final manuscript.

Conflicts of Interest: The authors declare no conflict of interest.

References

1.   Lvov, Y.M.; Shchukin, D.G.; Mohwald, H.; Price, R.R. Halloysite clay nanotubes for controlled release of protective agents. *ACS Nano* **2008**, *2*, 814–820. [CrossRef] [PubMed]

2.   Vergaro, V.; Abdullayev, E.; Lvov, Y.M.; Zeitoun, A.; Cingolani, R.; Rinaldi, R.; Leporatti, S. Cytocompatibility and uptake of halloysite clay nanotubes. *Biomacromolecules* **2010**, *11*, 820–826. [CrossRef] [PubMed]

3.   Yuan, P.; Southon, P.D.; Liu, A.; Green, M.E.R.; Hook, J.M.; Antill, S.; Kepert, C.J. Functionalization of halloysite clay nanotubes by grafting with γ-aminopropyltriethoxysilane. *J. Phys. Chem. C* **2008**, *112*, 15742–15751.

4.   Deepak, R.; Agrawal, Y. Multifarious applications of halloysite nanotubes: A review. *Rev. Adv. Mater. Sci.* **2012**, *30*, 282–295.

5.   Prashantha, K.; Lacrampe, M.; Krawczak, F. Processing and characterization of halloysite nanotubes filled polypropylene nanocomposites based on a masterbatch route: Effect of halloysites treatment on structural and mechanical properties. *eXPRESS Polym. Lett.* **2011**, *5*, 295–307.

6.   Wei, W.; Minullina, R.; Abdullayev, E.; Fakhrullin, R.; Mills, D.K.; Lvov, Y.M. Enhanced efficiency of antiseptics with sustained release from clay nanotubes. *RSC Adv.* **2014**, *4*, 488–485. [CrossRef]

7.   Wei, W.; Abdllayev, E.; Goeders, A.; Hollister, A.; Lvov, D.M.; Mills, D.K. Clay nanotube/poly(methyl methacrylate) bone cement composite with sustained antibiotic release. *Macromol. Mater. Eng.* **2012**, *297*, 645–653.

8.   Kommireddy, D.A.; Lvov, Y.; Mills, D.K. Nanoparticle multilayers: Surface modification for cell attachment and growth. *J. Biomed. Nanotech.* **2005**, *1*, 286–290. [CrossRef]

9.   Kommireddy, D.A.; Sriram, S.M.; Lvov, Y.M.; Mills, D.K. Stem cell attachment to layer-by-layer assembled TiO$_2$ nanoparticle thin films. *Biomaterials* **2006**, *27*, 4296–4303. [CrossRef] [PubMed]

10. Mills, D.K. Biocompatibility of halloysite clay nanotubes in a rat dermal model. *FASEB J.* **2014**, *28(1)* Supplement 87.4 (abstract).

11. Abdullayev, E.; Shchukin, D.G.; Mohwald, H.; Lvov, Y.M. Halloysite tubes as nanocontainers for anticorrosion coating with benzotriazole. *Appl. Mater. Interfaces* **2009**, *1*, 437–443. [CrossRef] [PubMed]

12. Qi, R.; Guo, R.; Shen, M.; Cao, X.; Zhang, L.; Xu, J.; Yu, J.; Shi, X. Electrospun poly(lactic-co-glycolic acid)/halloysite nanotube composite nanofibers for drug encapsulation and sustained release. *J. Mater. Chem.* **2010**, *20*, 10622–10629. [CrossRef]

13. Patel, S. Nanotechnology in Regenerative Medicine. Ph.D. Thesis, Louisiana Tech University, Ruston, LA, USA, 2011.

14. Xue, J.; Niu, Y.; Gong, M.; Shi, R.; Chen, D.; Zhang, L.; Lvov, Y. Electrospun microfiber membranes embedded with drug-loaded clay nanotubes for sustained antimicrobial protection. *ACS Nano* **2015**, *9*, 1600–1612. [CrossRef] [PubMed]

15. Dong, Y.; Chaudhary, D.; Haroosh, H.; Biackford, T. Development and characterisation of novel electrospun polylactic acid/tubular clay nanocomposites. *J. Mater. Sci.* **2011**, *46*, 6148–6153. [CrossRef]

16. Dong, Y.; Narshall, J.; Haroosh, H.; Mohammadzadehmoghadamam, S.; Liu, D.; Qi, X.; Lay, K.-T. Polylactic acid (PLA)/halloysite nanotube (HNT) composite mats: Influence of HNT content and modification. *Compos. A Appl. Sci. Manuf.* **2015**, *76*, 28–36. [CrossRef]

17. Sun, L.; Boyer, C.; Grimes, R.; Mills, D.K. Drug coated clay nanoparticles for delivery of chemotherapeutics. *Current Nanosci.* **2016**, in press. [CrossRef]

18. Sharma, J.; Lizu, M.; Stewart, S.; Zygula, K.; Lu, Y.; Rajat Chauhan, R.; Yan, X.; Guo, Z.; Wujcik, E.K.; Wei, S. Multifunctional nanofibers towards active biomedical therapeutics. *Polymers* **2015**, *7*, 186–219. [CrossRef]

19. Nitya, G.; Nair, G.; Mony, U.; Chennazhi, K.; Nair, S. *In vitro* evaluation of electrospun PCL/nanoclay composite scaffold for bone tissue engineering. *J. Mater. Sci. Mater. Med.* **2012**, *23*, 1749–1761. [CrossRef] [PubMed]

20. Vergaro, V.; Lvov, Y.; Leporatti, S. Halloysite clay nanotubes for resveratrol delivery to cancer cells. *Macromol. Biosci.* **2012**, *12*, 1265–1271. [CrossRef] [PubMed]

21. Lvov, Y.M.; Abdullayev, E. Polymer-clay nanotube composites with sustained release of bioactive agents. *Prog. Polym. Sci.* **2013**, *38*, 1690–1719. [CrossRef]

22. Du, M.; Guo, B.; Jia, D. Newly emerging applications of halloysite nanotubes: A review. *Polym. Int.* **2010**, *59*, 574–582. [CrossRef]

23. Jammalamadaka, U.; Tappa, K.; Mills, D.K. Osteoinductive calcium phosphate clay nanoparticle bone cements (CPCs) with enhanced mechanical properties. In Proceedings of the 36th Annual IEEE EMBS Conference, Chicago, IL, USA, 26–30 August 2014; Volume 759, pp. 1–4.

24. Karnik, S.; Mills, D.K. Bioactive hydrogels for TMJ repair. *FASEB J.* **2012**, *26*. [CrossRef]

25. Karnik, S.; Mills, D.K. Clay nanotubes as growth factor delivery vehicle for bone tissue engineering. *J. Nano Nanotech.* **2013**, *4*, 102. [CrossRef]

26. Sun, L.; Boyer, C.; Mills, D.K. Polyelectrolyte coated clay nanotubes with pH controlled release. *J. Nano Nanotech.* **2013**, *4*, 104. [CrossRef]

27. Zhou, W.Y.; Guo, W.; Liu, M.; Liao, R.; Bakr, M.; Rabie, A.B.M.; Jia, D. Poly(vinyl alcohol)/halloysite nanotubes bionanocomposite films: Properties and *in vitro* osteoblasts and fibroblasts response. *J. Biomed. Mater. Res.* **2010**, *93*, 1574–1587. [CrossRef] [PubMed]

28. De Volder, M.F.; Tawfick, S.H. Carbon nanotubes: Resent and future commercial applications. *Science* **2013**, *139*, 535–539. [CrossRef] [PubMed]

29. Grobert, N. Carbon nanotubes—Becoming clean. *Mater. Today* **2007**, *10*, 28–35. [CrossRef]

30. Firme, C.P.; Bandaru, P.R. Toxicity issues in the application of carbon nanotubes to biological systems. *Nanomedicine* **2010**, *6*, 245–256. [CrossRef] [PubMed]

31. El-Refaie, K.; Bowlin, G.; Mansfield, K.; Layman, J.; Simpson, D.G.; Sanders, E.H.; Wnek, G.E. Release of tetracycline hydrochloride from electrospun poly(ethylene-co-vinylacetate), poly(lactic acid), and a blend. *J. Control. Relat.* **2002**, *81*, 57–64.

32. Tappa, K.; Jammalamadaka, U.; Mills, D.K. Design and Evaluation of a Nanoenhanced Anti-Infective Calcium Phosphate Bone Cement. In Proceedings of the 36th Annual IEEE EMBS Conference, Chicago, IL, USA, 26–30 August 2014; Volume 759, pp. 1–4.

Transient Mechanical Response of Lung Airway Tissue during Mechanical Ventilation

Israr Bin Muhammad Ibrahim [1], Parya Aghasafari [1] and Ramana M. Pidaparti [2,*]

Academic Editor: Aldo Boccaccini

[1] Graduate Student, College of Engineering, University of Georgia, 597 DW Brooks Drive, Athens, GA 30602, USA; israr@uga.edu (I.B.M.I.); parya.aghasafari@uga.edu (P.A.)

[2] College of Engineering, University of Georgia, 132A Paul D. Coverdell Center, Athens, GA 30602, USA

[*] Correspondence: rmparti@uga.edu

Abstract: Patients with acute lung injury, airway and other pulmonary diseases often require Mechanical Ventilation (MV). Knowledge of the stress/strain environment in lung airway tissues is very important in order to avoid lung injuries for patients undergoing MV. Airway tissue strains responsible for stressing the lung's fiber network and rupturing the lung due to compliant airways are very difficult to measure experimentally. Multi-level modeling is adopted to investigate the transient mechanical response of the tissue under MV. First, airflow through a lung airway bifurcation (Generation 4–6) is modeled using Computational Fluid Dynamics (CFD) to obtain air pressure during 2 seconds of MV breathing. Next, the transient air pressure was used in structural analysis to obtain mechanical strain experienced by the airway tissue wall. Structural analysis showed that airway tissue from Generation 5 in one bifurcation can stretch eight times that of airway tissue of the same generation number but with different bifurcation. The results suggest sensitivity of load to geometrical features. Furthermore, the results of strain levels obtained from the tissue analysis are very important because these strains at the cellular-level can create inflammatory responses, thus damaging the airway tissues.

Keywords: mechanical strains; lung airway; mechanical ventilation; finite element analysis

1. Introduction

Patients with respiratory problems whose lungs are compromised are treated with Mechanical Ventilation (MV) to assist them with breathing. MV can cause injury to airway lung tissue resulting from air pressure, and may eventually also cause damage to the lungs [1,2]. In general, the incidence of respiratory failures resulting from ventilator-associated lung injury (VALI) [3–5] is increasing due to the fact that there are still unanswered questions with regards to the transmission of mechanical forces into lung tissues resulting from MV. Mechanical aspects of VILI have been reported in past works, such as [3–5]. From these studies, ventilator management practices have been developed, such as maintaining positive end-expiration pressure. Despite this development, MV-induced damage may still lead to systemic organ failure called biotrauma [6]. Mechanical strain caused by ventilators has been shown to cause cell signaling, leading to inflammation [7]. It has also been shown that the inflammation depends on the mode of MV [8,9].

Several studies have been conducted to model air pressure on lung airway wall with diseases, such as tumors [10,11] and chronic obstructive pulmonary disease [12,13]. Fluid–solid interaction modeling using the Finite Element Method (FEM) has been employed by Koombua, et al. [14] to study airflow characteristics and stress distribution on airway walls. The results in [14] were used in [15] and [16] to analyze stress and strain distribution occurring on a more detailed model of tissue.

However, the previous studies did not employ a full breathing cycle. The complex dynamics of lung tissue environment are still being investigated, and, although significant research has been done, there are still unanswered questions with regards to the transmission of mechanical forces into lung tissues resulting from MV. Even though there have been many studies investigating the effects of long term ventilation with respect to lungs, the connection between the global deformation of the whole lung and the strains reaching the lung tissue has not been studied. Capturing real-time progression of airway tissue stretch during breathing is still difficult to assess experimentally. Currently, there is a lack of data on real-time stretch progression of airway tissue during MV.

In this study, a scheme for studying real-time progression of airway tissue stretch is proposed. The scheme only incorporates fundamental dynamics of airway tissue structure to save on the computational cost of simulation. The scheme consists of two steps. First, airflow through a lung airway bifurcation (Generation 4–6) is modeled using Computational Fluid Dynamics (CFD) to obtain air pressure during MV breathing with a flow rate of 35 L/min, as a follow up on results from a previous study [17]. Next, the transient air pressure was used in structural analysis to obtain mechanical strain experienced by the airway tissue wall. Several simulations were carried out to investigate the stress/strain environment within the airway lung tissue, and the results are presented and discussed.

2. Materials and Methods

2.1. Airway Tissue Model

Human lung consists of bifurcating airways that begin at the trachea. The trachea is designated as Generation 0. The generation numbers are then counted from bifurcating airways that grow after the trachea. The total number of generations in the human lung is 23. In this study, airway bifurcation of Generation 4–6 is considered.

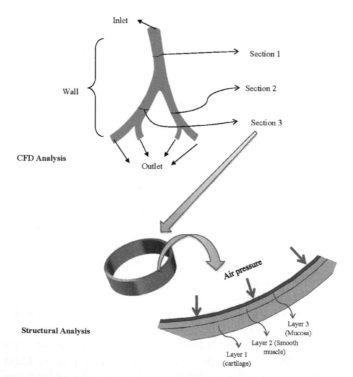

Figure 1. Multi-level analysis adopted for transient mechanical response of airway lung tissue. **Top**: model and boundary condition used for fluid flow analysis. **Bottom**: model for mechanical analysis.

A three-dimensional model of lung bifurcation Generation 4–6 was built based on approximation of a realistic human lung model. For the mechanical response analysis, the model was based on airway representation as discussed by Kamm [18]. The upper lung airway (Generation 1–8) is composed of three major layers based on the composition of each layer: mucosa, sub-mucosa and the area outside of the submucosa [19]. The submucosa includes the smooth muscle tissue. The area outside the submucosa consists of cartilage-fibrous layer and adventitia. In this study, a slice of airway model is considered for simplicity. The airway is assumed to stretch in radial and axial directions. The structure of the airway was simplified into a basic fundamental structure to capture the main dynamics of the airway. Figure 1 show the lung bifurcation along with the tissue layer model used in this study. Each layer is named as shown in Figure 1 for later discussion.

2.2. Computational Models and Boundary Conditions

In this study, two separate models were investigated using multi-level modeling. First, airflow through a lung airway bifurcation (Generation 4–6) is modeled using Computational Fluid Dynamics (CFD) to obtain air pressure (airflow model). Next, the transient air pressure was used in structural analysis to obtain mechanical strain experienced by the airway tissue wall (airway deformation model).

2.2.1. Airflow Model

The momentum of airflow through the lung bifurcation was assumed to follow the Navier-Stokes equations:

Conservation of mass:

$$\frac{\partial \rho}{\partial t} + \nabla.\left(\rho \mathbf{v}\right) = 0 \tag{1}$$

Conservation of momentum

$$\rho\left(\frac{\partial \mathbf{v}}{\partial t} + \mathbf{v}.\nabla \mathbf{v}\right) = -\nabla p + \nabla.(\mu(\nabla \mathbf{v} + (\nabla \mathbf{v})^T)) \tag{2}$$

where ρ is fluid density, $\mathbf{v}$ is velocity and p is pressure. The viscous force term in Equation (2) is in the form of incompressible viscous flow, which is the assumption of flow in airway Generation 4–6. The domain for the equations was the solid model as shown in Figure 1. Boundary conditions for Equations (1) and (2) were as follows: a velocity boundary conditions was imposed at the inlet ("top" part of airway as in Figure 1) and a pressure outlet boundary conditions at the four outlets ("bottom" part of airway as in Figure 1) assuming flow exits the bifurcation unobstructed. No-slip boundary conditions are imposed at the walls. For the pressure and velocity conditions occurring in the lung geometry, the flow is assumed to be incompressible. A rectangular MV flow rate waveform was considered in this study. A model was developed to approach the rectangular MV waveform [17,20]. The wave was modeled as a function of airway area and generation number. The equations are as follows:

$$v\left(t\right) = \frac{Q}{A\, 2^N - 1} \tag{3}$$

$$v\left(t\right) = TV.\frac{e^{-(t-t_i)/\tau}}{A\, 2^N - 1} \tag{4}$$

where t_i is total inspiration time, N is generation number, A is cross section of airway under consideration (in this case, the airway Generation 4 model), Q is averaged flow rate, TV is tidal volume of lung bifurcation, and τ is ratio of tidal volume and flow rate. The flow rate is equal to the ratio of tidal volume and inspiration time. Equation (3) was used as an approach for studying inspiration airflow, and Equation (4) was used to assess expiration airflow during MV breathing. The analysis was carried out for MV breathing with a flow rate of 35 L/min, following [17]. Hence, to maintain the flow rate, the tidal volume was chosen to be 14 mL, and inspiration and expiration times were chosen to be 0.4 s and 1.6 s, respectively.

Figure 1 shows three parts of the airway wall that are designated as Sections 1–3. Section 1 is the biggest airway in the model, followed by Sections 2 and 3. Pressures along these sections were obtained for a subsequent airway wall mechanical response analysis.

2.2.2. Airway Deformation

The lung airway tissue was assumed to be an elastic material. The stress developed in the tissue by pressure applied on it was modeled using the Cauchy momentum relationship:

$$\frac{\partial \sigma_{ij}}{\partial x_j} = \rho \frac{\partial^2 u_i}{\partial t^2} \tag{5}$$

The constitutive equation is a linear elastic material as follows:

$$\sigma_{ij} = E\epsilon_{ij} \tag{6}$$

where σ is stress tensor, ϵ is strain, u is displacement, E is elastic moduli, t is time and ρ is mass density. In this study, a cylindrical coordinate was used to define direction. Hence, the subscript refers to radial (r), rotational (θ) and (in three-dimensional case) axial (z) directions, respectively. The domain of Equation (5) is a slice of lung airway as shown in Figure 1. Radial contraction by smooth muscle is the dominant type of contraction when the lung airway inhales air. In this study, we are interested in the fundamental dynamics of the airway. Hence, the thickness of the model was determined to be one-third of its diameter to ensure the radial response is dominant over axial ones.

The diameters were chosen to approach three sections in the CFD analysis. These diameters are: 2.5 mm for Section 1, 1.86 mm for Section 2 and 1.45 mm for Section 3. The tissue model consists of three layers as shown in Figure 1. The thickness of layers for each section is shown in Table 1. The layers were modeled as an isotropic material. The Young's moduli are 75 KPa for smooth muscle and cartilage layer, and 115 KPa for mucosa layer. Poisson's ratio for all three layers is 0.45. The density is assumed to be 1365.6 kg/m^3, after the work of Koombua and Pidaparti [14]. Isotropic material relation was used in this study since the focus of our study was to analyze the basic dynamics (axial and radial structural response) of the tissue. A recent study by Weed et al. [21] also shows evidence of isotropy at least in lung parenchyma. Furthermore, Xia et al. [22] described in their work that an anisotropic effect appears when tissue strains reached 20%. In this study, small strain assumption was used for the tissue analysis. To compare the linear isotropic material relationship with non-linear ones, two non-linear material relationships were applied. The Neo-Hookean material relationship was used with parameters as outlined in [15]; C_{10} = 0.56 MPa, μ = 1.12 MPa and incompressibility parameter, $d = 2.14 \times 10^{-8}$. Ogden material model was also used, based on measurement by Zeng, et al. [23]. The parameters were as follows; μ_1 =177.93 Pa, A_1 = 5.0424, incompressibility parameter, D_1 = 0.

Table 1. Material model for each layer of the airway tissue.

Layer Name	Layer Number	Thickness at Each Section		
		Section 1	Section 2	Section 3
Mucosa	Layer 1	66.56 μm	49.59 μm	38.61 μm
Smooth Muscle	Layer 2	31.95 μm	22.17 μm	18.53 μm
Cartilage	Layer 3	15.29 μm	11.37 μm	8.86 μm

The boundary conditions were as follows: the model was subjected to transient pressures on the inner surface of the airway model. The transient pressures are obtained from airflow simulation described in the previous section. The displacement in the circular direction was constrained, following the assumption described above. The slice of tissue is supposed to be a part of the larger airway. It was assumed that one side of the slice gave better support than another. Hence, one side's face was constrained in the axial direction.

2.2.3. Numerical Solution to the Models

The model equations were all solved by Finite Element Analysis through ANSYS v.14 software. Both models were discretized using ANSYS meshing module. Fluent component of ANSYS was used to solve Equations (1) and (2) with boundary conditions mentioned earlier. A User-Defined Function (UDF) code was built for generating velocity inlet boundary conditions as expressed in Equations (3) and (4). The fluid domain was discretized with 428,834 tetrahedral elements. ANSYS Mechanical was used to solve Equation (7). The tissue domain was discretized with 277,440 tetrahedral elements.

3. Results and Discussion

3.1. Airflow Simulation

By applying CFD analysis to lung bifurcation 4–6 using the approach described above, velocity field and air pressure inside the lung airway were obtained. Figure 2 shows contour of velocity in the immediate bifurcation after the inlet of Generation 5 airway. It can be seen from Figure 2 that the velocity patterns during inspiration and expiration are different. The velocity patterns presented in Figure 2 followed a horseshoe pattern, as previously shown in studies by Sul *et al.* [24] and Isabey and Chang [25]. This finding provides preliminary validation that the present analysis is reasonable for estimating the fluid dynamics characteristics.

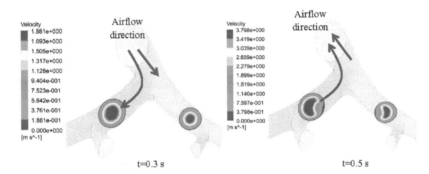

Figure 2. Velocity contour on bifurcation of Generation 5 lung airway, showing tendency to form horseshoe pattern.

The average pressure on three different regions of the bifurcations (Sections 1–3) is presented in Figure 3 in order to see how the MV air pressure affects the tissues. It can be seen from Figure 3 that the pressure variation with time follows the MV velocity waveform. Also, the pressure is higher in Section 1 in comparison to other bifurcation locations (Sections 2 and 3).

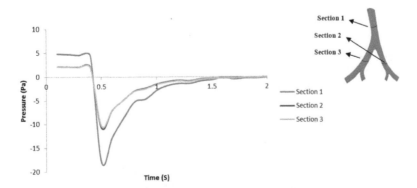

Figure 3. Average transient pressure for MV breathing in Sections 1–3 obtained from airflow simulation.

The pressure along the airway at several locations at various times during inspiration and expiration is shown in Figures 4 and 5. At the bifurcation junction, the pressure is concentrated in the center of the junction. In this area, the pressure on the left side of the junction is higher in value than on the right. The variations of pressure in all these location are shown to be less than 3 Pa. These results suggest that the MV air pressure is non-linear along the radial path of the airway. There are concentrations of pressure on the three junctions during inspiration.

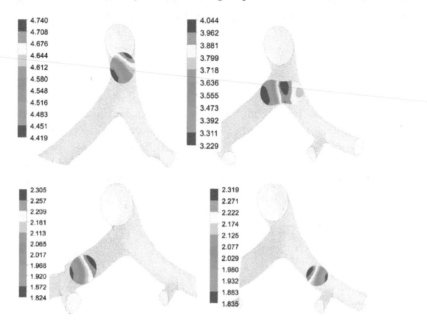

Figure 4. Pressure distribution in lung airway at several locations along airway at $t = 0.3$ s (inspiration) on four section, including junction. Color legend shows pressure value range in Pascal.

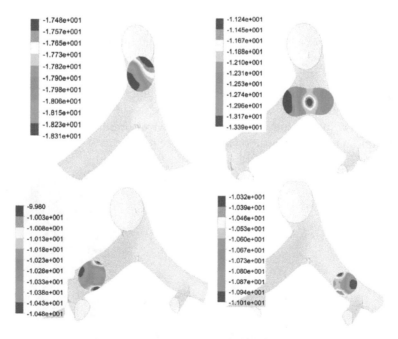

Figure 5. Pressure distribution in lung airway at several locations along airway at $t = 0.5$ s (peak expiration) on four section, including junction. Color legend shows pressure value range in Pascal.

3.2. Transient Mechanical Response

3.2.1. Mesh Convergence Study

Mesh convergence test was carried out before determining that the number of elements as mentioned in Section 2.2.2 was sufficient for this study. The tissue domain was discretized three times with an increasing number of elements as described in Figure 6. The Von Mises strains at two time values were chosen for comparison. The two time values will also be used in later sections. The inspiration time value of 0.3 s, which is close to start of expiration, was chosen. The second time value chosen was $t = 0.5$ s, which is the amplitude of the MV wave.

Figure 6 shows maximum and minimum Von Mises strain at these time values. It can be seen clearly from Figure 6 that the Von Mises strain has reached convergence at first discretization with 277,440 elements. As shown in Tables 2 and 3 a solution obtained with 784,992 elements was named Case 3. Data presented in Tables 2 and 3 also show that the differences between solution (Von Mises strain) of Case 3 and two other cases are negligible. Thus, discretization with 277,440 elements was chosen for this study.

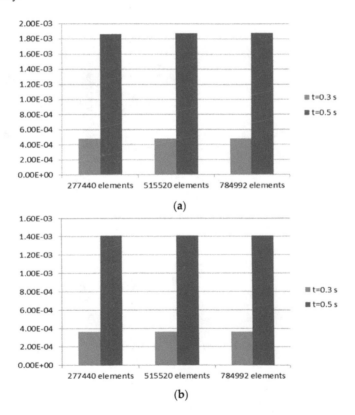

Figure 6. Mesh convergence test for tissue domain. Von Mises strain at two time values, *t*, was taken as criteria. (a) Maximum strain; (b) minimum strain.

Table 2. Difference of maximum Von Mises strain between solution of Case 3 and two other cases (*t* is time value).

Case Name	Number of Elements	$t = 0.3$ s	$t = 0.5$ s
Case 1	277,440	$3.45 \times 10^{-0.06}$	$1.40 \times 10^{-0.05}$
Case 2	515,520	$1.24 \times 10^{-0.06}$	$5.00 \times 10^{-0.06}$
Case 3	784,992		

Table 3. Difference of maximum Von Mises strain between solution of Case 3 and two other cases (t is time value).

Case Name	Number of Elements	$t = 0.3$ s	$t = 0.5$ s
Case 1	277,440	$-1.00 \times 10^{-0.08}$	$-1.00 \times 10^{-0.07}$
Case 2	515,520	0.00	0.00
Case 3	784,992		

3.2.2. Airway Strains

Figure 7 shows maximum Von Mises strain over the domain representing each bifurcation generation. Generally, the highest equivalent strain occurred in Section 3, compared to other sections of bifurcation. The highest equivalent strain is 0.711% and occurred in Layer 2 (smooth muscle) of Section 3 (mucosa). The maximum equivalent strain distribution from Section 3 also showed a significant difference (up to 0.24%) between Layer 1 (cartilage) and 3 (mucosa), compared to the same distribution between Layer 1 and 3 at Sections 1 and 2 (up to 0.015% and 0.0088%, consecutively).

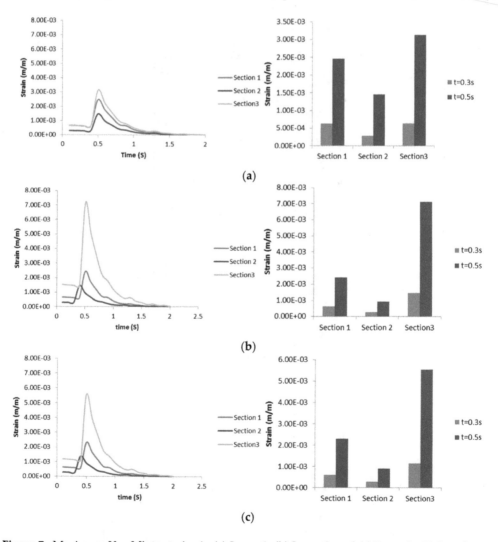

Figure 7. Maximum Von Mises strains in (**a**) Layer 1; (**b**) Layer 2; and (**c**) Layer 3. Right column shows strain progression over time, left column captures the Von Mises strain at a specific time for a clearer comparison.

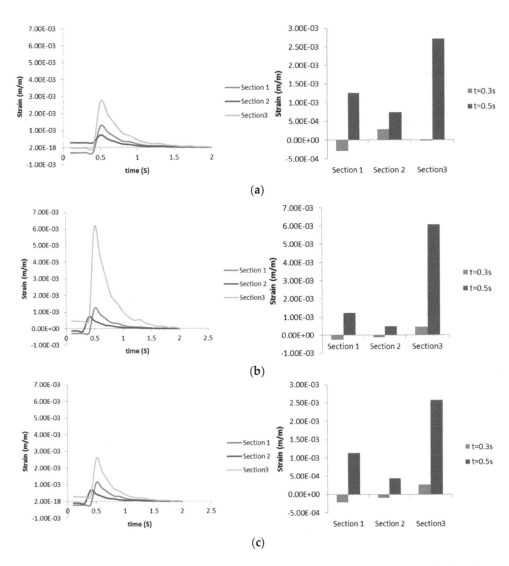

Figure 8. Maximum radial strains occurring in (a) Layer 1; (b) Layer 2; and (c) Layer 3. Right column shows strain progression over time, left column captures the Von Mises strain at specific time for a clearer comparison.

The maximum radial strains obtained from the structural analysis in each of the sections (Sections 1–3) are shown in Figure 8. It can be seen from Figure 8 that the radial strain is higher in Layer 2 at the location of Section 3. This trend is also true for axial strains (not shown). Also, the direction of radial strain can be different for each layer at the same time, revealing unusual dynamics between layers. Figure 9 shows there is no significant difference of Von Mises strain distribution between Layers, compared to strain distribution between Sections in Figure 7. In both Figure 10a,b, the maximum radial strains in Sections 1 and 2 during inspiration are in the negative direction. In Figure 10c, it is apparent that the direction of maximum radial strains is positive in Layers 2 (smooth muscle) and 3 (mucosa), while maximum axial strain in Section 3 is always in the positive direction during inspiration and expiration. The results presented in Figures 7–10 illustrate the distribution and magnitude of strains in different layers of the tissue during inspiration and expiration. This is very important as the tissue response depends on the composition of the layers making up the tissue.

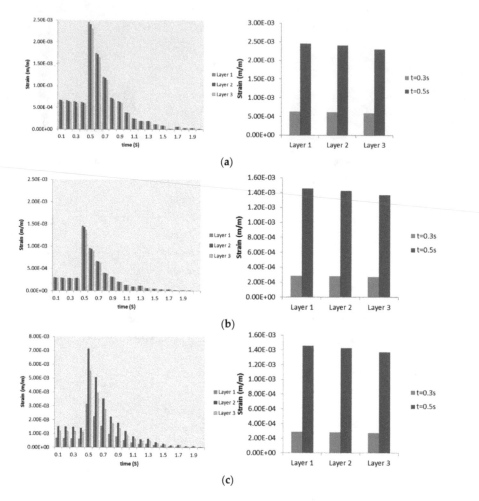

Figure 9. Maximum Von Mises strains occurring in (**a**) Section 1; (**b**) Section 2; and (**c**) Section 3. Right column shows strain progression over time, left column captures the Von Mises strain at specific time for a clearer comparison.

Figure 11 summarizes strains obtained for each section representing Generation 4–5 at time 0.5 s. Radial strain dominates over axial strain following our assumption as discussed in Section 2.2.2. It can be clearly seen that von-Mises strain in Section 3 (Generation 5 to the left of bifurcation defined by the illustration) is 2.9 times and 4.89 times the strains in Sections 1 and 2 respectively. The radial and axial components of the strain in Section 3 are also higher than in Section 1 (8.15 times for radial and 6.04 times for axial) and Section 2 (4.83 times for radial and 3.6 times for axial).

Figure 12 shows a comparison of strain occurring at Generation 4 (Section 1) at the same time as in Figure 9 ($t = 0.5$ s) by using different material models including non-linear, Neo-Hookean and Ogden models. The results were presented in a Log_{10} plot. The Neo-Hookean model produced the smallest range of strains. The von-Mises strain from assuming Neo-Hookean material is 8.16×10^{-4} smaller than that obtained from the linear isotropic material relationship. However, the von-Mises strain from applying the Ogden material relationship is 2.48 times higher than a linear isotropic relationship. Since non-linear deformation was ignored in this study, and only basic response (radial and axial strains) was the focus of this study, there was no need to continue using non-linear material relations in this case. However, the Ogden material parameters used in this study may lead to results that better match the linear isotropic ones.

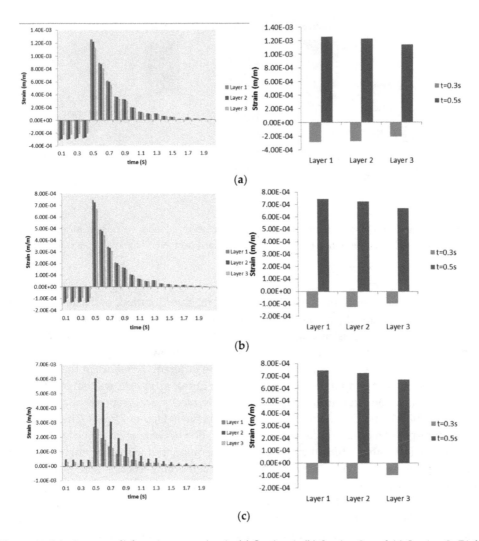

Figure 10. Maximum radial strains occurring in (**a**) Section 1; (**b**) Section 2; and (**c**) Section 3. Right column shows strain progression over time, left column captures the Von Mises strain at specific time for a clearer comparison.

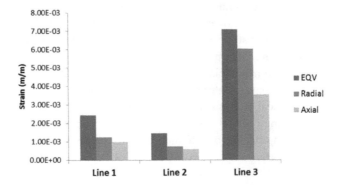

Figure 11. Maximum strains at peak expiration (0.5 s). Von Mises strain in Section 3 is 4.89 times and 2.9 times higher than in Sections 1 and 2 respectively. Radial strain in Section 3 is 8.15 times and 4.83 times higher than in Sections 1 and 2 respectively. Axial strain in Section 3 is 6.04 times and 3.6 times higher than in Sections 1 and 2 respectively.

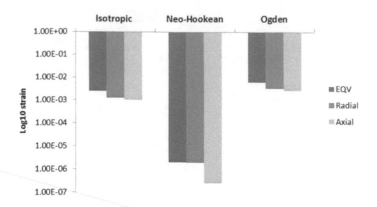

Figure 12. Comparison of maximum strains with different material model at peak expiration ($t = 0.5$ s). The plot is $\log_{10}$ plot to exaggerate the proportion of strains from Neo-Hookean material model.

3.3. Discussion

Figure 3 shows the average transient air pressure at the three locations of the airway. The average air pressure at Sections 2 and 3 (both are Generation 5 of airway) are almost the same, while Section 1 (Generation 4) shows the highest average air pressure distribution during inspiration and expiration. It can be concluded that the air pressure reduces as the air flows through higher generations of airway, as expected. However, this pressure could cause different dynamics at the tissue level that contain three layers. The simplified tissue model used in this study shows that location in one of the Generation 5 (Section 3) bifurcations generates more mechanical response than the location at Generation 4 (Section 1) and Generation 5 (Section 2), as shown in Figures 7 and 8. The Von Mises strain (as mechanical response measure) occurring in one bifurcation of Generation 5 (Section 1) is about four times higher than in another bifurcation of the same generation (but with different bifurcation).

The only physical difference between the two Generation 5 sections here is the radius (and consequently the thickness of each layer); generation in Section 2 (1.86 mm in diameter) is about 0.8 times bigger than in Section 3 (1.45 mm in diameter). When subjected to almost similar pressure cycles, these two sections behaved differently. The mechanical response can be up to four times higher for the bigger bifurcation as described in the previous section. The material properties of the two sections in Generation 5 are the same, and the loadings are similar as shown in Figure 3, however, the thickness is different. This suggests sensitivity to geometrical features. Further study is needed to determine factors that caused different mechanical responses in the same generation in this study.

The importance of investigating strain as a mechanical response lies in its ability to detect/assess inflammation and cell signaling. It is known that cyclic mechanical strain induces varied cell signaling, such as shown in [26]. Mechanical strain may change the properties of the airway [27] and lead to inflammation. The mechanism of inflammation associated with MV breathing is not well established and there are different proposed mechanisms, such as discussed in [28] and [7]. Additionally, some authors have conducted experiments to determine a strain threshold that leads to inflammation [29,30]. However, combining mechanical analysis as presented in this study with stochastic modeling of inflammation (such as [31]) may reveal important dynamics between organ level and cell level mechanics that lead to tissue injury and/or differences in airway properties.

4. Conclusions

In this study, a multi-level modeling scheme was employed to investigate mechanical responses caused by air pressure during MV breathing. A bifurcation at Generation 4–6 of a realistic human lung model was considered for this study. ANSYS software package v.14 was used to solve equations for modeling airflow and structural responses of the airway. One section in Generation 4 and two sections in Generation 5 were selected for structural analysis. Fluid flow simulation showed expected pressure distribution from higher generations to lower ones. Mechanical responses to the air pressure in three locations of bifurcation showed different dynamics between two sections at the same generation with similar transient load. Mechanical response at the same generation but with different bifurcation can differ by up to eight times (peak radial strain in Section 3 is eight times that of Section 2). This result suggests sensitivity to geometrical features. Further study is needed to quantify the sensitivity of mechanical responses to geometrical features.

Acknowledgments: The authors thank the NSF for support this research through a grant CMMI-1430379.

Author Contributions: Aghasafari conducted airflow simulation study and post-processing of the results. Bin M. Ibrahim conducted mechanical response analysis, post-processed the results and wrote the paper. Pidaparti designed the study and contributed to the writing of the paper.

Conflicts of Interest: The authors declare no conflict of interest.

References

1. Ricard, J.D.; Dreyfuss, D.; Saumon, G. Ventilator-induced lung injury. *Eur. Respir. J. Suppl.* **2003**, *42*, S2–S9. [CrossRef]
2. Pinhu, L.; Whitehead, T.; Evans, T.; Griffiths, M. Ventilator-associated lung injury. *Lancet* **2003**, *3613*, 32–40. [CrossRef]
3. Parker, J.C.; Townsley, M.I.; Rippe, B.; Taylor, A.E.; Thigpen, J. Increased microvascular permeability in dog lungs due to high peak airway pressures. *J. Appl. Physiol.* **1984**, *571*, 1809–1816.
4. Dreyfuss, D.; Saumon, G. Role of tidal volume, FRC, and end-inspiratory volume in the development of pulmonary edema following mechanical ventilation. *Am. Rev. Respir. Dis.* **1993**, *1481*, 1194–1203. [CrossRef] [PubMed]
5. Hernandez, L.A.; Peevy, K.J.; Moise, A.A.; Parker, J.C. Chest wall restriction limits high airway pressure-induced lung injury in young rabbits. *J. Appl. Physiol.* **1989**, *662*, 2364–2368.
6. Dos Santos, C.C.; Slutsky, A.S. The contribution of biophysical lung injury to the development of biotrauma. *Annu. Rev. Physiol.* **2006**, *68*, 585–618. [CrossRef] [PubMed]
7. Ning, Q.; Wang, X. Activations of mitogen-activated protein kinase and nuclear factor-kappaB by mechanical stretch result in ventilation-induced lung injury. *Med. Hypotheses* **2007**, *683*, 56–60. [CrossRef]
8. Matsuoka, T.; Kawano, T.; Miyasaka, K. Role of high-frequency ventilation in surfactant-depleted lung injury as measured by granulocytes. *J. Appl. Physiol.* **1994**, *765*, 539–544.
9. Sugiura, M.; McCulloch, P.R.; Wren, S.; Dawson, R.H.; Froese, A.B. Ventilator pattern influences neutrophil influx and activation in atelectasis-prone rabbit lung. *J. Appl. Physiol.* **1994**, *771*, 1355–1365.
10. Guan, X.; Segal, R.A.; Shearer, M.; Martonen, T.B. Mathematical Model of Airflow in the Lungs of Children II: Effects of Ventilatory Parameters. *J. Theor. Med.* **2000**, *3*, 51–62. [CrossRef]
11. Kleinstreuer, C.; Zhang, Z. Targeted drug aerosol deposition analysis for a four-generation lung airway model with hemispherical tumors. *J. Biomech. Eng.* **2003**, *125*, 197–206. [CrossRef] [PubMed]
12. Yang, X.L.; Liu, Y.; Luo, H.Y. Respiratory flow in obstructed airways. *J. Biomech.* **2006**, *39*, 2743–2751. [CrossRef] [PubMed]
13. Brouns, M.; Jayaraju, S.T.; Lacor, C.; de Mey, J.; Noppen, M.; Vincken, W.; Verbanck, S. Tracheal stenosis: A flow dynamics study. *J. Appl. Physiol.* **2007**, *102*, 1178–1184. [CrossRef] [PubMed]
14. Koombua, K.; Pidaparti, R.M. Inhalation Induced Stresses and Flow Characteristics in Human Airways through Fluid-Structure Interaction Analysis. *Model. Simul. Eng.* **2008**. Available online: http://www.hindawi.com/journals/mse/2008/358748/ (accessed on 28 September 2015). [CrossRef]
15. Rolle, T.; Pidaparti, R.M. Tissue strains induced in airways due to mechanical ventilation. In Proceedings of the ASME Early Career Technical Conference, Atlanta, Georgia, USA, 4–5 November 2011.
16. Pidaparti, R.M.; Koombua, K. Tissue strains induced in airways due to mechanical ventilation. *Mol. Cell Biomech.* **2011**, *81*, 149–168.
17. Koombua, K.; Pidaparti, R.M.; Longest, P.W.; Ward, K.R. Biomechanical Aspects of Compliant Airways due to Mechanical Ventilation. *Mol. Cell Biomech.* **2009**, *6*, 203–216. [PubMed]
18. Kamm, R.D. Airway wall mechanics. *Annu. Rev. Biomed. Eng.* **1999**, *1*, 47–72. [CrossRef] [PubMed]

19. Bai, A.; Eidelman, D.H.; Hogg, J.C.; James, A.L.; Lambert, R.K.; Ludwig, M.S.; Martin, J.; McDonald, D.M.; Mitzner, W.A.; *et al.* Proposed nomenclature for quantifying subdivisions of the bronchial wall. *J. Appl. Physiol.* **1994**, *72*, 1011–1014.

20. Arambakam, R.; Pidaparti, R.; Reynolds, R.M.; Heise, A. Computational Fluid Dynamics Anslysis of Typical Lower Lung Bifurcations under Mechanical Ventilation and Normal Breathing. In Proceedings of the Fourteenth Annual Early Career Technical Conference, Birmingham, UK, 1–2 November 2014.

21. Weed, B.; Patnaik, S.; Rougeau-Browning, M.; Brazile, B.; Liao, J.; Prabhu, R.; Williams, L. Experimental Evidence of Mechanical Isotropy in Porcine Lung Parenchyma. *Materials* **2015**, *82*, 2454–2466. [CrossRef]

22. Xia, G.; Tawhai, M.H.; Hoffman, E.A.; Lin, C.L. Airway wall stiffening increases peak wall shear stress: A fluid-structure interaction study in rigid and compliant airways. *Ann. Biomed. Eng.* **2010**, *38*, 1836–1853. [CrossRef] [PubMed]

23. Zeng, Y.J.; Yager, D.; Fung, Y.C. Measurement of the Mechanical Properties of the Human Lung Tissue. *J. Biomech. Eng.* **1987**, *109*, 169–174. [CrossRef] [PubMed]

24. Sul, B.; Wallqvist, A.; Morris, M.J.; Reifman, J.; Rakesh, V. A computational study of the respiratory airflow characteristics in normal and obstructed human airways. *Comput. Biol. Med.* **2014**, *52*, 130–143. [CrossRef] [PubMed]

25. Isabey, D.; Chang, H.K. A model study of flow dynamics in human central airways. Part II: Secondary flow velocities. *Respir. Physiol.* **1982**, *499*, 97–113. [CrossRef]

26. Shah, M.R.; Wedgwood, S.; Czech, L.; Kim, G.A.; Lakshminrusimha, S.; Schumacker, P.T.; Steinhorn, R.H.; Farrow, K.N. Cyclic stretch induces inducible nitric oxide synthase and soluble guanylate cyclase in pulmonary artery smooth muscle cells. *Int. J. Mol. Sci.* **2013**, *144*, 4334–4348. [CrossRef] [PubMed]

27. Tepper, R.S.; Ramchandani, R.; Argay, E.; Zhang, L.; Xue, Z.; Liu, Y.; Gunst, S.J. Chronic strain alters the passive and contractile properties of rabbit airways. *J. Appl. Physiol.* **2005**, *98*, 1949–1954. [CrossRef] [PubMed]

28. Chapman, K.E.; Sinclair, S.E.; Zhuang, D.; Hassid, A.; Desai, L.P.; Waters, C.M. Cyclic mechanical strain increases reactive oxygen species production in pulmonary epithelial cells. *Am. J. Physiol. Lung Cell Mol. Physiol.* **2005**, *289*, L834–L841. [CrossRef] [PubMed]

29. Chandel, N.S.; Sznajder, J.I. Stretching the lung and programmed cell death. *Am. J. Physiol. Lung Cell Mol. Physiol.* **2000**, *279*, L1003–L1004. Available online: http://ajplung.physiology.org/content/279/6/L1003.abstract (accessed on 15 March 2015).

30. Copland, I.B.; Post, M. Stretch-activated signaling pathways responsible for early response gene expression in fetal lung epithelial cells. *J. Cell. Physiol.* **2007**, *210*, 133–143. [CrossRef] [PubMed]

31. Reynolds, A.; Koombua, K.; Pidaparti, R.M.; Ward, K.R. Cellular Automata Modeling of Pulmonary Inflammation. *Mol. Cell Biomech.* **2012**, *91*, 141–156. [CrossRef]

3

DNA and RNA Extraction and Quantitative Real-Time PCR-Based Assays for Biogas Biocenoses in an Interlaboratory Comparison

Michael Lebuhn [1,*]**, Jaqueline Derenkó** [2,†]**, Antje Rademacher** [2,†]**, Susanne Helbig** [3,†]**, Bernhard Munk** [1,†]**, Alexander Pechtl** [4,†]**, Yvonne Stolze** [5,†]**, Steffen Prowe** [3]**, Wolfgang H. Schwarz** [4]**, Andreas Schlüter** [5]**, Wolfgang Liebl** [4] **and Michael Klocke** [2]

Academic Editor: Ali Khademhosseini

[1] Bavarian State Research Center for Agriculture, Department for Quality Assurance and Analytics, Lange Point 6, 85354 Freising, Germany; Bernhard.Munk@lfl.bayern.de
[2] Leibniz Institute for Agricultural Engineering Potsdam-Bornim, Department Bioengineering, Max-Eyth-Allee 100, 14469 Potsdam, Germany; jderenko@atb-potsdam.de (J.D.); arademacher@atb-potsdam.de (A.R.); mklocke@atb-potsdam.de (M.K.)
[3] Beuth University of Applied Sciences, Department of Life Sciences and Technology, Luxemburger Strasse 10, 13353 Berlin, Germany; shelbig@beuth-hochschule.de (S.H.); steffen.prowe@beuth-hochschule.de (S.P.)
[4] Department of Microbiology, Technische Universität München, Emil-Ramann-Str. 4, D-85354 Freising-Weihenstephan, Germany; alexander.pechtl@tum.de (A.P.); wschwarz@wzw.tum.de (W.H.S.); wliebl@mytum.de (W.L.)
[5] Institute for Genome Research and Systems Biology, CeBiTec, Bielefeld University, Bielefeld, Germany; ystolze@cebitec.uni-bielefeld.de (Y.S.); aschluet@cebitec.uni-bielefeld.de (A.S.)
*   Correspondence: michael.lebuhn@lfl.bayern.de
†   These authors contributed equally to this work.

Abstract: Five institutional partners participated in an interlaboratory comparison of nucleic acid extraction, RNA preservation and quantitative Real-Time PCR (qPCR) based assays for biogas biocenoses derived from different grass silage digesting laboratory and pilot scale fermenters. A kit format DNA extraction system based on physical and chemical lysis with excellent extraction efficiency yielded highly reproducible results among the partners and clearly outperformed a traditional CTAB/chloroform/isoamylalcohol based method. Analytical purpose, sample texture, consistency and upstream pretreatment steps determine the modifications that should be applied to achieve maximum efficiency in the trade-off between extract purity and nucleic acid recovery rate. RNA extraction was much more variable, and the destination of the extract determines the method to be used. RNA stabilization with quaternary ammonium salts was an as satisfactory approach as flash freezing in liquid N_2. Due to co-eluted impurities, spectrophotometry proved to be of limited value for nucleic acid qualification and quantification in extracts obtained with the kit, and picoGreen® based quantification was more trustworthy. Absorbance at 230 nm can be extremely high in the presence of certain chaotropic guanidine salts, but guanidinium isothiocyanate does not affect (q)PCR. Absolute quantification by qPCR requires application of a reliable internal standard for which correct PCR efficiency and Y-intercept values are important and must be reported.

Keywords: ring-trial; fermenter sludge; DNA purity; PCR suitability; extraction efficiency; RNA preservation; reverse transcription; methanogens; *Bacteria*; absolute quantification

1. Introduction

Since its invention in 1983, polymerase chain reaction (PCR) and PCR based techniques for analyzing DNA started a triumphal march in many segments of life science research and biotechnological applications, and represented a break-through in culture independent analysis of biological specimens from diverse ecosystems. The same was true for the quantitative PCR variant, the Real-Time quantitative PCR (qPCR) [1], which was introduced in the mid-1990s. Several technically different qPCR variations were developed [2], of which systems using a fluorescent DNA binding dye or a hydrolysis probe in 5′-nuclease assays are most widely applied. By integration of a reverse-transcription (RT) step after digestion of genomic DNA, it is intended to quantify RNA species by RT-qPCR.

Providing quantitative data in the analysis of biogas biocenoses is a major task not only for research but particularly for practice. For example, quantitative assessment of functionally important microbial guilds and their activity can be used for process diagnosis and control. As recently shown, novel early warning systems of biogas process failure based on qPCR and RT-qPCR analyses announced microbial metabolic stress weeks before the response of conventional chemical parameters [3,4]. However, the quality of (RT-)qPCR analyses is directly dependent on the quality of the nucleic acid extract which depends on the applied sample processing measures and the extraction procedure. Extract quality thus predetermines suitability for the intended analysis. Some steps of these sample processing and analysis pipelines entail uncertainty. This is expressed, e.g., in the enormous number on "the best" DNA or RNA extraction routine for PCR based approaches for challenging samples such as from biogas fermenter sludges or similar matrices, e.g., [5–8]. However, the suggestions made are partly contradictory. These challenges and uncertainties are more detailed in the following:

If the concentration of certain organisms, genes or transcripts is to be specifically determined in environmental samples or other specimens, absolute quantification of nucleic acid copies by qPCR or RT-qPCR currently is the method of choice. Since qPCR and, particularly, RT-qPCR approaches present some pitfalls, details of crucial analysis steps and data of control points should be reported [9,10]. This is particularly important if it is intended to compare the results with those of other laboratories. The efficiency of the qPCR reaction can be calculated and should not be below 90% and above 110% [11]. Optimum PCR efficiency is usually obtained if the concentration of inhibitors in the extract is minimized (see below). PCR inhibition can be detected in a dilution assay, allowing the elimination of inhibited reactions from further evaluations.

Processivity, robustness and reaction environment of DNA-polymerases have greatly been improved since their first introduction. Much less is known about the efficiency of RT reactions with RNA extracts. Different RT enzymes and chemistries are available on the market but information on RT inhibitors and modes of RT inhibition and respective research is scarce, although the RT reaction is located upstream of PCR and is thus more prone to co-extracted inhibitors. Absolute quantification of RNA by RT-qPCR is thus a prominent challenge for the future.

Reliable nucleic acid quantification requires not only efficient PCR- (and RT-) reactions, meaning that the concentration of inhibitory compounds in the DNA and RNA extracts must have been sufficiently removed. The extraction routine should also provide high extraction efficiency conferring high assay sensitivity. Moreover, different nucleic acid isolation procedures for biogas biocenoses from fermenter samples can considerably impact the DNA composition in the extract [6,7], e.g., by selective lysis of different microorganisms. This can result in biased detection of taxonomic groups. Nucleic acid extraction is thus a tradeoff between sample purity on the one hand and maximum recovery rate on the other hand.

Extraction of nucleic acids has a longer history than PCR applications. Phenol/chloroform-based extraction techniques were mainly developed to yield nucleic acids as pure as possible, e.g., for DNA/DNA hybridization studies. Nucleic acid purity is conventionally judged by spectro-photometrical absorption at 260 nm, 280 nm and 230 nm and benchmark thresholds or intervals for the corresponding ratios A260/280 and A260/230 [12]. However, several drawbacks are associated with phenol/chloroform-based extraction techniques such as working with hazardous chemicals in sometimes many repetitive steps resulting often in phenol contamination of the extract and (partial) inhibition of subsequent PCR, as well as poor extraction efficiency (see below), precluding work with small sample volumes. Moreover, recent findings challenge the above mentioned thresholds for nucleic acid purity if extracts are intended to be used for PCR approaches. For example, Munk *et al.* [13] reported very low A260/230 ratios in DNA extracts, probably due to the presence of guanidine (iso)thiocyanate (GITC) which is used in several more user-friendly nucleic acid extraction kits. GITC absorbs strongly at 230 nm, but no inhibition of qPCR reactions was observed in spite of its presence. Since such kits are becoming more and more popular, their suitability and efficiency for (RT-)qPCR-based approaches had to be evaluated in the present work. Kit-based systems, however, mostly use small sample volumes, resulting in relatively low nucleic acid yields. This can be limiting for downstream applications requiring higher nucleic acid amounts. Moreover, harsh sample disruption can result in too strong nucleic acid fragmentation for analyses requiring longer template strands. In the case of total RNA extraction, this can result in compromised RNA quality as indicated by low RIN (RNA integrity number) values [14], impeding e.g., metatranscriptome analysis, as is addressed in this paper. The destination of the extract must therefore determine the extraction methodology used.

A further problem associated with RNA and, particularly, mRNA is its instability. If samples are taken from biologically active environments such as biogas fermenters, RNA degradation must be obviated at once to avoid biased results. They should thus either immediately be processed or treated by physical or chemical means preventing RNA decomposition. Several approaches of RNA preservation have been developed with snap-freezing in liquid nitrogen as an effective and widespread method. More recently, quaternary ammonium salts have been reported to be efficient RNA preservatives in some [15,16] but not all [17] studies. It should be evaluated in the present study, which RNA preservation method may be suitable for samples from biogas fermenter sludge.

For absolute quantification by (RT-)qPCR, the extraction efficiency, the DNA or RNA recovery rate, must be known to account for losses during the extraction procedure. The recovery rate can be determined, e.g., by sample spiking with pre-quantified standards [18]. Without these data, absolute concentrations of target genes or transcripts cannot be calculated reliably in environmental samples. However, respective information is only rarely presented in studies reporting quantitative results, and, particularly, data on RNA extraction efficiencies are almost inexistent. This paper will show an example of how DNA extraction efficiencies can be assessed.

The generation of a reliable standard is a further prerequisite for absolute quantification, and the few papers dealing with this subject indicate that this is obviously a neglected issue. It is strongly suggested to apply an internal standard. Dilution series of the standard should result in a standard curve with qPCR efficiency close to 100% (slope close to −3.3), indicating absence of PCR inhibitors, and with most qPCR platforms, a Y-intercept between Cq (cycle of quantification) values of about 34–39 [19]. Lower Y-intercept values of the standard will result in considerable underestimation and higher values in substantial overestimation of the target in the sample.

With this background and considering the challenges associated with nucleic acid extraction from difficult matrices such as biogas digester sludge and subsequent (RT-)qPCR analysis, this communication will address the following questions and issues:

- Does a given kit-based DNA extraction protocol (with small modifications) yield similar qPCR results in a ring-trial in an interlaboratory comparison? Do modifications that were introduced by some partners impact the results, and which differences can be seen if a traditional chloroform-based routine is used?
- How efficient is the extraction considering that maximum assay sensitivity, *i.e.*, lowest possible method detection limit should be provided?
- How suitable are the extracts for qPCR? If PCR inhibition is observed, which qualities do the extracts have?

- Do the different qPCR protocols that were applied in the ring-trial yield different results? Are there any peculiarities explaining the differences?
- Which are the most important process steps, control points and benchmarks that should be respected in the qPCR analysis pipeline, and are traditional absorption ratio thresholds helpful?
- Which methods of RNA preservation can be recommended for biogas fermenter sludge samples?
- Which pitfalls do the different RNA extraction protocols present for different purposes such as RT-qPCR and metatranscriptome analysis?

2. Experimental Section

2.1. General Outline of the Experiments

Five partners (A–E) participated in the interlaboratory comparison of nucleic acid extraction and RNA preservation methods.

Partner A operated four two-stage two-phase biogas fermenter systems and partner B four single-stage biogas fermenters digesting grass silage, as specified below. All biogas systems were fed with grass silage originating from the same stock. The silage was provided as big-packs by partner B.

In an initial analysis, partner B compared different DNA extraction kits in a standard-spiking approach which was described to be suitable for the given plant-based fermenter sludge matrix. A kit-based routine was optimized yielding a DNA recovery rate of about 90%. In the following DNA extraction ring-trial, all five partners participated. Four of these (partners A–D) served as analysts performing quantitative Real-Time PCR (qPCR) with each of the DNA extracts provided by the five partners. The analysts used their established in-house routines (see Section 2.7) targeting *Bacteria* (protocols i, iv and v), *Archaea* (protocol ii) or methanogenic *Archaea* (protocol iii).

In order to compare RNA preservation methods, digestate from the "hydrolysis" stage of partner A was treated by four different preservation methods as specified below and sent to partner B for RNA extraction, reverse transcription and cDNA synthesis following two different protocols (see below). cDNA was quantified for methanogenic *Archaea* directly by partner B (protocol iii) and after shipping cDNA to partner A for *Bacteria* (protocol i) and *Archaea* (protocol ii).

Flow charts of the experimental setup of the interlaboratory DNA extraction and the RNA preservation experiment are shown in Figure 1.

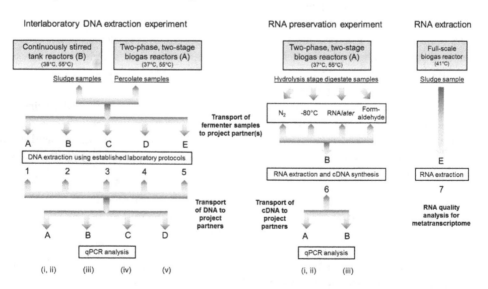

Figure 1. Experimental flow chart of the interlaboratory DNA extraction and RNA preservation experiments; numbers refer to laboratory protocols, letters to partners.

2.2. Substrate, Reactor Set-Up and Sampling

Grass silage was provided by partner B in big-packs. It was fed to the single-stage fermenters operated by partner B (see below) and shipped to partner A as substrate for its two-stage two-phase fermenter systems. The grass silage showed the following properties: organic dry matter (ODM) = 32.57% fresh matter (FM); NH_4^+-N = 0.63 g·kg_{FM}^{-1}; pH = 5.11; volatile organic acids (VOA) = 8.26 g·kg_{FM}^{-1}. There was no hint for spoilage or minor quality. Partner A chopped the silage to approximately 3 cm, froze aliquots and fed them after defrosting in batches to the "hydrolysis" stages. Partner B chopped the silage to about 1 cm after freezing, prepared daily portions and fed aliquots to the single-stage fermenters.

Partner A operated two mesophilic (37 °C) and two thermophilic (55 °C) two-stage two-phase biogas fermenter systems. The principle of the reactor set up was described in detail by Schönberg and Linke [20]. Each system consisted of one hydrolysis reactor (HR), one percolate (process fluid) storage tank and one anaerobic filter compartment. The HR reactors were fed with grass silage (see above) in batches every 28 days. After one batch, the digested substrate (digestate) was replaced by fresh grass silage. The subsequent percolate storage tank supplied process fluid for the semi-continuously percolated HRs and the anaerobic filter compartment which contained biofilm carriers. The substrate load of the HRs was successively increased from 250 g to 500 g FM, which was applied in a start-up procedure of five batch fermentations to ensure adaptation of the microbial community. Digestate was taken from the HRs after four batches (112 days) of the start-up procedure at high methane yield (356.5 ± 8.5 L_{STP}·kg_{VS}^{-1}). It was treated instantly with four different preservation methods to evaluate the most efficient one for RNA stability (Figure 1). Percolate samples were taken from the HRs and immediately frozen for interlaboratory comparison of DNA extraction and qPCR analysis (Figure 1) after five batch runs (140 days) at high methane yield (370.2 ± 4.2 L_{STP}·kg_{VS}^{-1}).

Partner B operated two mesophilic (38 °C) and two thermophilic (55 °C) continuously stirred tank reactors (CSTRs) with the grass silage in parallel. At the time of sampling, the organic loading rate was relatively low (0.5 g_{VS}·L^{-1}·d^{-1}), and the process ran stably and very efficiently with a methane yield of about 400 L_{STP}·kg_{VS}^{-1} at both temperatures. The concentration of short-chain fatty acids (SCFA) was low, all SCFA were below 500 mg·L^{-1}. Free ammonia-nitrogen (FAN), however, was continuously rising, and amounted to about 500 mg·L^{-1} and 1500 mg·L^{-1} in the meso- and thermophilic processes, respectively. Although there were no symptoms of process disturbance yet, the rising FAN and, particularly, the already high FAN values in the thermophilic process suggested precaution in the further operation strategy [3].

2.3. Comparison of DNA Extraction Kits and DNA Recovery Rates

It has been shown in a previous study [5] that initial repeated washing of cattle manure and samples from fermenters operated with renewable resources greatly reduced co-elution of PCR-inhibitors in DNA extraction. The washing procedure was consequently integrated in DNA and RNA extraction protocols [13,21]. With this initial modification, different DNA extraction kits (FastDNA™ SPIN Kit for Soil, MP Biomedicals (FSKS); UltraClean® Soil DNA Isolation Kit, MoBio Laboratories Inc. (USIK); QIAamp DNA Stool Mini Kit, QIAGEN (QSMK)) were compared. The comparison was based on a standard spiking approach:

A pre-quantified *Escherichia coli* suspension (10 µL) was spiked to 390 µL of an autoclaved cattle manure sample. The absence of *E. coli* DNA in the sample was confirmed using the DNA extraction protocol of Lebuhn *et al.* [5] and the qPCR protocol given below. After mixing thoroughly and washing with PBS and 0.85% KCl (centrifugation steps 5 min at ca. 12,000 g), DNA was extracted from 40 µL sample aliquots in parallel with the three DNA extraction kits in accordance with the manufacturers' recipes. 5'-nuclease qPCR assays specific for *E. coli* were carried out for undiluted extracts and serial (1:10) extract dilutions in triplicate in 25 µL reaction volumes containing 0.6 µM primers Ec_mur_27f (GATAAATTTCGTGTTCAGGGGCCA) and Ec_mur_177r (CTTTCGCACCCAGCTGGCTT), 0.2 µM hydrolysis probe Ec_mur_126S (5'-FAM-AGAACCGGTAGAGATCCAGAACGTCC-TAMRA-3'),

0.75 U Platinum *Taq* polymerase (Invitrogen) and 1 × supplied PCR buffer (without MgCl$_2$), 200 µM dNTPs, 6 mM MgCl$_2$ and 2.5 µL DNA template. The relatively high MgCl$_2$ concentration was generally used by partner and analyst B because it was shown in a previous experiment with a DNA extract from cattle manure spiked with *E. coli* that 6 mM MgCl$_2$ in the given qPCR assays provided optimum conditions for the DNA polymerase (Supplementary Figure S1). We assume that co-eluted compounds scavenge or mask Mg^{2+} ions, impeding the processivity of the polymerase and leading to increased Cq-values if Mg^{2+} was not supplemented.

For comparison of the DNA extraction kits and calculation of the achieved DNA recovery rates, the Cq-values of the dilution series were plotted against the calculated gene copy numbers in the qPCR reactions along with a 1:10 dilution series of the spiked *E. coli* suspension as the reference.

2.4. DNA Extraction Ring-Trial

In a ring-trial of five laboratories (partners A–E), DNA was extracted from a set of biogas fermenter samples with different physicochemical properties following their established adapted DNA isolation routine protocols (1–5 in Figure 1). The extracts were used to study the impact of the DNA extraction protocol on the results of qPCR-based DNA analysis. All five project participants isolated genomic DNA from eight biogas reactor samples (two mesophilic and two thermophilic "hydrolysis" reactors, and two mesophilic and two thermophilic single-stage CSTRs). Four partners (A–D) used DNA extraction protocols involving the FastDNA™ SPIN Kit for Soil (protocols 1–4) due to previous experience that this system combines ease of operation and optimum DNA recovery rates for this type of samples [5], (see also Section 3.1). These protocols mainly differed in applied sample volume, sample treatment and lysis conditions (Table 1). In DNA extraction from the CSTRs, the applied sample volume was limited to 40 or 50 µL (Table 1) due to the previous experience with cattle manure [5], and similar CSTR sludges (not shown) that a higher load can result in co-elution of PCR inhibitors resulting in compromised PCR efficiency or even PCR failure, particularly with undiluted extracts. DNA concentration and quality of the extracts were determined by examination of absorbance at 230, 260, 280 and partially 320 nm, and by the ratios A260/280 and A260/230, as indicated in Supplementary Table S2. Partners A and D used the Quant-iT™ PicoGreen® system (Invitrogen Ltd., Paisley, UK) for DNA quantification and calibration [22].

Table 1. DNA extraction protocols for fermenter samples.

Protocol		1	2 *	3 *	4	5
extracted sample volume #	percolate (HR)	2 mL	100 µL	40 µL	2 mL	26 g
	sludge (CSTR)	50 µL	40 µL		50 µL	26 g
pretreatment of samples			60 s Ultra-Turrax, centrifuged, 2× 0.85% KCl-washed		centrifuged, resuspended in ultrapure H$_2$O	washed with PBS
cell lysis		2× 20 s @ 5.0 bead beater (MP Biome-dicals, Lysing Matrix E Tubes)	40 s @ 6.0 bead beater (MP Bio-medicals, Lysing Matrix E Tubes)		disruptor (Scientif. industr.); MP Bio-medicals Lysing Matrix E Tubes	CTAB/pronase ε/SDS/65 °C
involvement of a kit		FastDNA Spin kit for Soil				no
DNA-purification		kit components, silica matrix based, no post-purification				chloroform:iso-amylalcohole, isopropanol
elution volume #		100 µL		75 µL		100 µL

* Following the extraction routine described in [5,13] with modifications as indicated; # all calculations are for 1 mL or 1 g of fresh fermenter sludge or percolate sample.

Extraction procedure 5 (Table 1) was included as a widely applied reference since it followed a CTAB-chloroform:isoamylalcohol routine for DNA isolation: 26 g of sample were mixed with 50 mL of

1 M phosphate buffered saline solution (PBS, 137 mM NaCl, 2.7 mM KCl, 10 mM Na_2HPO_4, 1.8 mM KH_2PO_4). The mixture was centrifuged at $9000\times g$ for 5 min. Pellet resuspension in 50 mL PBS (4 °C), shaking at 400 rpm, centrifugation at $200\times g$ for 5 min and supernatant collection followed and was repeated two more times. The collected supernatant was centrifuged at $9000\times g$ for 5 min, and the pellet was resuspended in 40 mL PBS and centrifuged again at $5000\times g$ for 15 min. The supernatant was discarded. For cell lysis, the pellet was resuspended in CTAB containing DNA extraction buffer (DEP, described previously in Henne *et al.*, 1999) with 5 mg Pronase ε (Serva Electrophoresis GmbH, Heidelberg, Germany) and 2 mg RNase (Qiagen, Germantown, MD, USA) and shaken at 180 rpm and 37 °C for one hour. The suspension was incubated at 65 °C for 2 h after adding 3 mL of 10% SDS solution, while inverting every 15 min. Centrifugation at $3900\times g$ for 10 min, filtering through a folded filter (pore size 15–18 μm) and mixing the filtrate 1:1 (v/v) with a 24:1 chloroform/isoamylalcohol (v/v) mixture were followed by centrifugation at $8000\times g$ and 4 °C for 5 min. The upper phase was mixed 1:0.7 (v/v) with isopropyl alcohol and incubated at room temperature for one hour. DNA was pelleted by centrifugation at $9000\times g$ and 4 °C for 20 min. For DNA purification, NucleoBond AX-G (Macherey-Nagel, Düren, Germany) ion exchange columns and solutions were used in accordance with the manufacturer's protocol. The DNA pellet was resuspended in 2 mL N2-buffer, incubated overnight at 60 °C, and after following the instructions, DNA was eluted with 100 μL TE buffer (10 mM Tris, 1 mM EDTA, pH 8.0) overnight at 4 °C. DNA was quantified using a NanoDrop 2000 Spectrophotometer (Thermo Scientific, Waltham, MA, USA) and was used additionally for DNA qualification and evaluation of contaminants.

Noticeably, DNA preparation routine 5 is typically used to obtain long and unfragmented double-stranded DNA e.g., for whole microbial genome and metagenome analysis. This preparation method must thus confer other qualities than protocols 1–4 and 6 (see Section 2.4) which were developed for representative community composition analysis using a combination of physical cell disruption and chemical lysis, and for sensitive PCR-based quantification of relatively short nucleic acid targets. For this purpose, DNA loss during the extraction procedure and separation of PCR inhibitors are minimized, and the most important goal is to provide the maximum possible recovery rate of PCR-suitable DNA.

The five DNA preparations were sent to four project partners (A–D, Figure 1) who compared DNA quality by laboratory specific qPCR-based quantification of bacterial 16S rDNA, archaeal 16S rDNA and *mcr*A/*mrt*A gene copy numbers of methanogenic *Archaea* (see Section 2.7). 16S rDNA is the target of choice if the diversity of a specified microbial community is the focus, and *mcr*A/*mrt*A is used to specifically determine methanogenic *Archaea*. As a gene encoding subunit A of methyl-CoM-reductase, the key enzyme of methanogenesis, its transcription and translation are a prerequisite for methane production. It is intimately linked with energy production and therefore essential for activity and viability of methane producing *Archaea*.

2.5. Preservation Methods for RNA Stability

In a preliminary experiment, samples from the four CSTRs operated by partner B (see Section 2.2) were taken in repetition, and RNA was extracted immediately (see Section 2.6), or after treatment with quaternary ammonium salts as supplied in commercial RNA*later*® (Ambion, Carlsbad, CA, USA) in accordance with the manufacturer's protocol and after treatment with "homemade" RNA preservative (http://sfg.stanford.edu/RNAbuffer.pdf, http://www.protocol-online.org/prot/Protocols/RNAlater-3999.html; final concentrations: 25 mM sodium citrate, 10 mM EDTA, 70 g ammonium sulfate/100 mL solution, pH 5.2) and storage at −20 °C overnight to simulate a "worst case scenario". Since commercial RNA*later*® performed slightly better than "homemade" RNA preservative in RT-qPCR evaluation of extracts by partner/analyst B targeting *mcr*A/*mrt*A (see Section 2.7), and almost 60% of *mcr*A/*mrt*A cDNA was quantified relative to the fresh samples (see Section 3.3.1, Figure 6), preservation by commercial RNA*later*® was chosen for sample shipping.

In order to identify possibly even more suitable preservation methods for RNA stabilization in biogas fermenter samples, four methods were tested with digestate from the HR of the two-stage systems (see Section 2.2) operated by partner A (Figure 1); (a) flash freezing in liquid N_2 and (b) freezing at $-80\,°C$; (c) treatment with RNA*later*®; and (d) treatment with formaldehyde. Digestates treated with RNA*later*® stabilization solution (c) were incubated at $4\,°C$ overnight, according to the protocol of the supplier, and then frozen at $-80\,°C$. In method (d), samples were treated with 1.25 mL 1x phosphate buffered solution (PBS) and 3.75 mL of 3.7% formaldehyde. After incubation overnight at $4\,°C$, the solution was removed subsequently, and samples were washed twice with $1\times$ PBS and stored at $-80\,°C$ in a $1\times$ PBS/96% ethanol (1:1, v/v) solution. The frozen samples were sent in dry ice to project partner B for RNA extraction (see Section 2.6).

2.6. RNA Extraction and Reverse Transcription

After two months of storage at $-80\,°C$, RNA extraction was carried out (protocol 6 in Figure 1) by partner B with 40 µL fermenter samples that were washed twice using the FastRNA Pro™ Soil–Direct Kit (MP Biomedicals, Santa Ana, CA, USA) with a second washing step, as reported by Munk and coworkers [21]. The TURBO DNA-*free*™ Kit (Ambion, Carlsbad, CA, USA) was used to digest co-extracted DNA in 20 µL eluate in accordance with the manufacturer's instructions.

Two different reverse transcription (RT) reactions were performed by partner B (Figure 1). The first reaction aimed at transcribing total RNA with random hexamer primers provided in the AffinityScript™ Multiple Temperature Reverse Transcriptase (Agilent Technologies, Santa Clara, CA, USA) kit. Complementary DNA (cDNA) was dedicated to partner A for assays targeting 16S rRNA of *Bacteria* and *Archaea*. The second reaction was specific for *mcr*A/*mrt*A mRNA using reverse primer MeA-i 1435r (Table 1). RT reactions were performed with 5 µL RNA, 200 ng hexamer primers or 600 nM MeA-i 1435r, 2 µL AffinityScript RT buffer (10x), 2 µL DTT (100 mM), 2 µL dNTPs (10 mM) and 1 µL AffinityScript RT filled up with RNase free H_2O to a final volume of 20 µL. The RT reaction was performed at $42\,°C$ or $45\,°C$ for 60 min for the reactions with random hexamers and *mcr*A/*mrt*A, respectively, and stopped by heat inactivation for 15 min at $70\,°C$. cDNA was kept on ice for subsequent *mcr*A/*mrt*A qPCR analysis by partner B (see Section 2.7) or was frozen at $-20\,°C$ and shipped to partner A on ice for quantification of 16S rRNA transcripts (see Section 2.7).

Protocol 6 was used for the RNA preservation experiment because it was developed for representative community analysis using combined mechanical and biochemical cell disruption and for the sensitive quantification of relatively short RNA fragments by RT-qPCR. This requires minimization of RNA loss during the extraction procedure and should provide a maximum recovery rate of RT-qPCR-suitable RNA. However, preliminary results (see Section 3.3.2) showed that protocol 6 may not fulfill the requirements for metatranscriptome studies since RIN values were unsatisfactory (6.4, 5.4 after rRNA depletion) and because the amount of RNA recovered by the miniprep-procedure from the single-stage fermenters operated by partner B was insufficient. For this reason, a classical acid phenol based RNA extraction routine (protocol 7 in Figure 1) is described here with which the required RNA quality and quantity from biogas fermenter sludge were obtained:

In RNA extraction protocol 7, a full-scale biogas plant (BGP_WF in [23]) was sampled. Fermenter sludge sample was taken by partner E and immediately processed according to a sample volume adapted version of the protocol described by Zoetendal *et al.* [24], followed by application of the RNeasy Midi kit (Qiagen, Hilden, Germany) in accordance with the manufacturer's protocol. In brief, RNA extraction included the following steps (between steps samples were always kept on ice): 4.4 g of fermenter sludge were mixed with 2.5 mL TE buffer ($4\,°C$), applied on a nylon filter (40 µm nylon BD Biosciences, Heidelberg, Germany) and centrifuged at $400\times g$ and $4\,°C$ for 2 min. The filtrate was mixed $1+1$ (v/v) with acid phenol ($4\,°C$), filled into a 15 mL tube with approximately 0.5 g glass beads (0.1 mm) and vortexed for 4 min at highest speed. After centrifugation at $5000\times g$ for 5 min, the upper phase was mixed 1:4 (v/v) with RLT buffer (RNeasy Kit, with 2-mercaptoethanol). The following steps were done in accordance with the RNeasy Midi bacteria protocol starting at step 5, with the upper phase/ethanol ratio being 1:2.8 (v/v, without RLT). Finally, RNA was eluted in 50 µL RNase-free water. DNA was digested using the RNase-Free DNase Set (Qiagen, Hilden, Germany) following the manufacturer's instructions. DNA removal was verified by gel electrophoresis. Quality and quantity of the extracted RNA were evaluated using a Prokaryote RNA 6000 Pico chip and the Agilent 2100 Bioanalyzer (Agilent, Santa Clara, CA, USA). For metatranscriptome sequencing, depletion of rRNA from the whole RNA sample [25] was done once using the Ribo-Zero rRNA Removal Kit (Bacteria) (Epicentre, Chicago, IL, USA) and measured again using the Agilent 2100 Bioanalyzer.

2.7. Real-Time Quantitative PCR Assays

For evaluation of the DNA extraction ring-trial (Section 2.4) and of the RNA preservation methods (Section 2.5), DNA or cDNA was quantified applying different qPCR-assays. These were targeting bacterial or archaeal 16S rRNA encoding genes and transcripts (measured as cDNA), or the methanogenic archaeal community and its activity by analysis of the functional *mcrA/mrtA* genes and transcripts. One Cq value was generally subtracted from RT-qPCR results for cDNA to compensate for the first PCR cycle transforming DNA–RNA hybrids into double-stranded DNA. qPCR reactions were mostly carried out in triplicates.

Partner (and analyst) A applied primers and probes published by Yu and coworkers [26] for the *Bacteria* (Bac, protocol i) and the *Archaea* assays (Arc, protocol ii) with temperature profiles as described in Supplementary Table S1. qPCR reactions were run on a CFX96 Touch™ Real-Time PCR Detection System (BioRad, Sundbyberg, Sweden) using the DyNAmo Flash Probe qPCR Kit (Thermo Scientific, Waltham, MA, USA) in a 20 µL reaction setup. This included 900 nM primers and 200 nM labeled probes (Supplementary Table S1) as final concentration, 10 µL 2× master mix with Mg^{2+} as supplied, 2 µL template, and 4 µL DNase free H_2O. For the qPCR standards, plasmids containing a target sequence (16S rDNA of *Archaea* [27] and of *Pectobacterium carotovorum* ssp. *carotovorum* DSM 30168 for *Bacteria*) were prepared as described by Klocke *et al.* [27]. The concentration of the linearized plasmid was determined with a NanoDrop™ 3300 fluorospectrometer (Thermo Fisher Scientific Inc., Waltham, MA, USA), and the copy number of 16S rDNA per defined volume was calculated [28]. A tenfold dilution series of these plasmids served for the standard curves (presented as joint projects in Figure 2) with PCR efficiencies of 93.6% and 95.4% for the Bac (i) and Arc (ii) assays, respectively. The single slopes covered a range from −3.4 to −3.6. The Y-intercept ranged between 41.6 and 44.2 for the Bac and 38.4–41.1 for the Arc assays, indicating overestimation particularly of the *Bacteria* target sequences (see also Section 3.2.2). Template (c)DNA was added 10-fold or 100-fold diluted and determined in triplicate for qPCR analyses.

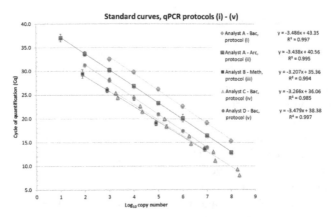

Figure 2. Standard curves applied for the qPCR determination of *Bacteria* (Bac), *Archaea* (Arc) and methanogenic *Archaea* (Meth) in the interlaboratory DNA extraction experiment; partners/analysts A, C and D used plasmid preparations and spectrophotometry to calculate target gene copy numbers, and partner/analyst B applied MPN-qPCR for lysed transformed *E. coli* suspensions.

The assay applied by partner/analyst B (protocol iii, Meth) quantifies the methanogenic archaeal community, targeting the functional $mcrA/mrtA$ genes or their transcripts. Assay design, temperature profile, reactants, preparation of standards and generation of standard curves are described in [13]. Primers MeA-i 1046f/MeA-i 1435r [29] and further details on the assay composition are listed in Supplementary Table S1. Due to their high degeneracy, a higher concentration of 600 nM primers and 6 mM $MgCl_2$ were used in qPCR reactions because the DNA polymerase coped best at higher Mg^{2+} concentrations with this type of extracts (see Section 2.3, Supplementary Figure S1). For the standard, an $mcrA$ PCR-fragment of a $Methanosarcina$ sp. from a CSTR sample was cloned in $E. coli$, and the transformed $E. coli$ suspension was propagated and quantified after washing by most probable number (MPN) qPCR analysis of a dilution series of the lysed clone suspension resulting in achieving the target gene concentration in the stock suspension. No signs of qPCR inhibition were seen even for the undiluted stock suspension. More details are found in [13]. The joint project $mcrA/mrtA$ standard curve obtained with serial clone suspension dilutions (Figure 2) yielded reasonable values for the Y-intercept (Cq 35.4) and the slope (-3.21) resulting in 105% PCR efficiency. Template DNA or cDNA was added undiluted, tenfold and for some extracts 100-fold diluted to reveal PCR inhibition in $mcrA/mrtA$ qPCR. In –RT reactions (DNase digested RNA extract without Reverse Transcriptase) serving as negative controls, only weak qPCR signals were obtained occasionally. The corresponding amount of residual DNA was negligible, ranging between 0 and 2.8 double-stranded cDNA copies per qPCR assay, and was subtracted from the results to specify the cDNA pool.

Partner C used a SYBR® Green based qPCR system as initially described by Fierer $et al.$ [29] for the analysis of $Bacteria$ (qPCR protocol iv). For the standard, a PCR product obtained with bacterial 16S rDNA primers AGAGTTTGATYMTGGCTC and CAKAAAGGAGGTGATCC and $Clostridium$ $thermocellum$ ATCC 27405 gDNA was cloned into a pJet1.2 blunt end vector. Correct insertion was examined by sequencing. The copy number of inserted nucleotides in the linearized vector was determined by molecular weight calculation [30] and NanoDrop 1000 spectrophotometry. Template (c)DNA was added 10-fold or 100-fold diluted. The joint project standard curve obtained with serial dilutions of the standard displayed reasonable values for the Y-intercept (Cq 36.1) and the slope (-3.27) resulting in 102% PCR efficiency (Figure 2). qPCR assays were performed with the Biorad SsoAdvanced™ Universal SYBR® Green Supermix in a 17 µL setup with 200 nM primers and 2 µL DNA template and run on a CFX96 Touch™ Real-Time PCR Detection System (BioRad, Sundbyberg, Sweden). The temperature profile and more details on the qPCR assay are found in Supplementary Table S1.

Partner (and analyst) D used the same primers and probes for $Bacteria$ as partner/analyst A [26]. The LightCycler® 480 (Roche Diagnostics, Basel, Switzerland) with the temperature profile listed in Supplementary Table S1 was used for analysis. Each quantitative real-time PCR assay was carried out in 20 µL of a reaction mixture containing 5 µL of extracted DNA as template, 10 µL of 2× LightCycler® 480 Probes Master (Roche Diagnostics, Basel, Switzerland) with dUTP, 0.1 U Uracil-N-glycosylase (UNG), 500 nM of each primer and 100 nM of the labeled probes (Supplementary Table S1). DNA extracts were diluted 1:100 and 1:1000 and measured in duplicate per dilution. The 16S rDNA plasmid standard was kindly provided by partner/analyst A (see above). The concentration of the linearized plasmid was determined with the fluorescent nucleic acid stain PicoGreen (Quant- iT™ PicoGreen dsDNA Assay Kit, Thermo Fisher Scientific, Waltham, MA, USA) on a SPECTRAmax® GEMINI XS Microplate spectrofluorometer (Molecular Devices, Sunnyvale, CA, USA). After calculation of the copy number of 16S rDNA per defined volume [28], a 10-fold dilution series from 10^7–10^2 copies of 16S rDNA per PCR reaction was established and used as the standard. For each qPCR experiment, a plasmid dilution series was prepared and measured in triplicate. For experiments of partner D, Y-intercept and slope of the standard were 38.4 and -3.48, respectively, resulting in 93.8% PCR efficiency (Figure 2).

3. Results and Discussion

3.1. Comparison of DNA Extraction Kits and DNA Recovery Rates

In a previous experiment, partner/analyst B spiked cattle manure devoid of *Escherichia coli* DNA with *E. coli* cells and evaluated the efficiencies obtained in parallel DNA extractions with the FastDNA™ SPIN Kit for Soil (FSKS), the UltraClean® Soil DNA Isolation Kit (USIK) and the QIAamp DNA Stool Mini Kit (QSMK) applying *E. coli* specific qPCR. Figure 3 shows the results. Serial extract dilutions yielded similar slopes in *E. coli murA* qPCR (efficiencies 99.30% for the *E. coli* spike, 99.97% for variant FSKS, 89.30% for variant USIK and 89.92% for variant QSMK) with reasonable Y-intercept values (Figure 3). Extraction efficiencies were calculated as means of DNA recovery rates, relating the results for the different extracts obtained at Cq 30 and Cq 20 to the respective whole cell qPCR values for the *E. coli* spike suspension dilutions. Variant FSKS was the most efficient with 88.24%, followed by variant USIK (60.58%) and variant QSMK (37.48%). For comparison, conventional phenol-chloroform-isoamylalcohol extraction of *E. coli* spiked cattle manure did not yield PCR amplifiable DNA (not shown) and was not considered further. Due to the results, DNA extraction procedure FSKS was integrated in the sample processing pipeline of partner/analyst B.

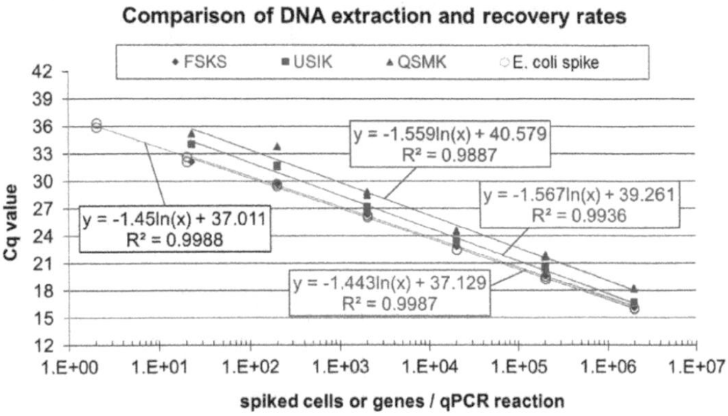

Figure 3. Comparison of DNA extraction efficiencies from treated cattle manure spiked with *Escherichia coli* cells using the FastDNA™ SPIN Kit for Soil (FSKS), the UltraClean® Soil DNA Isolation Kit (USIK) and the QIAamp DNA Stool Mini Kit (QSMK); Cq: cycle of quantification.

Using the FSKS DNA extraction routine and the Platinum *Taq* polymerase based qPCR assay setup (see Section 2.3), addition of polyvinylpyrrolidone (PVP, 0.85%, 1.7%, 3.4%) and cetyltrimethylammoniumbromide (CTAB, 5%, 10%, 20%, in 1 M NaCl) did not enhance but diminished the extraction efficiency [5], and similarly, addition of T4-gene-32-protein was not found to improve the DNA recovery rate (not shown). These results suggest that such additives as adjuvants might only be helpful in cases of suboptimal nucleic acid extraction or (partially) inhibited PCR but not at an already high recovery rate.

3.2. Interlaboratory Comparison of DNA Extraction

Data tracing back to obvious (partially) inhibited PCR reactions, as evidenced e.g., by serial dilution qPCR assays, are consequently excluded from calculations and data processing. However, since identifying and preventing PCR inhibition is extremely important to avoid false negative results, the occurrence of PCR inhibition is discussed in relation to identified or suspected factors.

DNA extracts of all laboratories yielded consistent results in the qPCR assays of the analysts confirming high reproducibility of the analytical procedure. For analytical (qPCR), technical (DNA extraction) and biological (parallel fermenters) replicates, coefficients of variation (CV) were similar (Table 2).

Table 2. Coefficients of variation (CV) of analytical, technical and biological replicates. Data represent means of all Bac, Arc and Meth qPCR results over all DNA extracts.

qPCR Assay, Partner/Analyst	Bac (i), A	Arc (ii), A	Meth (iii), B	Bac (iv), C	Bac (v), D
			CV (%)		
analytical	7.56	7.20	7.47	10.10	13.38
technical	20.92	20.80	24.85	n.d.	16.33
biological	28.49	28.70	32.24	25.45	23.27

n.d.: not determined; Bac: *Bacteria*; Arc: *Archaea*; Meth: methanogenic *Archaea*.

3.2.1. Differences between Reactor Types and Operation

Total 16S rDNA copy numbers of *Bacteria* (Bac-PCR protocols, Figure 4) and of *Archaea* (Arc-PCR protocol, Figure 5A) differed between reactor types. 16S rDNA copy numbers per milliliter of fermenter material obtained from the CSTR reactors were approximately 10-fold higher than from the HR reactors, independently of the extraction protocol used, no matter if extracts 5 (suboptimal performance, see Section 3.2.3) were considered or not. Only minor differences were detected between 16S rDNA copy numbers at thermophilic or mesophilic conditions in one reactor type, indicating that the tested temperatures (see Section 2.2) did not influence considerably the presence of total *Bacteria* and total *Archaea* in the examined HR and CSTR reactors.

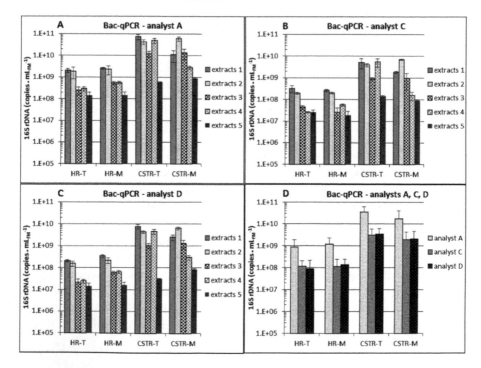

Figure 4. 16S rDNA copy numbers for *Bacteria* (Bac) assessed by qPCR and calculated per fresh sample matter as performed by partner/analyst A (**A**); partner/analyst C (**B**) and partner/analyst D (**C**). Means and standard deviations of two biological replicates of four different biogas reactor samples are shown. For each biogas reactor sample, DNA was extracted in different laboratories (partners/analysts A–E) in two technical replicates (1 exception) using their established protocols, and each extract was analyzed in (one exception) two, mostly six and occasionally 10 qPCR replicates. Extracts with the given numbers were obtained with the corresponding DNA extraction protocol. Figure 4D shows the Bac-qPCR data as overall means of extraction protocols 1–5, and the corresponding standard deviations. HR: "hydrolysis" reactor; CSTR: continuously stirred tank reactor; T: thermophilic; M: mesophilic.

However, the concentration of methanogenic *Archaea* was higher in the mesophilic than in the thermophilic CSTR and HR reactors (Figure 5B, inconsistent data of extracts 5 not considered), although methane productivity rates were almost identical (see Section 2.2). One reason can be a higher richness of methanogens at mesophilic conditions [31–33] with lower mean specific activity. In principle, the higher temperature allows for higher activity, and a narrowed spectrum of methanogens can be capable of increased specific metabolic performance [4,34]. Higher specific methanogenic activity at thermophilic conditions can thus explain the observed differences.

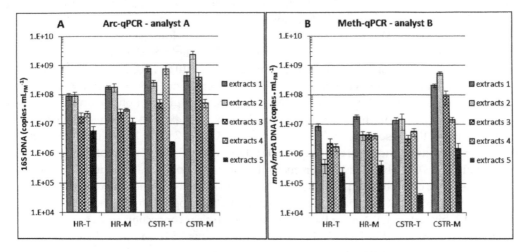

Figure 5. 16S rDNA copy numbers for *Archaea* (Arc) and *mcr*A/*mrt*A copy numbers for methanogens assessed by qPCR and calculated per fresh sample matter as performed by partner/analyst A (**A**) and partner/analyst B (**B**). Means and standard deviations of two biological replicates of four different biogas reactor samples are shown. For each biogas reactor sample, DNA was extracted in different laboratories (partners/analysts A–E) in two technical replicates (one exception) using their established protocols, and each extract was analyzed in six, or eight to ten qPCR replicates in case of partner/analyst A or B, respectively. Extracts with the given numbers were obtained with the corresponding DNA extraction protocol. HR: "hydrolysis" reactor; CSTR: continuously stirred tank reactor; T: thermophilic; M: mesophilic.

The concentration of methanogenic *Archaea* was high in the percolates of the HRs (Figure 5B), and the approximately 10-fold higher concentration of *Archaea* (Figure 5A) supported these data. Since significant amounts of methane were produced not only in the CSTRs but also in the HRs at both temperatures, this confirmed that the methanogens were not only present but displayed high activity also in the HRs.

3.2.2. Differences between Analytical Routines

Figure 4D shows that overall means of bacterial 16S rDNA copy numbers quantified by partners/analysts C and D (protocols iv and v, respectively) for each of the four reactors were similar whereas the overall means determined by partner/analyst A (protocol i) was consistently elevated by one order of magnitude even though the same assay system was used by partners/analysts A and D (Supplementary Table S1). This results from a significant difference in the Y-intercept value of the standard curve of partner/analyst A for *Bacteria* leading to a higher copy number count deduced from the obtained Cq values (Figure 2). The corresponding Y-intercepts of partners/analysts C and D were in a normal range whereas the value of partner/analyst A was unusually high [19]. The high value can be due to DNA degradation or overestimation of the DNA standard template, both of which result in overestimation of calculated copy numbers in samples. No such difference could be observed for archaeal 16S rDNA copy numbers of partner/analyst A (Figure 5A), and the corresponding standard curve Y-intercept value was in the upper normal range (Figure 2). These findings emphasize the necessity of including all necessary data demanded by the MIQE guidelines [9,10,35] to be able to evaluate the comparability of different studies when absolute gene or transcript copy numbers are compared.

A critical issue is also the sensitivity of the assay. For any quantitative assay, the method detection limit (MDL) should be provided. However, the optimum MDL is not achieved with compromised samples, e.g., if DNA is too fragmented or extracts are containing PCR inhibitors [18,36]. PCR inhibition is typically detected in dilution-PCR. Adding e.g., 100-fold diluted DNA template to the reaction mixture will reduce the possibility of PCR inhibition by co-extracted compounds but lower the assay sensitivity by a factor of 100 at the same time. Negative results (below the detection limit) may thus be obtained in case of low target organism abundance in the sample. Such compromised sensitivity relativizes the applicability of the assay. Analyst A used 2 µL of 1:10 or 1:100 diluted DNA template in 20 µL reaction volume (cf. Supplementary Table S1) and observed no PCR inhibition. Analyst B used 1 µL of undiluted, 1:10 and 1:100 of diluted DNA template in 25 µL reaction volume and observed PCR inhibition only with undiluted template for three of four extracts obtained with protocol 4 (by partner D) from the mesophilic CSTR reactors. Analyst C used 2 µL of 1:10 diluted DNA template in 17 µL reaction volume and observed three single inhibited qPCR reactions in total for extracts from the thermophilic CSTR1 obtained with protocols 1, 2 and 5 (by partners A, B and E, respectively). Analyst D used 5 µL of 1:100 and 1:1000 diluted DNA template in 20 µL reaction volume and observed one possibly inhibited PCR reaction with an extract from a thermophilic hydrolysis reactor prepared with protocol 3 (by partner C). Among the tested variants, no clear connection between PCR inhibition and analytical partner could be identified, but the highly viscous CSTR samples obviously presented the greatest challenge for the tested DNA extraction systems.

3.2.3. Differences between DNA Extraction Protocols

Four of the five tested DNA extraction variants (Table 1) involved the FastDNA™ SPIN Kit for soil whereas procedure 5 followed a CTAB-chloroform:isoamylalcohol routine. When mean values over the different analysts and reactor types were calculated per milliliter (or g) fresh sample matter for the five extracts (Table 3), protocol 2 yielded the highest copy numbers followed by protocols 1, 4 or 3, and, finally, protocol 5 in Bac, Arc- and Meth-qPCR.

Table 3. Performance of five DNA extracts in different qPCR assays, mean for all analysts and fermenters.

DNA Extracts Obtained with	Bac-qPCR (i, iv, v)	Arc-qPCR (ii)	Meth-qPCR (iii)
	Mean over Analysts and Fermenters (copies $\cdot$ mL$_{FM}$$^{-1}$)		
protocol 1	8.79×10^9	3.64×10^8	6.09×10^7
protocol 2	1.07×10^{10}	6.99×10^8	$1,38 \times 10^8$
protocol 3	2.54×10^9	1.18×10^8	2.47×10^7
protocol 4	5.26×10^9	2.08×10^8	6.38×10^6
protocol 5	3.52×10^8	7.17×10^6	5.42×10^5

Bac: *Bacteria*; Arc: *Archaea*; Meth: methanogenic *Archaea*; FM: fresh matter.

The highest microbial DNA yield was thus obtained for all analysts and fermenters with extraction protocol 2 but it appeared to be best suited for qPCR analysis only with the extracts from the more viscous CSTR samples. DNA extraction protocols 1, 3, 4 and 5 yielded only 43%, 19%, 19% and 1% in means of the copies obtained with protocol 2 for CSTR sludge samples, respectively. Protocol 2 provided about 90% DNA recovery in previous spiking experiments with cattle manure (see Section 3.1) and can be modified to warrant similar DNA recovery rates for recalcitrant organisms or mixed communities in a ramping approach with pooling of extracted sub-fractions [37,38]. However, DNA extracts 1 (obtained with protocol 1, provided by partner A) yielded the highest copy numbers for percolate samples from HR reactors by all four analysts (Figure 4A–C and Figure 5). DNA extraction protocols 2, 3, 4 and 5 yielded only 69%, 17%, 18% and 6% in means of the copies obtained with protocol 1 for HR percolates, respectively. One of the reasons can be that it was possible to apply a higher sample volume (2 mL, Table 1, except for protocol 5, see below) of the more liquid HR percolates without risking PCR inhibition. However, 2 mL percolate was also applied in protocol 4 (Table 1) which yielded only 18% of the copies obtained by protocol 1. This might be due to suboptimal cell disruption obtained with the different disruptor device (Table 1). Another reason for the minor performance of protocol 2 as compared to protocol 1 with the HR samples is that analyst B determined relatively low concentrations of methanogens for all four protocol 2 extracts from the two thermophilic HRs (Figure 5B). Since these extracts yielded the highest numbers of archaeal 16S rDNA (Figure 5A) and were among the best in the analysis of *Bacteria* (Figure 4), a problem leading to underestimation may have occurred during the analytical process of analyst B but probably not at the extraction level. It can be recorded that applying a higher volume of HR percolates in extraction protocols 1–4 was of advantage by improving the method detection limit and statistics, since these percolates were more liquid and contained less organic matter and microbial biomass than the more viscous CSTR samples for which protocol 2 provided the best trade-off between optimum qPCR assay sensitivity (method detection limit) and risk of PCR inhibition.

Protocol 5 yielded the lowest copy numbers in all analyses (Figure 4A–C and Figure 5, Table 3). Because of insufficient DNA yields (see also Supplementary Figure S2), partner E sent only one of four planned extracts from the thermophilic CSTRs to the partners. Protocol 5 was thus found to be the least suited for DNA extraction from biogas reactor samples if high DNA extraction efficiency for sensitive downstream qPCR analysis is the goal. However, it must be recalled that this method has been developed to obtain relatively long, unfragmented DNA double-strands, suitable, e.g., for whole microbial genome and metagenome analysis (see Section 2.4). Whether protocols 1–4 are suitable for these applications is currently being tested more exhaustively and is discussed in Section 3.3.2.

DNA concentration in extracts and suitability of extracted DNA for downstream applications such as PCR are traditionally quantified and qualified by spectrophotometric analysis of absorbance at 260 nm and ratios A260/280, A260/230 [12]. Although the respective benchmarks undoubtedly qualify nucleic acid extracts as (almost) pure, some application such as qPCR with specialized novel enzymes obviously do not necessarily require reaching all of these high standards, particularly an A260/230 ratio of >1.8, and the considerable labor investment to reach these levels. However, this depends on the extraction system used: high A230 values are obtained in the presence of co-eluted GITC but do not compromise PCR [13] in contrast to the presence of PCR inhibitors such as phenol. One of the tasks of this work was thus to substantiate this finding in the interlaboratory ring-trial and to figure out important characteristics for PCR-suitable DNA extracts.

As expected, all of the DNA extracts, except for those provided by partner E (extracts/protocol 5) which did not use the FastDNA Spin Kit for Soil with GITC but a CTAB/pronase ε/SDS system instead (Table 1), showed consistently very low A260/230 ratios with a maximum of 0.146, whereas the extracts provided by partner E yielded considerably higher A260/230 ratios, but only one extract exceeded 1.8 (Supplementary Table S2). About half of the extracts showed A260/280 ratios between 1.7 and 2.0, several were lower, probably due to suboptimal exclusion of proteins, and some were higher, even reaching values above 3 (Supplementary Table S2). However, there was no obvious evidence that A260/230 ratios lower than 1.8 or with A260/280 ratios below 1.7 or above 2.0 were responsible for qPCR inhibition: clear qPCR inhibition was only observed by analyst/partner B for four undiluted extracts (CSTR1-M2 by partner A; CSTR1-M1, CSTR2-M1, CSTR2-M2 by partner D) but many other extracts with worse A260/280 and A260/230 ratios did not inhibit the qPCRs. None of the A320 values were above 0.05, the level at which co-eluted humic compounds appeared to (partially) inhibit qPCR [13]. The extract with the highest value (0.044, Supplementary Table S2) did not compromise qPCR. There was thus no indication that extracts as prepared by one of the proposed protocols and missing the mentioned benchmarks of sample purity will fail or are prone to fail in qPCR. However, A260/280 values >2.5 cannot simply be explained by the presence of RNA and pH shifts. Most abnormally high A260/280 values and most inhibited qPCR reactions were observed for the CSTR extracts (Supplementary Table S2). The inhibitor may particularly be produced in mesophilic anaerobic digestion of grass silage and have an absorbance maximum at 260 nm. It remains to be noted here that even very low A260/230 ratios originating from co-elution of GITC are no indicator of qPCR inhibition.

The results of DNA quantification were in general agreement with those of the qPCR approaches, both confirmed higher DNA concentrations in the CSTR sludges than in the HR percolates (Supplementary Figures S2 and S3). However, DNA quantification yielded partially variable results. For several extracts, differences up to a factor of ca. 10 between the results obtained by the different analysts were not rare (Supplementary Figure S2). Strikingly, all DNA concentrations determined for the CSTR extracts 1–4 by partners/analysts B, C and E which applied the extinction coefficient for double-stranded DNA were considerably higher than those determined by partners/analysts A and D which used the PicoGreen® system (Supplementary Figure S2). This was not the case for both, CSTR and HR extracts 5 from partner E, who applied a different extraction system and did not use GITC. Differences between the DNA quantification methods were less evident for the HR extracts, probably due to the low recovered DNA concentrations (Supplementary Figure S2). Besides strong absorption at 230 nm, GITC shows weak absorbance at 260 nm. We assume therefore that DNA concentrations calculated with the extinction coefficient (at 260 nm) were slightly biased for extracts obtained with the FastDNA™ Spin Kit for Soil. However, this cannot explain the considerable overestimation observed with the extinction coefficient for the extracts that were derived from the CSTRs. We assume that an impurity, possibly the inhibitor mentioned above, particularly present in the CSTR extracts operated with grass silage was responsible for the high A260 values.

In conclusion, DNA quantification with the PicoGreen® system appeared to produce more reliable results. Considering only these data (Supplementary Figure S3), DNA concentrations calculated per milliliter or gram of fermenter percolate or sludge were by far higher for the CSTRs. For both, HRs and CSTRs, they were lowest in extract 5 provided by partner E, as discussed above, and highest at almost the same level in extracts 1 and 2 (by partners A and B). Extracts 3 and 4 (by partners C and D) were intermediary. For extract 4, a reason may be that a disruptor was used (Table 1) with which microbial cells may not have been broken up as efficiently as with the bead beater. The low applied percolate volume of only 40 μL (Table 1) resulting in signals partly below the detection limit may explain the relatively low values for HR, but since no such reason could be identified for the CSTR extract 3, we assume that a systemic problem in the extraction procedure of partner C may have occurred.

3.3. RNA Experiments

3.3.1. RNA Preservation

RNA preservation was initially tested with commercial RNA*later*® and "homemade" RNA preservative in relation to immediate RNA extraction for samples from the four CSTRs operated by partner B (see Section 2.2). The results are shown in Figure 6.

In comparing the medians of the eight extracts relative to the immediate extraction (fresh, control), preservation by RNA*later*® yielded 57% of the control *mcrA/mrtA* cDNA with a SD of 19%, whereas preservation by "homemade" RNA preservative produced only 33% of the control *mcrA/mrtA* cDNA with a high SD of 74%. The RNA preservation efficiency with nearly 60% as compared to direct sample processing was considered satisfactory. Due to these results, further evaluations were carried out with commercial RNA*later*®. The comparison with flash freezing with N_2 as the widely applied reference method was of particular interest.

The appropriate method for stabilization of RNA was examined for digestates derived from a HR of the two-stage systems operated by partner A by comparing four different preservation methods, formaldehyde treatment, RNA*later*®, flash freezing in liquid N_2 and freezing at −80 °C (see Section 2.5).

Figure 7 shows the results of the corresponding RT-qPCR analyses performed by partners and analysts A and B for the RNA extracts prepared by partner B (Figure 1), as described in Section 2.6.

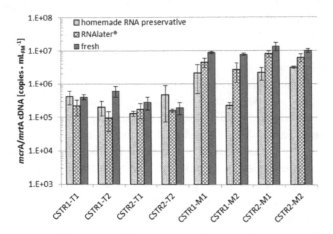

Figure 6. RT-qPCR analysis of *mcr*A/*mrt*A mRNA (analyst B) extracted from two mesophilic (M1,2) and thermophilic (T1,2) biogas CSTR sludge samples in parallel (CSTR1,2). RNA was preserved with commercial RNA*later*® and "homemade" RNA preservative or immediately extracted (fresh). All qPCR analyses were accomplished in triplicate.

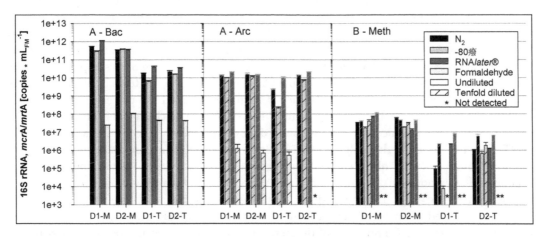

Figure 7. RT-qPCR analysis (analyst A) of bacterial (Bac) and archaeal (Arc) 16S rRNA and of *mcr*A/*mrt*A mRNA (analyst B, Meth) to evaluate the most efficient preservation method for RNA stability: digestates (D) of two (1, 2) mesophilic (M) and thermophilic (T) biogas fermentations ("hydrolysis" stages of partner A) were preserved by flash freezing in liquid N_2, at $-80\,^{\circ}$C, treatment with RNA*later*® or formaldehyde. All qPCR analyses were accomplished in triplicate.

The formaldehyde treatment led to the lowest detection of RNA copy numbers of all preservation methods, and some RT-qPCR signals were even below the corresponding detection limits. The other preservation methods resulted in much higher and mostly similar RNA copy numbers between the variants (Figure 7). Among the three high valued preservation methods, slightly lower RNA copy numbers were obtained for samples that were frozen at $-80\,^{\circ}$C while RNA*later*® treated samples yielded comparable or higher copy numbers than flash freezing with liquid N_2 (Figure 7). This is also indicated by Cq factors which were calculated in relation to liquid N_2 fixation as the reference. Taking into account the data of all RT-qPCR assays, Cq factor medians were 1.5 for RNA*later*® and 0.5 for freezing at $-80\,^{\circ}$C.

Previous reports on RNA extraction using different sample preservation methods are contradictory: Wang and coworkers [39] obtained equivalent or better RNA quality of cervical tissues preserved by flash freezing with N_2 than with RNA*later*®-preserved tissues. Riesgo and coworkers [40] similarly found for sponge samples that freezing in liquid N_2 outperformed RNA*later*® in quality and quantity and showed no RNA loss during two months of storage. However, RNA*later*® was favored as a preservation method in studies analyzing tissues, yeast or prokaryotic cultures [15,16,41].

In contrast, there are only very few studies dealing with preservation of RNA from environmental samples. Rissanen and coworkers [17] recommend a phenol-chloroform solution as preservation method for environmental sediment samples containing humic acids. They found that RNA*later*® preservation decreased nucleic acid yields drastically and slightly biased community analysis towards certain microbial types. The authors argued that the high content of humic compounds could precipitate nucleic acids by ammonium salts contained in RNA*later*®. However, in spite of the fact that the digestates investigated in the current study did contain humic acids, RNA recovery from RNA*later*® preservation was at least equivalent with recovery from preservation by liquid N_2. Moreover, although physical, chemical and structural differences in the mesophilic and thermophilic reactor samples were expected, no major difference between preservation by liquid N_2 and RNA*later*® was observed. Thus, RNA*later*® appears to be useful to preserve samples e.g., from agricultural biogas plants where laboratory sample processing chains for RNA are missing.

The bacterial 16S RNA cDNA/DNA ratios for the RNA*later*® preservation method indicated very low transcriptional activity, ranging between 0.45 and 1.47 for mesophilic samples, and, in particular, for the thermophilic digestate samples, between 0.06 and 0.1. Lebuhn *et al.* [34] reported higher cDNA/DNA ratios for *Bacteria*, about 10 for mesophilic and about 5 for thermophilic flow-through hydrolytic/acidogenic CSTRs operated with a straw/hay mixture, in spite of the much shorter retention time and lower VS degradation efficiency in the CSTRs and the lower energy content in the straw/hay mixture. Major RNA degradation can be ruled out because almost 60% of the RNA was recovered after preservation by RNA*later*® relative to immediate RNA extraction from fresh samples. The lower transcriptional activity can originate from the particular conditions in the "hydrolysis" stages of partner A. Taken together, the results support the efficacy of the highly valued RNA preservation methods.

However, reliable quantitative assessment of RNA species is a challenge in its current state, and results must remain uncertain because solid data on the (m)RNA recovery rate (see below) and RT efficiency are difficult to obtain [42,43]. However, RT efficiency data are an integral part in the assessment of the RT-qPCR efficiency and both data are essential to determine the method detection limit. Quantitative determination of RNA species is not reliable without solid assessment of the RT-qPCR efficiency. It is pointed out that if relative comparisons are intended and one methodology is used in one single experiment, data on the RT-qPCR efficiency are welcome but not absolutely necessary. Comparison of different experiments, however, especially if performed in different laboratories, is critical. If the RT efficiency is not known and absolute quantification of RNA molecules is based on the possibly erroneous assumption that all (100%) of the targeted RNA molecules in the extract are transcribed to cDNA, respective values are speculative and most probably underestimated. Using a pre-quantified enterovirus RNA spike in similar approaches as described in Section 2.3 with the FastRNA® Pro Soil-Direct Kit (see Section 2.6), RNA recovery rates of 70% [4], 30% [21] and 20% [44] were reported, and other RNA extraction systems (RNeasy® Plant Mini Kit (QIAGEN), Strep Thermo-Fast® Plates (ABgene), RNaid® Plus Kit with SPIN™ (Bio 101), QIAamp® MinElute™ Virus Spin Kit (QIAGEN) and QIAamp® Viral RNA Mini Kit (QIAGEN)) yielded only up to 5% RNA recovery [44]. This illustrates the current difficulty of absolute RNA quantification. The scientific community is encouraged to invest more effort in this field.

3.3.2. RNA Qualification and Quantification for Metatranscriptome Sequencing

Metatranscriptome sequencing allows the taxonomic and functional analysis of metabolically active microbial communities on transcript level [45–47]. For this purpose, environmental samples are taken and total RNA is extracted. DNA removal, rRNA depletion and cDNA synthesis follow prior to sequencing [45,47–49]. Complete DNA removal is a prerequisite because left-over-DNA would be co-sequenced and falsify the sequencing results.

In contrast to RNA extraction protocol 6 aiming at maximizing RNA recovery to provide maximum RT-qPCR sensitivity and to perform representative community analysis, RNA extraction protocol 7 has been designed to extract longer and maximum quality RNA strands, for (meta)transcriptome analysis predominantly consisting of mRNA. Consequently, some RNA loss during the extraction procedure is taken into account. Alternative systems targeting poly-A tails are excluded since a large fraction of prokaryotic mRNA is devoid of poly-A tails.

rRNA removal is essential for metatranscriptome sequencing if considerably deeper sequencing and extensive data filtering is not considered. Messenger RNA (mRNA) in prokaryotic cells only has a share of 1%–20% of the total RNA, while the rest is mostly comprised of rRNA, tRNA and small RNAs. Due to its low percentage, specific depletion particularly of rRNA is important to increase the number of mRNA molecules that harbor the metabolic information of the microbial communities up to one order of magnitude [47,48,50]. The two most frequently used depletion approaches are either based on the enzymatic digestion of rRNAs or on hybridization of these. The enzymatic approach has its drawback in the depletion of degraded mRNAs because this can cause transcriptomic information loss. By contrast, the approach based on hybridization is reported to produce better results and to be more suitable for quantitative analysis [46,47].

For this purpose, hybridization based rRNA depletion using the Ribo-Zero rRNA Removal Kit (Bacteria) (Epicentre, Chicago, IL, USA) was chosen (see Section 2.6). Quantity and quality control of the RNA samples was done prior to and after rRNA depletion using the Agilent 2100 Bioanalyzer, resulting in electropherograms as shown in Figure 8. Figure 8A depicts an electropherogram of total RNA extracted from a biogas fermenter and Figure 8B the rRNA-depleted RNA. 23S rRNA has a peak at about 2900 nt and 16S rRNA at about 1540 nt, while mRNAs are typically shorter. The intensity of fluorescence (fluorescence units, FU) is positively correlated with the RNA concentration. For 16S rRNA, the intensity was approximately 67-fold higher prior to depletion than thereafter (approximately 200 FU compared to 3), while for the 23S rRNA fragments, the FU value dropped from approximately 100 to 3. This shows that the depletion was successful for this sample which was prepared following protocol 7.

For metatranscriptome sequencing using whole-community RNA, the RNA integrity number (RIN-value) is important. It indicates the integrity of the RNA and ranges from 1 (completely degraded) to 10 (high integrity, intact) [14]. A challenge is that the RIN value should not be lower than 7 even after rRNA depletion for metatranscriptome sequencing because library preparation and sequencing can be difficult in the presence of considerable amounts of short RNA fragments [14,48].

In contrast, 16S and 23S rRNA peaks were hardly visible in RNA preparations following protocol 6. The RIN value was 6.4 before (Figure 8A) and 5.4 after rRNA depletion (Figure 8B), and RNA yield was not sufficient for metatranscriptome analysis (not shown). If RNA fragmentation or other factors were reasons for the absence of rRNA peaks, precluding rRNA depletion with the given system (see Section 2.6), is a matter of current research. The small sample volume used (see Section 2.6) can explain the low RNA yield. Anyway, protocol 6 cannot be recommended for RNA extraction in the current form with rRNA depletion if metatranscriptome analysis is the aim. As noted by Pascault et al. [48], a convenient solution may be, considering the steadily falling sequencing price, to increase the sequencing depth and separate rRNA by bioinformatic tools instead of removing rRNA by technical means.

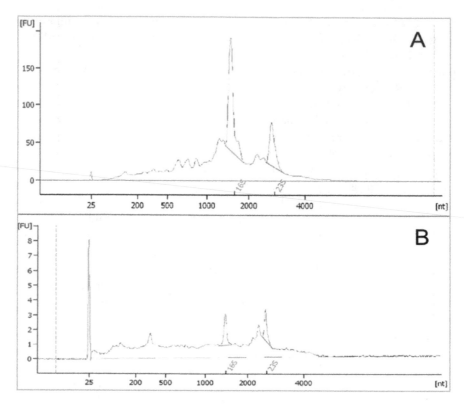

Figure 8. Electropherograms of whole-community RNA extracted from biogas fermenter sludge after DNase digestion. Whole-RNA before (**A**) and after (**B**) rRNA depletion using the Ribo-Zero rRNA Removal Kit (Bacteria) (Epicentre). Fluorescence units (FU) are positively correlated with the RNA concentration. nt: nucleotides.

4. Conclusions

An interlaboratory comparison of nucleic acid extraction, RNA preservation and quantitative Real-Time PCR (qPCR) based assays was conducted. Extracts compared in the ring-trial were derived from differently operated biogas fermenters digesting grass silage as difficult matrices. A kit based DNA extraction system with modifications such as upstream sample washing showed excellent DNA recovery rates. In order to optimize the trade-off between PCR suitability and DNA loss during multiple purification steps, less material could be processed with highly viscous fermenter sludge samples, but no post-purification was necessary. qPCR inhibition was observed only in very few cases in dilution assays as an exception. The system produced qPCR-suitable extracts and highly reproducible qPCR results in the ring-trial between five institutional partners, whereas a traditional chloroform-CTAB based protocol performed considerably worse in qPCR analyses of *Bacteria*, *Archaea* and methanogens. Classical absorption coefficients were of limited use to predict qPCR-suitability of nucleic acid extracts from highly viscous biogas fermenter sludges. Particularly, the ratio A260/230 was of no significance when GITC was present in the extracts. As compared to quantification of total DNA with the PicoGreen® system, using the extinction coefficient appeared to generate overestimated results particularly for the CSTR extracts obtained with the kit system, possibly due to interference of a co-extracted impurity. Quantitative data on total DNA contents in extracts from highly viscous fermenter sludges must thus be regarded with precaution, particularly if spectrophotometry is used.

Besides optimal primer design, special attention must be paid to the generation of a reliable standard with a plausible slope and an appropriate Y-intercept value. If this is not the case, considerable over- or underestimation in absolute quantification is the consequence. Respective data for the standard should be provided along with data for the extraction efficiency of the nucleic acid extraction system used.

Several methods were tested in an experiment to evaluate RNA preservation methods. Sample storage in RNA*later*® was at least as efficient as flash freezing and storage in liquid N_2. As compared to immediate processing of fresh samples, RNA loss of about 40% was observed if samples were stored in RNA*later*®. Although there is potential to optimize the system, RNA preservation with RNA*later*® can thus be recommended for field applications where sample shipping is an issue.

Considerable variation and poor reproducibility were observed for the tested total RNA extraction systems with the given fermenter sludge samples. It is not clear if the deficiencies are due to suboptimal extraction efficiency, compromised reverse transcription (RT) or both. Although plausible RT-qPCR results could be generated for distinct experiments, RNA yield and quality was insufficient after rRNA depletion for metatranscriptome analysis. The scientific community is encouraged to invest more effort in this field.

Acknowledgments: The following persons were supported by the German Federal Ministry of Education and Research (BMBF; grant 03SF0440) through Project Management Jülich (PTJ): Jaqueline Derenkó, Antje Rademacher, Bernhard Munk, Alexander Pechtl, Susanne Helbig and Yvonne Stolze. Parts of the work were supported by the Bavarian State Ministry for Nutrition, Agriculture and Forestry, grant K/08/06. The authors thank Carsten Jost and Jessica Behrend for their excellent technical and laboratory support, Elena Madge-Pimentel for excellent technical assistance and Irena Maus for her laboratory support.

Author Contributions: Jaqueline Derenkó, Bernhard Munk, Susanne Helbig, Alexander Pechtl and Yvonne Stolze performed laboratory experiments and wrote manuscript sections. Antje Rademacher, Michael Klocke, Steffen Prowe, Andreas Schlüter, Wolfgang H. Schwarz and Wolfgang Liebl were involved in conceptual work, coordination of individual experiments, supervision and proof reading of the manuscript. Michael Lebuhn performed laboratory experiments, wrote manuscript sections, coordinated the manuscript and was responsible for final editing.

Conflicts of Interest: The authors declare no conflict of interest.

References

1. Heid, C.A.; Stevens, J.; Livak, K.J.; Williams, P.M. Real time quantitative PCR. *Genome Res.* **1996**, *6*, 986–994. [CrossRef] [PubMed]

2. Josefsen, M.H.; Löfström, C.; Hansen, T.; Reynisson, E.; Hoorfar, J. Instrumentation and Fluorescent Chemistries Used in qPCR. In *Quantitative Real-Time PCR in Applied Microbiology*; Filion, M., Ed.; Caister Academic Press: Norfolk, UK, 2012; pp. 27–52.

3. Lebuhn, M.; Munk, B.; Effenberger, M. Agricultural biogas production in Germany—From practice to microbiology basics. *Energy Sustain. Soc.* **2014**, *4*. [CrossRef]

4. Munk, B.; Lebuhn, M. Process diagnosis using methanogenic *Archaea* in maize-fed, trace element depleted fermenters. *Anaerobe* **2014**, *29*, 22–28. [CrossRef] [PubMed]

5. Lebuhn, M.; Effenberger, M.; Gronauer, A.; Wilderer, P.A.; Wuertz, S. Using quantitative real-time PCR to determine the hygienic status of cattle manure. *Water Sci. Technol.* **2003**, *48*, 97–103. [PubMed]

6. Bergmann, I.; Mundt, K.; Sontag, M.; Baumstark, I.; Nettmann, E.; Klocke, M. Influence of DNA isolation on Q-PCR-based quantification of methanogenic *Archaea* in biogas fermenters. *Syst. Appl. Microbiol.* **2010**, *33*, 78–84. [CrossRef] [PubMed]

7. Henderson, G.; Cox, F.; Kittelmann, S.; Miri, V.H.; Zethof, M.; Noel, S.J.; Waghorn, G.C.; Janssen, P.H. Effect of DNA extraction methods and sampling techniques on the apparent structure of cow and sheep rumen microbial communities. *PLoS ONE* **2013**, *8*. [CrossRef]

8. Li, A.; Chu, Y.N.; Wang, X.; Ren, L.; Yu, J.; Liu, X.; Yan, J.; Zhang, L.; Wu, S.; Li, S. A pyrosequencing-based metagenomic study of methane-producing microbial community in solid-state biogas reactor. *Biotechnol. Biofuels* **2013**, *6*. [CrossRef] [PubMed]

9. Huggett, J.; Nolan, T.; Bustin, S.A. MIQE: Guidelines for the Design and Publication of a Reliable Real-time PCR Assay. In *Real-Time PCR: Advanced Technologies and Applications*; Caister Acad. Press: Norfolk, UK, 2013; pp. 247–258.

10. Johnson, G.; Nour, A.A.; Nolan, T.; Huggett, J.; Bustin, S. Minimum information necessary for quantitative real-time PCR experiments. In *Quantitative Real-Time PCR*; Springer: New York, NY, USA, 2014; pp. 5–17.

11. Quantitative Real-Time PCR: Methods and Protocols. *Methods in Molecular Biology*; Biassoni, R., Raso, A., Eds.; Springer Science & Business Media: New York, NY, USA, 2014; Volume 1160.

12. Sambrook, J.; Russel, D.W. *Molecular Cloning. A Laboratory Manual*, 3rd ed.; Cold Spring Harbor Laboratory Press, Cold Spring Harbor: New York, NY, USA, 2001; Volume 3.

13. Munk, B.; Bauer, C.; Gronauer, A.; Lebuhn, M. Population dynamics of methanogens during acidification of biogas fermenters fed with maize silage. *Eng. Life Sci.* **2010**, *10*, 496–508. [CrossRef]

14. Schroeder, A.; Mueller, O.; Stocker, S.; Salowsky, R.; Leiber, M.; Gassmann, M.; Lightfoot, S.; Menzel, W.; Granzow, M.; Ragg, T. The RIN: An RNA integrity number for assigning integrity values to RNA measurements. *BMC Mol. Biol.* **2006**, *7*. [CrossRef] [PubMed]

15. Bachoon, D.S.; Chen, F.; Hodson, R.E. RNA recovery and detection of mRNA by RT-PCR from preserved prokaryotic samples. *FEMS Microbiol. Lett.* **2001**, *201*, 127–132. [CrossRef] [PubMed]

16. Van Eijsden, R.G.E.; Stassen, C.; Daenen, L.; van Mulders, S.E.; Bapat, P.M.; Siewers, V.; Goossens, K.V.Y.; Nielsen, J.; Delvaux, F.R.; van Hummelen, P.; *et al.* A universal fixation method based on quaternary ammonium salts (RNA*later*) for omics-technologies: *Saccharomyces cerevisiae* as a case study. *Biotechnol. Lett.* **2013**, *35*, 891–900. [CrossRef] [PubMed]

17. Rissanen, A.J.; Kurhela, E.; Aho, T.; Oittinen, T.; Tiisola, M. Storage of environmental samples for guaranteeing nucleic acid yields for molecular microbiological studies. *Appl. Microbiol. Biotechnol.* **2010**, *88*, 977–984. [CrossRef] [PubMed]

18. Lebuhn, M.; Effenberger, M.; Garces, G.; Gronauer, A.; Wilderer, P.A. Evaluating real-time PCR for the quantification of distinct pathogens and indicator organisms in environmental samples. *Water Sci. Technol.* **2004**, *50*, 263–270. [PubMed]

19. Scott Adams, P. Data analysis and reporting. In *Real-time PCR*; Dorak, M.T., Ed.; Taylor & Francis Group: Abingdon, UK, 2006; pp. 39–62.

20. Schönberg, M.; Linke, B. The influence of the temperature regime on the formation of methane in a two-phase anaerobic digestion process. *Eng. Life Sci.* **2012**, *12*, 279–286. [CrossRef]

21. Munk, B.; Bauer, C.; Gronauer, A.; Lebuhn, M. A metabolic quotient for methanogenic *Archaea*. *Water Sci. Technol.* **2012**, *66*, 2311–2317. [CrossRef] [PubMed]

22. Nettmann, E.; Bergmann, I.; Mundt, K.; Linke, B.; Klocke, M. Archaea diversity within a commercial biogas plant utilizing herbal biomass determined by 16S rDNA and mcrA analysis. *J. Appl. Microbiol.* **2008**, *105*, 1835–1850. [CrossRef] [PubMed]

23. Stolze, Y.; Zakrzewski, M.; Maus, I.; Eikmeyer, F.; Jaenicke, S.; Rottmann, N.; Siebner, C.; Pühler, A.; Schlüter, A. Comparative metagenomics of biogas-producing microbial communities from production-scale biogas plants operating under wet or dry fermentation conditions. *Biotechnol. Biofuels* **2015**, *8*. [CrossRef] [PubMed]

24. Zoetendal, E.G.; Booijink, C.C.; Klaassens, E.S.; Heilig, H.G.; Kleerebezem, M.; Smidt, H.; de Vos, W.M. Isolation of RNA from bacterial samples of the human gastrointestinal tract. *Nat. Protoc.* **2006**, *1*, 954–959. [CrossRef] [PubMed]

25. Urich, T.; Lanzén, A.; Qi, J.; Huson, D.H.; Schleper, C.; Schuster, S.C. Simultaneous assessment of soil microbial community structure and function through analysis of the meta-transcriptome. *PLoS ONE* **2008**, *3*. [CrossRef] [PubMed]

26. Yu, Y.; Lee, C.; Kim, J.; Hwang, S. Group-specific primer and probe sets to detect methanogenic communities using quantitative real-time polymerase chain reaction. *Biotechnol. Bioeng.* **2005**, *89*, 670–679. [CrossRef] [PubMed]

27. Klocke, M.; Nettmann, E.; Bergmann, I.; Mundt, K.; Souidi, K.; Mumme, J.; Linke, B. Characterization of the methanogenic *Archaea* within two-phase biogas reactor systems operated with plant biomass. *Syst. Appl. Microbiol.* **2008**, *31*, 190–205. [CrossRef] [PubMed]

28. Fey, A.; Eichler, S.; Flavier, S.; Christen, R.; Höfle, M.G.; Guzmán, C.A. Establishment of a real-time PCR-based approach for accurate quantification of bacterial RNA targets in water, using *Salmonella* as a model organism. *Appl. Environ. Microbiol.* **2004**, *70*, 3618–3623. [CrossRef] [PubMed]

29. Bauer, C.; Korthals, M.; Gronauer, A.; Lebuhn, M. Methanogens in biogas production from renewable resources—A novel molecular population analysis approach. *Water Sci. Technol.* **2008**, *58*, 1433–1439. [CrossRef] [PubMed]

30. Fierer, N.; Jackson, J.; Vilgalys, R.; Jackson, R.B. Assessment of soil microbial community structure by use of taxon-specific quantitative PCR assays. *Appl. Environ. Microbiol.* **2005**, *71*, 4117–4120. [CrossRef] [PubMed]

31. Karakashev, D.; Batstone, D.J.; Angelidaki, I. Influence of environmental conditions on methanogenic compositions in anaerobic biogas reactors. *Appl. Environ. Microbiol.* **2005**, *71*, 331–338. [CrossRef] [PubMed]

32. Pervin, H.M.; Dennis, P.G.; Lim, H.J.; Tyson, G.W.; Batstone, D.J.; Bond, P.L. Drivers of microbial community composition in mesophilic and thermophilic temperature-phased anaerobic digestion pre-treatment reactors. *Water Res.* **2013**, *47*, 7098–7108. [CrossRef] [PubMed]

33. Pap, B.; Györkei, Á.; Boboescu, I.Z.; Nagy, I.K.; Bíró, T.; Kondorosi, É.; Maróti, G. Temperature-dependent transformation of biogas-producing microbial communities points to the increased importance of hydrogenotrophic methanogenesis under thermophilic operation. *Bioresour. Technol.* **2015**, *177*, 375–380. [CrossRef] [PubMed]

34. Lebuhn, M.; Hanreich, A.; Klocke, M.; Schlüter, A.; Bauer, C.; Marín Pérez, C. Towards molecular biomarkers for biogas production from lignocellulose-rich substrates. *Anaerobe* **2014**, *29*, 10–21. [CrossRef] [PubMed]

35. Bustin, S.; Benes, V.; Garson, J.; Hellemans, J.; Huggett, J.; Kubista, M.; Mueller, R.; Nolan, T.; Pfaffl, M.W.; Shipley, G.L.; *et al.* The MIQE guidelines: Minimum information for publication of quantitative real-time PCR experiments. *Clin. Chem.* **2009**, *55*, 611–622. [CrossRef] [PubMed]

36. Hospodsky, D.; Yamamoto, N.; Peccia, J. Accuracy, precision, and method detection limits of quantitative PCR for airborne bacteria and fungi. *Appl. Environ. Microbiol.* **2010**, *76*, 7004–7012. [CrossRef] [PubMed]

37. Garcés, G.; Effenberger, M.; Najdrowski, M.; Gronauer, A.; Wilderer, P.A.; Lebuhn, M. Quantitative real-time PCR for detecting *Cryptosporidium parvum* in cattle manure and anaerobic digester samples—Methodological advances in DNA extraction. In Proceedings of the 8th Latin American Workshop and Symposium on Anaerobic Digestion, Punta del Este, Uruguay, 2–5 October 2005; pp. 68–73.

38. Garcés, G.; Effenberger, M.; Najdrowski, M.; Wackwitz, C.; Gronauer, A.; Wilderer, P.A.; Lebuhn, M. Quantification of *Cryptosporidium parvum* in anaerobic digesters treating manure by (reverse-transcription) quantitative real-time PCR, infectivity and excystation tests. *Water Sci. Technol.* **2006**, *53*, 195–202. [CrossRef] [PubMed]

39. Wang, S.S.; Sherman, M.E.; Rader, J.S.; Carreon, J.; Schiffman, M.; Baker, C.C. Cervical Tissue collection Methods for RNA Preservation Comparison of Snap-frozen, Ethanol-fixed and RNA*later*-fixation. *Diagn. Mol. Pathol.* **2006**, *15*, 144–148. [CrossRef] [PubMed]

40. Riesgo, A.; Pérez-Porro, A.R.; Carmona, S.; Leys, S.P.; Giribet, G. Optimization of preservation and storage time of sponge tissues to obtain quality mRNA for next-generation sequencing. *Mol. Ecol. Res.* **2012**, *12*, 312–322. [CrossRef] [PubMed]

41. Mutter, G.L.; Zahrieh, D.; Liu, C.; Neuberg, D.; Finkelstein, D.; Baker, H.E.; Warrington, J.A. Comparison of frozen and RNA*later* solid tissue storage methods for use in RNA expression microarrays. *BMC Genom.* **2004**, *5*. [CrossRef] [PubMed]

42. Gallup, J.M. qPCR inhibition and amplification of difficult templates. In *PCR Troubleshooting and Optimization, the Essential Guide*; Kennedy, S., Oswald, N., Eds.; Caister Academic Press: Norfolk, UK, 2011; pp. 23–65.

43. Lebuhn, M.; Weiß, S.; Munk, B.; Guebitz, G.M. Microbiology and Molecular Biology Tools for Biogas Process Analysis, Diagnosis and Control. In *Biogas Science and Technology: Advances in Biochemical Engineering/Biotechnology*; Guebitz, G.M., Ed.; Springer: Cham, Switzerland, 2015; Volume 151, pp. 1–40.

44. Lebuhn, M.; Wilderer, P. Abschlussbericht des StMUGV-Projekts "Biogastechnologie zur umweltverträglichen Flüssigmistverwertung und Energiegewinnung in Wasserschutzgebieten: wasserwirtschaftliche und hygienische Begleituntersuchung, Projektteil: Mikrobiologische, parasitologische und virologische Untersuchungen". Technische Universität München, Lehrstuhl für Siedlungswasserwirtschaft. 2006. Available online: http://www.sww.bgu.tum.de/fileadmin/w00bom/www/_migrated_content_uploads/StMUGV-Abschlussbericht_2006_Lebuhn.pdf (accessed on 28 December 2015).

45. Su, C.; Lei, L.; Duan, Y.; Zhang, K.Q.; Yang, J. Culture-independent methods for studying environmental microorganisms: Methods, application, and perspective. *Appl. Microbiol. Biotechnol.* **2012**, *93*, 993–1003. [CrossRef] [PubMed]

46. Filiatrault, M.J. Progress in prokaryotic transcriptomics. *Curr. Opin. Microbiol.* **2011**, *14*, 579–586. [CrossRef] [PubMed]

47. Sorek, R.; Cossart, P. Prokaryotic transcriptomics: A new view on regulation, physiology and pathogenicity. *Nat. Rev. Genet.* **2010**, *11*, 9–16. [CrossRef] [PubMed]

48. Pascault, N.; Loux, V.; Derozier, S.; Martin, V.; Debroas, D.; Maloufi, S.; Humbert, J.F.; Leloup, J. Technical challenges in metatranscriptomic studies applied to the bacterial communities of freshwater ecosystems. *Genetica* **2014**, *143*, 157–167. [CrossRef] [PubMed]

49. Mäder, U.; Nicolas, P.; Richard, H.; Bessières, P.; Aymerich, S. Comprehensive identification and quantification of microbial transcriptomes by genome-wide unbiased methods. *Curr. Opin. Biotechnol.* **2011**, *22*, 32–41. [CrossRef] [PubMed]

50. Moran, M.A. Metatranscriptomics: Eavesdropping on complex microbial communities. *Microbe* **2009**, *4*, 329–335.

4

Rebalancing Redox to Improve Biobutanol Production by *Clostridium tyrobutyricum*

Chao Ma [1], Jianfa Ou [1], Ningning Xu [1], Janna L. Fierst [2], Shang-Tian Yang [3] and Xiaoguang (Margaret) Liu [1,*]

Academic Editors: Mark Blenner and Michael D. Lynch

[1] Department of Chemical and Biological Engineering, The University of Alabama, 245 7th Avenue, Tuscaloosa, AL 35401, USA; cma3@crimson.ua.edu (C.M.); jou4@crimson.ua.edu (J.O.); nxu2@crimson.ua.edu (N.X.)

[2] Department of Biological Science, The University of Alabama, 300 Hackberry Lane, Tuscaloosa, AL 35487, USA; janna.l.fierst@ua.edu

[3] Department of Chemical and Biomolecular Engineering, The Ohio State University, 151 West Woodruff Avenue, Columbus, OH 43210, USA; yangst.osu.2013@gmail.com

* Correspondence: mliu@eng.ua.edu

Abstract: Biobutanol is a sustainable green biofuel that can substitute for gasoline. Carbon flux has been redistributed in *Clostridium tyrobutyricum* via metabolic cell engineering to produce biobutanol. However, the lack of reducing power hampered the further improvement of butanol production. The objective of this study was to improve butanol production by rebalancing redox. Firstly, a metabolically-engineered mutant CTC-*fdh-adhE2* was constructed by introducing heterologous formate dehydrogenase (*fdh*) and bifunctional aldehyde/alcohol dehydrogenase (*adhE2*) simultaneously into wild-type *C. tyrobutyricum*. The mutant evaluation indicated that the *fdh*-catalyzed NADH-producing pathway improved butanol titer by 2.15-fold in the serum bottle and 2.72-fold in the bioreactor. Secondly, the medium supplements that could shift metabolic flux to improve the production of butyrate or butanol were identified, including vanadate, acetamide, sodium formate, vitamin B12 and methyl viologen hydrate. Finally, the free-cell fermentation produced 12.34 g/L of butanol from glucose using the mutant CTC-*fdh-adhE2*, which was 3.88-fold higher than that produced by the control mutant CTC-*adhE2*. This study demonstrated that the redox engineering in *C. tyrobutyricum* could greatly increase butanol production.

Keywords: *Clostridium tyrobutyricum*; butanol production; redox engineering; metabolic cell engineering; metabolic shift

1. Introduction

n-Butanol is a potential substitute for gasoline, a raw material to generate bio-jet fuel and an important industrial chemical. Biobutanol has been produced by solventogenic Clostridia, such as *C. acetobutylicum* and *C. beijerinckii*, in acetone-butanol-ethanol (ABE) fermentation [1,2]. Extensive metabolic engineering and fermentation development have improved butanol production [3–6]. However, ABE fermentation still suffers from the relatively low butanol yield, titer and productivity due to the butanol inhibition and the complicated metabolic pathway involved in acidogenesis, solventogenesis and sporulation.

One alternative strategy for biobutanol production is to synthesize a heterologous butanol biosynthesis pathway from solventogenic Clostridia in other microbes, including *Escherichia coli*, *Saccharomyces cerevisiae*, *Pseudomonas putida*, *Bacillus subtilis* and *Lactobacillus brevis* [7–12]. For instance, integrated synthetic biology, carbon engineering and redox engineering have been performed in

E. coli, which produced 14 g/L of n-butanol [13]. Despite these achievements, the inability to ferment low-cost feedstock, the low efficiency to catabolize the hydrolysate of biomass, the low butanol tolerance or the low butanol productivity by these mutants has limited their industrial applications in biobutanol production.

C. tyrobutyricum, an anaerobic acidogenic strain naturally producing butyrate, acetate, CO_2 and H_2 [14,15], is a promising bacterium for biobutanol production due to its advantages of high butanol tolerance, relatively simple metabolic pathway and high conversion rate of butyryl-CoA from sugars [16–18]. Recently, the butanol-producing mutants CTC-*adhE2* and ACKKO-*adhE2* were developed by expressing a bifunctional aldehyde/alcohol dehydrogenase (*adhE2*) from *C. acetobutylicum* ATCC 824 in wild-type and high butyrate-producing mutants of *C. tyrobutyricum,* respectively. The free-cell fermentation produced 2.5–6.5 g/L of butanol by CTC-*adhE2* and 14.5–16 g/L of butanol by ACKKO-*adhE2* from glucose [16,17]. Advanced fermentation processes, such as using mannitol as a substrate or immobilizing mutant cells in a fibrous bed bioreactor supplemented with methyl viologen hydrate, were also developed to further improve biobutanol titer to 20 g/L [16,18].

Although rebalancing carbon resulted in high butanol production by *C. tyrobutyricum,* the lack of reducing power is still a challenge to further improve biobutanol production [17,19]. Redox engineering has been applied to improve the accumulation of NADH in *E. coli, S. cerevisiae* and *Klebsiella pneumonia* [10,13,20–22]. For example, the overexpression of formate dehydrogenase (*fdh*) improved biobutanol titer from 0.2 g/L–0.52 g/L by *E. coli* and from ~0.6 g/L–0.9 g/L by *K. pneumonia* [10,22]. The redox engineering via introducing Fdh into *C. tyrobutyricum* has not been evaluated so far [19].

The objective of this study was to improve biobutanol production via rebalancing redox in *C. tyrobutyricum* (Figure 1). Redox engineering was performed to construct a new mutant CTC-*fdh-adhE2,* and the components that boosted its butanol production were also identified. The free-cell butanol fermentation indicated that redox rebalance, in addition to carbon rebalance, is an efficient approach to significantly improve butanol production.

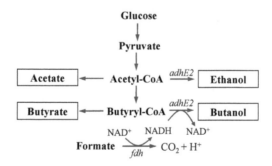

Figure 1. Butanol producing metabolic pathway in *C. tyrobutyricum.* The wild-type CTC produces butyrate and acetate; mutant CTC-*adhE2* produces butanol and ethanol via carbon rebalance by overexpressing the *adhE2* gene [16]; and mutant CTC-*fdh-adhE2* constructed in this study shows improved butanol production via redox rebalance by overexpressing the *fdh* gene.

2. Materials and Methods

2.1. Strains and Media

As listed in Table 1, wild-type, mutant CTC-*adhE2* and mutant CTC-*fdh-adhE2* of *C. tyrobutyricum* ATCC 25755 were used. The wild-type strain was purchased from ATCC (ATCC, Manassas, VA, USA). The control mutant (CTC-*adhE2*) was obtained from Yang's lab [23]. In this study, the mutant CTC-*fdh-adhE2* was constructed by simultaneously expressing the heterologous formate dehydrogenase gene (*fdh*) and bifunctional aldehyde/alcohol dehydrogenase (*adhE2*) in wild-type *C. tyrobutyricum.*

Table 1. Strains and plasmids used in this study.

Plasmid/Strain	Relevant Characteristics	Reference/Source
Plasmids		
pMTL-adhE2	*adhE2* overexpression with *thl* promoter	This study
pMTL-fdh-adhE2	*fdh* and *adhE2* overexpression with *thl* promoter	This study
Strains		
C. tyrobutyricum	Clostridium, ATCC 25755, wild-type	ATCC
CTC-adhE2	Clostridium with *adhE2* overexpression, thiamphenicol resistant	[23]
CTC-fdh-adhE2	Clostridium with *fdh* and *adhE2* overexpression, thiamphenicol resistant	This study
E. coli CA434	E. coli HB101 with plasmid R702, kanamycin resistant	[24]

The seed culture of *C. tyrobutyricum* was maintained anaerobically in reinforced Clostridial medium (RCM; Difco, Kansa City, MO, USA). The 30 μg/mL of thiamphenicol (Tm, Alfa Aesar, Ward Hill, MA, USA) was used to select and maintain mutants. In serum bottle and bioreactor fermentations, the modified Clostridial growth medium (CGM) containing 40 g/L of glucose was used [25]. Antibiotics were added to the baseline study in the serum bottle, but not added to media study and bioreactor fermentation. *E. coli* was grown aerobically in Luria–Bertani (LB) media supplemented with 30 μg/mL of chloramphenicol (Cm, Alfa Aesar) or 50 μg/mL of kanamycin (Kan, Alfa Aesar). The chemical reagents were purchased from Sigma-Aldrich (St. Louis, MO, USA), unless otherwise specified.

2.2. Plasmid Construction

The flowchart of plasmid construction is described in Figure 2. Specifically, the pMTL007 plasmid obtained from Minton's lab [26] was used as an expression vector. The *fdh* gene was amplified from *Moorella thermoacetica* ATCC 39073 (ATCC, Manassas, VA, USA) using Q5 High-Fidelity DNA Polymerase (NEB, Ipswich, MA, USA). The plasmid pMTL-*adhE2* was constructed following a previous publication [23]. The promoter of the homologous *thl* gene (Pthl) cloned from *C. tyrobutyricum*, the *fdh* gene and pMTL-*adhE2* backbone were assembled to generate the plasmid pMTL-*fdh-adhE2* using the Gibson Assembly kit (NEB) following the manufacture's instruction.

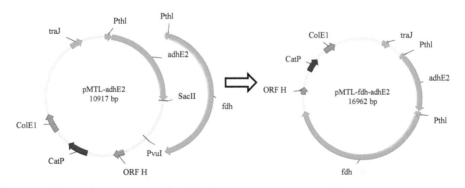

Figure 2. Plasmid construction. *ORF H*: *pCB102* replicon from *Clostridium butyricum*; *CatP*: chloramphenicol- and thiamphenicol-resistant gene; *ColE1*: *E. coli* replicon; *fdh*: formate dehydrogenase from *Moorella thermoacetica*; *Pthl*: promoter of thiolase from *Clostridium tyrobutyricum*; *traJ*: regulator of the F plasmid transfer operon; *adhE2*: bifunctional aldehyde/alcohol dehydrogenase (*adhE2*) from *Clostridium acetobutylicum*.

2.3. Mutant Construction

The transformation of plasmid pMTL-*fdh-adhE2* into the wild-type *C. tyrobutyricum* was performed via conjugation in an anaerobic chamber according to a previous publication [23,24] with the following modifications. The *E. coli CA434* [24] was transformed with pMTL-*fdh-adhE2* and cultivated in LB medium containing 30 µg/mL of Cm and 50 µg/mL of Kan. The *E. coli CA434*/pMTL-*fdh-adhE2* was harvested as donor cells when the optical density at 600 nm (OD_{600}) reached ~1.5. The wild-type cells were grown in RCM medium and collected as recipient cells when OD_{600} reached 1.5–3.0. The transformed clostridial cells were spread on RCM selection plates that contained 30 µg/mL of Tm and 250 µg/mL of D-cycloserine (Cy, Alfa Aesar). The selection plates were incubated at 37 °C for 72–96 h or until colonies appeared. Twenty colonies were picked and evaluated in a 50-mL serum bottle culture to screen the clone with the highest butanol production, which was named CTC-*fdh-adhE2*.

2.4. Medium Supplements Screening

To engineer the butanol fermentation process, medium optimization was performed in the small-scale fermentations of CTC-*fdh-adhE2* in serum bottles. In this study, ten components at two levels (*i.e.*, concentrations) were analyzed using the new mutant CTC-*fdh-adhE2*. Only the representative conditions that enhanced the production of butyrate or butanol are summarized in Table 2. Fresh seed culture with an OD_{600} of ~1.5 was used to inoculate the 50 mL CGM medium containing 40 g/L of glucose and different medium supplements. The cultures with inoculation OD_{600} of ~0.04 were anaerobically incubated at 37 °C without pH adjustment. The samples were taken daily to monitor cell growth, glucose consumption and product formation. All conditions were carried out in duplicate, and data are reported as means with standard deviations.

Table 2. Medium component screening to rebalance redox and carbon.

No.	Sodium Formate (1 g/L)	Vitamin B12 (0.001 g/L)	Methyl Viologen Hydrate (0.1 g/L)	Vanadate (1 g/L)	Acetamide (1 g/L)
C	−	−	−	−	−
1	+	−	−	−	−
2	−	+	−	−	−
3	−	−	+	−	−
4	+	+	−	−	−
5	+	−	+	−	−
6	+	+	+	−	−
7	+	−	−	+	−
8	+	−	−	−	+
9	+	+	−	+	−
10	+	+	−	−	+

2.5. Butanol Fermentation

The kinetics studies of the fed-batch fermentation by CTC-*adhE2* (control) and CTC-*fdh-adhE2* mutants were performed in a stirred-tank bioreactor (FS-01-A; Major science, Saratoga, CA, USA). Fermentation setup, operation and sampling were performed as reported previously [17]. Fermentations were run in duplicate, and the means and standard deviations were calculated.

2.6. Activity Assay of Formate Dehydrogenase

The preparation of cell extract and the measurement of formate dehydrogenase activity were performed as reported with modification [27,28]. The cells of CTC-*adhE2* and CTC-*fdh-adhE2* were collected when the OD_{600} reached ~2.0, centrifuged, washed and cooled in an anaerobic chamber. The suspended cells on ice were sonicated for 15 min using a sonifier (Models 250; Branson Ultrasonics,

Danbury, CT, USA) with 40% power and a 30 s interval for cooling each minute. The concentration of total protein in cell extract was titrated using Bradford's kit (Bio-Rad, Hercules, CA, USA) with bovine serum albumin as the standard.

The activity assay of formate dehydrogenase was performed at 37 °C in an anaerobic chamber. Specifically, 0.1 mL of cell extract was added to 1 mL of pre-mixed potassium phosphate buffer containing 1.67 mM of NAD^+ and 167 mM of sodium formate. The kinetics of NADH generation was monitored by measuring the absorbance at 340 nm at the reaction times of 10, 20, 30, 60, 90 and 120 s using a spectrophotometer (Biomate3; Thermo Fisher Scientific, Hudson, NH, USA). The concentration of NADH was calculated using a specific absorbance coefficient of $6220 \ M^{-1} \cdot cm^{-1}$. One unit of the activity of formate dehydrogenase was defined as 1 µmol of NADH produced per minute at pH 7.5 and 37 °C. The specific activity of Fdh was calculated from the measured activity divided by total protein in the cell extract.

2.7. Butanol Tolerance

To evaluate the effect of redox engineering on butanol tolerance, the wild-type, CTC-*adhE2* and CTC-*fdh-adhE2* were cultivated in serum bottles with a seeding OD_{600} of 0.3. Butanol at final concentrations of 0, 5, 10, 15 and 20 g/L was added to the basal CGM medium. The cell growth was analyzed by sampling serum bottles every 3 h. The butanol inhibition was modeled by the equation of $\mu = \mu_{max} Kp/(Kp + P)$, where μ is the specific growth rate (h^{-1}), Kp is the inhibition rate constant (g/L) and P is the butanol concentration (g/L).

2.8. NADH Assay

An NAD/NADH quantitation colorimetric kit (BioVision, San Francisco, CA, USA) was used to measure the intracellular concentration of NADH and NAD^+ in the CTC, CTC-*adhE2* and CTC-*fdh-adhE2*. The cells were cultivated in 50 mL CGM medium containing 1 g/L of sodium formate in serum bottles with duplication. The cultures were sampled in mid-log phase, and 1-mL samples were centrifuged at 10,000 rpm for 10 min. The cell pellets were re-suspended in 400 µL of NADH/NAD extraction buffer provided in the kit. The NADH and NAD^+ were extracted by freeze/thaw for two cycles (20 min on dry-ice followed by 10 min at room temperature in each cycle). To titrate NADH, NAD^+ was decomposed by heating the cell lysate at 60 °C by 30 min. After cooling on ice for 10 min, the NADH samples were centrifuged at 4000 rpm for 2 min. The 50 µL of supernatant were mixed with 100 µL of NAD^+ cycling enzyme provided in the kit. The reaction mixture was incubated at room temperature for 4 h before the value of OD_{450} was read. To measure the concentration of NAD^+ and NADH, 4 µL of cell lysate was mixed with 46 µL of NADH/NAD extraction buffer and 100 µL of NAD cycling enzyme. The NADH concentration was calculated using the equation "Concentration$_{NADH}$ (mM) = Mole$_{NADH}$ /Volume$_{wet \ cell}$ = Mole$_{NADH}$ /(OD_{600}/mL × 0.38 g-dry cell/L × 10^{-3} L/mL × 4.27 mL/g-wet cell × 10^{-3} g/mg)".

3. Analytical Methods

The cell density was analyzed by measuring the OD_{600} of the cell suspension. The concentrations of fermentation substrate and products, including glucose, butanol, butyrate, acetate and ethanol, were analyzed using high performance liquid chromatography (HPLC, Shimadzu, Columbia, MD, USA) following previously-published methods [17]. The glucose concentration was also analyzed daily using a YSI 2700 Select Biochemistry Analyzer (YSI Life Sciences, Yellow Springs, OH, USA) to determine the feeding strategy.

4. Results

4.1. Construction and Evaluation of Redox-Engineered Mutant

In this study, the heterologous formate dehydrogenase gene (*fdh*) was overexpressed together with *adhE2* in wild-type *C. tyrobutyricum* to generate the new mutant CTC-*fdh-adhE2*. The enzyme AdhE2 catalyzed the formation of butanol from butyryl-CoA, which had been demonstrated in the previous study [23]. In this study, the enzyme Fdh was expressed to rebalance redox by generating more reducing power, *i.e.*, NADH.

The mutant CTC-*fdh-adhE2* was characterized using PCR, an enzyme activity assay and a NADH assay. The heterologous *fdh* gene was successfully amplified from CTC-*fdh-adhE2* in the PCR reaction, indicating that the pMTL-*fdh-adhE2* plasmid was transformed into *C. tyrobutyricum*. The enzyme assay showed that the specific enzymatic activity of the NAD^+-dependent Fdh was 990 U/g in CTC-*fdh-adhE2* and zero in CTC-*adhE2*. These results confirmed that the heterologous fdh was expressed properly in the CTC-*fdh-adhE2*. In addition, the NADH assay demonstrated that the intracellular NADH concentration was increased from 0.11 mM in CTC-*adhE2* to 0.28 mM in CTC-*fdh-adhE2*, as shown in Figure 3. The mole ratio between NADH and NAD^+ was also enhanced in CTC-*fdh-adhE2*. These results confirmed that the synthesized Fdh enzyme boosted the intracellular accumulation of NADH.

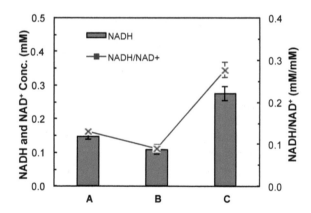

Figure 3. NADH assay. A: CTC wild-type; B: CTC-*adhE2*; and C: CTC-*fdh-adhE2*. ■: mole ratio between NADH and NAD^+. Conc.: concentration.

The baselines of butanol production of wild-type (Control 1), CTC-*adhE2* (Control 2), and CTC-*fdh-adhE2* were created using a 100-mL culture in serum bottles. The modified CGM medium was used without the addition of formate. As presented in the kinetics profiles in Figure 4, all three cells grew immediately after inoculation and reached a similar maximum OD_{600}. The fermentation data showed that the wild-type produced 9.02 g/L of butyrate and no butanol, as expected. The CTC-*adhE2* produced 8.78 g/L of butyrate and 2.51 g/L of butanol, and the CTC-*fdh-adhE2* produced 6.40 g/L of butyrate and 5.41 g/L of butanol. The CTC-*fdh-adhE2* also produced 2.15-fold higher butanol than CTC-*adhE2*. These results indicated that the overexpression of the heterologous gene *fdh* in CTC-*fdh-adhE2* shifted carbon flux from butyrate to butanol.

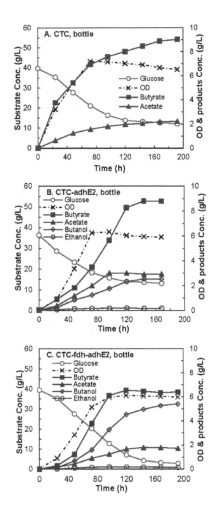

Figure 4. Kinetics butanol fermentation by *C. tyrobutyricum* in serum bottles. **(A)** CTC wild-type; **(B)** CTC-*adhE2*; and **(C)** CTC-*fdh-adhE2*. ○: glucose; ×: OD; ■: butyrate; ▲: acetate; ◇: butanol; ○: ethanol.

The butanol production consumes reducing power NAD(P)H in *C. tyrobutyricum*, and it is important to evaluate the feasibility of improving butanol production via rebalancing redox. In this study, we introduced an NADH-producing pathway by overexpressing an NAD$^+$-dependent enzyme Fdh in CTC-*adhE2*. NADH can be produced from formate through the synthesized reaction "Formate + NAD$^+$ ⇌ NADH + CO$_2$ + H$^+$". The increase of butanol production by CTC-*fdh-adhE2* indicated that the expression of Fdh generated more NADH and, thus, benefited butanol production.

In addition, it was noted that the butanol production was improved by CTC-*fdh-adhE2* even though no additional formate was added to the medium. Our proteomics study showed that pyruvate-formate lyase (*pfl*) was expressed in *C. tyrobutyricum* (data not shown). The enzyme *pfl* catalyzes the reversible reaction "pyruvate + CoA ⇌ formate + acetyl-CoA". Because pyruvate was produced from glucose through the glycolysis pathway, formate could be intracellularly produced from pyruvate. The generated formate could be consumed by Fdh to boost the NADH and improve butanol production.

4.2. Evaluation of Medium Supplements

Medium optimization, an important process engineering strategy, was carried out to screen the components that could increase the formation of butanol. The average data of the final titers of acetate, butyrate, ethanol and butanol by CTC-*fdh-adhE2* are described in Figure 5. The control culture without any medium supplement produced 5.35 g/L of butanol. Some components increased butanol up to 8.00 g/L, but some components increased butyrate up to 9.00 g/L. The ten experimental conditions were classified into two groups (Table 2 and Figure 5), including one that improved butanol production and a second that enhanced butyrate production.

In the first group, either sodium formate (SF, 1 g/L, Condition 1) or methyl viologen hydrate (MVH, 0.1 g/L, Condition 3) improved the butanol production. As discussed above, formate increased butanol production because the Fdh enzyme synthesized more NADH. MVH is an artificial electron carrier that can boost electron supply from ferredoxin, so it increased butanol production by providing more reducing power. MVH also used to improve butanol production by immobilized *C. tyrobutyricum* in a recent study [18]. A previous study reported that B12 improved the cell growth of *C. acetobutylicum* [29], but did not investigate the effect of B12 on butanol production. This study showed that vitamin B12 (B12, 0.001 g/L) did not change butanol production (Condition 2), although it improved the maximum OD by 36% compared to the control.

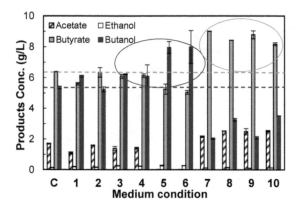

Figure 5. Screening of medium components to rebalance carbon and redox flux in CTC-*fdh-adhE2*. C: control without medium supplement; 1: + sodium formate (formate); 2: + vitamin B12 (B12); 3: + methyl viologen hydrate (MVH); 4: + formate + B12; 5: + formate + MVH; 6: + formate + B12 + MVH; 7: + formate + vanadate; 8: + formate + acetamide; 9: + formate + B12 + vanadate; and 10: + formate + B12 + acetamide. □: acetate; □: ethanol; ■: butyrate; ■: butanol. The experimental design is given in Table 2. The data are the means of the replicated fermentations in serum bottles.

The combination of formate, B12 and MVH (Condition 6) improved the butanol titer to 8.00 g/L and reduced the butyrate titer to 5.05 g/L compared to the control condition. The selectivity of butanol was also improved from 0.39 g/g (control) to 0.60 g/g. Interestingly, the supplement of these three components significantly reduced the acetate production to 0.01 g/L and improved C4 (butyrate and butanol) selectivity to 0.97 g/g, while the control culture produced acetate with a titer of 1.70 g/L and C4 (butyrate and butanol) with a selectivity of 0.87 g/g. Taken together, the addition of formate, B12 and MVH to CGM medium not only shifted more carbon flux from butyrate to butanol, but also redirected carbon from C2 to C4. We also observed that the addition of any of these three components (Conditions 1–3) did not reduce the acetate formation, but the combination of formate and MVH (Conditions 5 and 6) blocked the production of acetate. We hypothesized that a synergistic effect between formate and MVH shifted carbon flux from C2 to C4.

In the second group, the addition of vanadate and acetamide increased the production of butyrate and acetate. Acetamide could provide an extra nitrogen source, and vanadate could affect the metabolism of sugar phosphate. Both components shifted carbon from butanol to butyrate. Other key findings that are not described in Table 2 or Figure 5 include: (1) sodium pyruvate (1 g/L) did not increase the production of solvents and acids; and(2) manganese sulfate (0.1 g/L) increased acetate production (2.55 *vs.* 1.70 g/L by the control) because the high activity of acetate kinase needs manganese [30].

Very interestingly, both butanol and ethanol formation pathways consumed NADH, but neither the expression of Fdh nor the medium supplements improved the production of ethanol, *i.e.*, 0.28 g/L in Condition 6 supplemented with formate, B12 and MVH *vs.* 0.14 g/L in the control. Our previous proteomics study of *C. tyrobutyricum* showed that the expression of alcohol dehydrogenase (*adh*) was very low, while the expression of butanol dehydrogenase (*bdh*) was pretty high [17]. Therefore, the redox engineering increased the production of butanol, but not ethanol due to the low expression of *adh* and the low metabolic flux distributed to ethanol.

4.3. Butanol Fermentation

Both CTC-*adhE2* and CTC-*fdh-adhE2* were evaluated in 2-L fed-batch fermentations in a stirred-tank bioreactor at 37 °C and pH 6.0 using CGM medium supplemented with the identified components. As shown in Figure 6, four conditions were investigated, including CTC-*adhE2* supplemented with SF (Figure 6A), CTC-*adhE2* with sodium formate (SF), vitamin B12 (B12) and methyl viologen hydrate (MVH) (Figure 6B), CTC-*fdh-adhE2* with SF (Figure 6C) and CTC-*fdh-adhE2* with SF, B12 and MVH (Figure 6D). The kinetics profiles in Figure 5 showed that there was no obvious lag phase in all fermentations. As presented in Table 3, all four conditions had similar specific growth rates of 0.15–0.18 h^{-1} and a maximum OD$_{600}$ of 7.5–9.0.

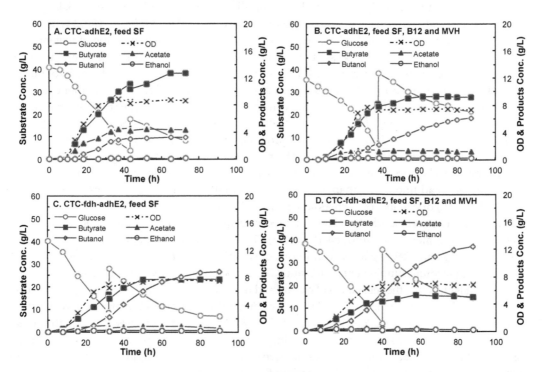

Figure 6. Kinetics of butanol fermentation by *C. tyrobutyricum* in 2-L bioreactors at pH 6.0, 37 °C and 100 rpm. (**A**) CTC-*adhE2* with sodium formate (SF); (**B**) CTC-*adhE2* with sodium formate, vitamin B12 (B12) and methyl viologen hydrate (MVH); (**C**) CTC-*fdh-adhE2* with SF; and (**D**) CTC-*fdh-adhE2* with SF, B12 and MVH. ○: glucose; ×: OD; ■: butyric acid; ▲: acetic acid; ◇: butanol; ○: ethanol.

Table 3. Butanol fermentations using metabolically-engineered *C. tyrobutyricum* by rebalancing redox.

Products		CTC-*adhE2* (Control)		CTC-*fdh-adhE2*	
		SF	SF, B12 and MVH	SF	SF, B12 and MVH
Cell growth (h^{-1})		0.18 ± 0.01	0.16 ± 0.0004	0.15 ± 0.001	0.15 ± 0.001
Biomass yield (g/g)		0.10 ± 0.01	0.09 ± 0.003	0.09 ± 0.001	0.08 ± 0.002
Concentration (g/L)	Butanol	3.18 ± 0.09	6.14 ± 0.05	8.83 ± 0.02	12.34 ± 0.02
	Butyrate	13.22 ± 0.73	9.32 ± 0.03	7.72 ± 0.05	5.05 ± 0.04
	Acetate	4.15 ± 0.25	1.42 ± 0.07	0.69 ± 0.01	0.26 ± 0.03
	Ethanol	0.11 ± 0.04	0.25 ± 0.02	0.24 ± 0.01	0.28 ± 0.07
Yield (g/g-glucose)	Butanol	0.05 ± 0.01	0.11 ± 0.003	0.16 ± 0.002	0.23 ± 0.002
	Butyrate	0.27 ± 0.02	0.20 ± 0.004	0.19 ± 0.01	0.10 ± 0.001
	Acetate	0.06 ± 0.01	0.02 ± 0.002	0.02 ± 0.001	0.003 ± 0.0001
	Ethanol	0.001 ± 0.0005	0.003 ± 0.0004	0.004 ± 0.0002	0.004 ± 0.0003
Productivity (g/L/h)	Butanol	0.12 ± 0.01	0.14 ± 0.01	0.17 ± 0.002	0.26 ± 0.01
	Butyrate	0.33 ± 0.01	0.26 ± 0.003	0.22 ± 0.004	0.15 ± 0.004
	Acetate	0.14 ± 0.01	0.07 ± 0.004	0.05 ± 0.01	0.01 ± 0.002
	Ethanol	0.006 ± 0.001	0.008 ± 0.0001	0.010 ± 0.0003	0.011 ± 0.0004

Notes: The means in this table were calculated from the duplicated fermentations. Yield = g-product/g-glucose consumed; and selectivity = g-product/g-total products. SF: sodium formate; B12: vitamin B12; MVH: methyl viologen hydrate.

As summarized in Table 3, the butanol concentrations were 3.18 g/L, 6.14 g/L, 8.83 g/L and 12.34 g/L by CTC-*adhE2* with SF, CTC-*adhE2* with SF, B12 and MVH, CTC-*fdh-adhE2* with SF and CTC-*fdh-adhE2* with SF, B12 and MVH, respectively. It is obvious that the integration of metabolic cell engineering (Fdh synthesis) and process engineering (medium optimization) significantly increased the butanol production from 3.18 g/L–12.34 g/L, with 3.88-fold improvement. The yield of butanol through integrated cell-process engineering (0.23 g/g) was also much higher than other conditions (0.05–0.16 g/g). The butanol production by redox engineering was slightly higher than that by medium supplement addition, *i.e.*, titer of 8.83 g/L and yield of 0.16 g/g (Figure 5C) *vs.* titer of 6.14 g/L and yield of 0.11 g/g (Figure 5B). However, butanol productivity was 0.26 g/L/h by CTC-*fdh-adhE2* with medium supplements, which was much higher than other conditions (0.12–0.17 g/L/h). These results confirmed that the rebalance of redox directed more carbon to the production of butanol that consumed reducing power. In addition, the ethanol concentrations in these four fermentations were fairly low (0.11 g/L–0.28 g/L) due to the low expression of alcohol dehydrogenase (*adh*).

To better understand the effect of carbon and redox rebalance on butanol production, we compared the selectivity of fermentation products (Figure 7). The selectivity of butanol was 0 g/g, 0.15 g/g, 0.51 g/g, 0.41 g/g and 0.69 g/g at the conditions of control, CTC-*adhE2* (C), CTC-*fdh-adhE2* (C-R1), CTC-*adhE2* with medium supplement (C-R2) and CTC-*fdh-adhE2* with medium supplement (C-R1-R2), respectively. It is clear that the rebalances of carbon and redox via integrating metabolic cell and process engineering could effectively boost the selectivity of butanol and improve the production of butanol. In addition; the conditions of C, C-R1, C-R2 and C-R1-R2 produced butyrate with concentrations of 13.22 g/L, 7.72 g/L, 9.32 g/L and 5.05 g/L and the acetate with concentrations of 4.15 g/L, 0.69 g/L, 1.42 g/L and 0.26 g/L, respectively. These results revealed that rebalancing redox not only redistributed carbon between C4, but also redirected the carbon flux from C2 to C4.

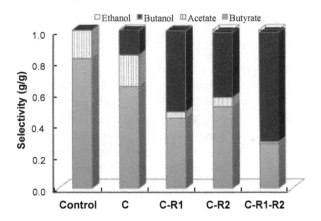

Figure 7. Improvement of n-butanol selectivity by rebalancing carbon and redox flux. Control: wild-type CTC without sodium formate; C: CTC-*adhE2* without sodium formate; C-R1: CTC-*adhE2* with sodium formate, vitamin B12 and methyl viologen hydrate; C-R2: CTC-*fdh-adhE2* with sodium formate; and C-R1-R2: CTC-*fdh-adhE2* with sodium formate, vitamin B12 and methyl viologen hydrate.

4.4. Butanol Tolerance

Butanol is a toxic solvent that could change the cell membrane fluidity and inhibit cell growth; butanol tolerance is therefore an important factor to evaluate in the new mutant. In this study, the effect of butanol on the cell growth of wild-type CTC, mutant CTC-*adhE2* and mutant CTC-*fdh-adhE2* was investigated (Figure 8). At 0 g/L of butanol, both mutants and the wild-type showed the highest specific growth rates, while higher butanol concentrations inhibited cell growth. At 20 g/L of butanol, less than 20% of the maximum growth rate was retained. The growth inhibition by butanol followed the noncompetitive inhibition kinetics with K_P (inhibition rate constant) of 3.32 g/L. 2.70 g/L and 3.37 g/L for CTC, CTC-*adhE2* and CTC-*fdh-adhE2*, respectively. Therefore, this tolerance study showed that metabolic engineering in the new mutant did not reduce butanol tolerance and cell growth.

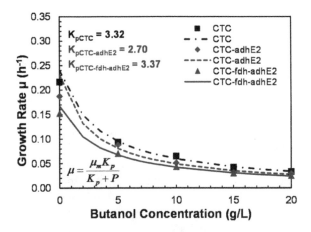

Figure 8. Butanol tolerance study of *C. tyrobutyricum* in serum bottles. ■: WT; -.-.-.-: WT; ◆: CTC-*adhE2*; - - - - -: CTC-adhE2; ▲: CTC-*fdh-adhE2*; ———: CTC-*fdh-adhE2*.

5. Discussion

5.1. Improvement of Butanol Production by Rebalancing Redox

In previous studies, it was found that the production of solvents involved a global remodeling of metabolism in *C. acetobutylicum* [31], and different NADH/NAD$^+$ ratios could redistribute the metabolic flux in *E. coli* and *C. acetobutylicum* [20,32]. Our previous omics study also showed a strong correlation between the expression level of NAD(P)H-dependent enzymes and the production of butanol in *C. tyrobutyricum*, indicating that high butanol production required more reducing power [17]. Another study showed that the addition of methyl viologen hydrate increased the butanol production in fermentation [18]. Therefore, it was hypothesized that the integration of redox rebalance in the engineered mutant of *C. tyrobutyricum* with carbon rebalance could greatly improve the butanol production.

To examine the above hypothesis, we performed redox engineering using both cell engineering and process engineering in this study. The control mutant is CTC-*adhE2*, i.e., the wild-type *C. tyrobutyricum* containing synthesized *adhE2* [23]. Compared to CTC-*adhE2*, the redox engineering by cell engineering or process engineering (*i.e.*, medium optimization in this study) doubled the butanol production, while the integration of cell-process engineering quadrupled the butanol production (12.34 g/L *vs.* 3.18 g/L). Our results suggest that the integration of these two engineering methods effectively improved reducing power supply and rebalanced redox. Our results also demonstrated that redox engineering was essential to achieve high butanol production in addition to carbon engineering in *C. tyrobutyricum*.

Another interesting observation from this study is that redox rebalance significantly reduced the production of acids (butyrate and acetate) by *C. tyrobutyricum*. Because butanol formation consumed NAD(P)H, the synthesis of the NADH-producing pathway and the medium supplements resulted in carbon redistribution from acetate and butyrate to butanol. Therefore, the selectivity of butanol was significantly increased in the fermentation of CTC-*fdh-adhE2* with medium supplements. Moreover, ethanol production was still low even though the butanol production was increased by redox rebalance. The high selectivity of butanol and low production of byproducts could benefit the separation of biobutanol and reduce the production cost of biofuel from bacterial fermentation.

5.2. Comparison with Previous Studies

As a promising host cell for butanol production, *C. tyrobutyricum* has the desired features, such as unassociated sporulation and cell autolysis, high butanol tolerance and the ability to convert low-value lignocellulosic biomass feedstock. Recently, two studies have sought to improve biobutanol production by *C. tyrobutyricum* through metabolic engineering or process engineering. As summarized in Table 4, the mutant CTC-*adhE2* with overexpressed *adhE2* and the pCB12 replicon in wild-type *C. tyrobutyricum* produced butanol with a titer of 2.5 g/L from glucose in a serum bottle in a previous study [23], and a similar result was obtained in this study. The pBP1 replicon improved butanol titer to 6.5 g/L from glucose in the serum bottle and further improved butanol titer to 20.0 g/L from mannitol in the bioreactor [16]. These results elucidated that the replicon optimization and substrate were very important for biobutanol production. Replicon optimization could be performed in our future metabolic engineering study. However, mannitol is an expensive raw material for butanol production, with a price of ~2000 U.S. dollars per ton, which prevents it from becoming an economically-competitive biobutanol in the fuel market. In this study, the redox was rebalanced by synthesizing a NADH-producing pathway in addition to carbon redistribution, resulting in a doubled butanol titer. The reducing power supplement further improved butanol production by quadrupling the titer to reach 12.3 g/L. Therefore, we can conclude that redox rebalance is an effective approach to improve the biobutanol production by *C. tyrobutyricum*.

Table 4. Recent progresses of butanol production with *C. tyrobutyricum*.

Strain	Characteristics	Mode	Carbon	Titer (g/L)	Yield (g/g)	Productivity (g/L/h)	Reference
CTC-*adhE2*	+*adhE2*, pCB102 replicon	Bottle	Glucose	2.5	0.15 *	0.04 *	[16]
		Bottle	Glucose	2.6	0.14	0.03	This study
CTpM2	+*adhE2*, pBP1 replicon	Bottle	Glucose	6.5	0.24	0.20 *	[16]
		Bioreactor	Mannitol #	20.0 #	0.33 #	0.32 #	[16]
CTC-*fdh-adhE2*	+*fdh*+*adhE2*, pCB102 replicon	Bottle	Glucose	6.9	0.21	0.20	This study
		Bioreactor	Glucose	12.3	0.23	0.26	This study

Notes: * These data were estimated from the fermentation kinetic profile. # Mannitol is an expensive substrate that significantly increases the production cost of biobutanol.

Our recent butyrate fermentation using *C. tyrobutyricum* showed that the fermentation pH significantly affected the production of butyrate and cell growth [33]. Specifically, the immobilized-cell fermentations at pH 6.0, 6.5 and 7.0 produced butyrate with concentrations of 50.11 g/L, 63.02 g/L and 61.01 g/L. The metabolic flux analysis demonstrated that carbon flux from acetyl-CoA to acetoacetyl-CoA and the carbon flux from butyryl-CoA to butyrate were increased at pHs of 6.5 and 7.0 as compared to pH 6.0. In addition, the biomass yield of *C. tyrobutyricum* was also decreased at a higher pH. The pH values of the butanol fermentations in bioreactors were maintained at 6.0 in this study. Because butanol is produced from butyryl-CoA directly, the improved carbon flux from C2 to C4 at higher pH could benefit the formation of butanol. In the future study, we will optimize butanol fermentation by the constructed mutant CTC-*fdh-adhE2* in order to further improve butanol production.

Other metabolic engineering strategies were used to increase butanol production by redistributing the metabolic flux from C2 to C4. For example, the ACKKO-*adhE2* achieved the highest concentration of butanol (14.5 g/L) in free-cell fermentation, and process engineering in immobilized-cell fermentation generated >20 g/L of butanol [18]. To achieve a higher butanol production, redox engineering could be applied in a high butanol-producing mutant, e.g., ACKKO-*adhE2*, in the future. To effectively integrate cell engineering and process engineering, metabolomics should be developed and applied to *C. tyrobutyricum*. Metabolic flux analysis modeling is also an effective tool to guide the design of metabolic engineering [33,34]. In addition, the medium components can be rationally designed to assist in butanol production post cell engineering.

6. Conclusions

In this study, we demonstrated that redox engineering was essential to improve butanol production in *C. tyrobutyricum*. Both the titer and selectivity of butanol were significantly improved by rebalancing redox in addition to rebalancing carbon. We also found that the integration of metabolic cell and process engineering was an efficient strategy to improve biofuel production.

Acknowledgments: The authors would like to thank Professor Nigel P. Minton at the University of Nottingham for providing us with plasmid construction material. This work was supported by a research grant from the National Science Foundation (Award No. 1342390) and the Research Grants Committee (RGC) Level 2 fund (Award No. RG14657) from The University of Alabama. The professional writing center at The University of Alabama polished the language of this manuscript.

Author Contributions: Chao Ma completed most of the lab work and also wrote most of this paper. Jianfa Ou and Ningning Xu contributed to plasmid construction and transformation optimization. Janna L. Fierst contributed to data analysis. Shang-Tian Yang provided the control mutant CTC-*adhE2* and was involved in mutant development. X. Margaret Liu led the project and finalized the manuscript as corresponding author.

Conflicts of Interest: The authors declare no conflict of interest. The founding sponsors had no role in the design of the study; in the collection, analyses or interpretation of data; in the writing of the manuscript; nor in the decision to publish the results.

References

1. Zheng, Y.N.; Li, L.Z.; Xian, M.; Ma, Y.J.; Yang, J.M.; Xu, X.; He, D.Z. Problems with the microbial production of butanol. *J. Ind. Microbiol. Biotechnol.* **2009**, *36*, 1127–1138. [CrossRef] [PubMed]
2. Milne, C.B.; Eddy, J.A.; Raju, R.; Ardekani, S.; Kim, P.J.; Senger, R.S.; Jin, Y.S.; Blaschek, H.P.; Price, N.D. Metabolic network reconstruction and genomescale model of butanol-producing strain *Clostridium beijerinckii* NCIMB 8052. *BMC Syst. Biol.* **2011**, *5*, 130–144. [CrossRef] [PubMed]
3. Lee, S.Y.; Park, J.H.; Jang, S.H.; Nielsen, L.K.; Kim, J.; Jung, K.S. Fermentative butanol production by Clostridia. *Biotechnol. Bioeng.* **2008**, *101*, 209–228. [CrossRef] [PubMed]
4. Nicolaou, S.A.; Gaida, S.M.; Papoutsakis, E.T. A comparative view of metabolite and substrate stress and tolerance in microbial bioprocessing: From biofuels and chemicals, to biocatalysis and bioremediation. *Metab. Eng.* **2010**, *12*, 307–331. [CrossRef] [PubMed]
5. Papoutsakis, E.T. Engineering solventogenic clostridia. *Curr. Opin. Biotechnol.* **2008**, *19*, 420–429. [CrossRef] [PubMed]
6. Al-Hinai, M.A.; Fast, A.G.; Papoutsakis, E.T. Novel system for efficient isolation of Clostridium double-crossover allelic exchange mutants enabling markerless chromosomal gene deletions and DNA integration. *Appl. Environ. Microbiol.* **2012**, *78*, 8112–8121. [CrossRef] [PubMed]
7. Atsumi, S.; Cann, A.F.; Connor, M.R.; Shen, C.R.; Smith, K.M.; Brynildsen, M.P.; Chou, K.J.; Hanai, T.; Liao, J.C. Metabolic engineering of *Escherichia coli* for 1-butanol production. *Metab. Eng.* **2008**, *10*, 305–311. [CrossRef] [PubMed]
8. Berezina, O.V.; Zakharova, N.V.; Brandt, A.; Yarotsky, S.V.; Schwarz, W.H.; Zverlov, V.V. Reconstructing the clostridial n-butanol metabolic pathway in *Lactobacillus brevis*. *Appl. Microbiol. Biotechnol.* **2010**, *87*, 635–646. [CrossRef] [PubMed]
9. Blombach, B.; Eikmanns, B.J. Current knowledge on isobutanol production with *Escherichia coli*, *Bacillus subtilis* and *Corynebacterium glutamicum*. *Bioeng. Bugs.* **2011**, *2*, 346–350. [CrossRef] [PubMed]
10. Nielsen, D.R.; Leonard, E.; Yoon, S.H.; Tseng, H.C.; Yuan, C.; Prather, K.L. Engineering alternative butanol production platforms in heterologous bacteria. *Metab. Eng.* **2009**, *11*, 262–273. [CrossRef] [PubMed]
11. Shen, C.R.; Liao, J.C. Metabolic engineering of *Escherichia coli* for 1-butanol and 1-propanol production via the keto-acid pathways. *Metab. Eng.* **2008**, *10*, 312–320. [CrossRef] [PubMed]
12. Steen, E.J.; Chan, R.; Prasad, N.; Myers, S.; Petzold, C.J.; Redding, A.; Ouellet, M.; Keasling, J.D. Metabolic engineering of *Saccharomyces cerevisiae* for the production of n-butanol. *Microb. Cell. Fact.* **2008**, *7*, 36. [CrossRef] [PubMed]
13. Shen, C.R.; Lan, E.I.; Dekishima, Y.; Baez, A.; Cho, K.M.; Liao, J.C. Driving forces enable high-titer anaerobic 1-butanol synthesis in *Escherichia coli*. *Appl. Environ. Microbiol.* **2011**, *77*, 2905–2915. [CrossRef] [PubMed]
14. Jiang, L.; Wang, J.; Liang, S.; Wang, X.; Cen, P.; Xu, Z. Production of butyric acid from glucose and xylose with immobilized cells of *Clostridium tyrobutyricum* in a fibrous-bed bioreactor. *Appl. Biochem. Biotechnol.* **2010**, *160*, 350–359. [CrossRef] [PubMed]
15. Liu, X.; Yang, S.T. Butyric acid and hydrogen production by *Clostridium tyrobutyricum* ATCC 25755 and mutants. *Enzyme Microb. Technol.* **2006**, *38*, 521–528. [CrossRef]
16. Yu, M.; Du, Y.; Jiang, W.; Chang, W.L.; Yang, S.T.; Tang, I.C. Effects of different replicons in conjugative plasmids on transformation efficiency, plasmid stability, gene expression and n-butanol biosynthesis in *Clostridium tyrobutyricum*. *Appl. Microbiol. Biotechnol.* **2012**, *93*, 881–889. [CrossRef] [PubMed]
17. Ma, C.; Kojima, K.; Xu, N.; Mobley, J.; Zhou, L.; Yang, S.T.; Liu, X.M. Comparative proteomics analysis of high n-butanol producing metabolically engineered *Clostridium tyrobutyricum*. *J. Biotechnol.* **2015**, *193*, 108–119. [CrossRef] [PubMed]
18. Du, Y.; Jiang, W.; Yu, M.; Tang, I.C.; Yang, S.T. Metabolic process engineering of *Clostridium tyrobutyricum Deltaack-adhE2* for enhanced n-butanol production from glucose: Effects of methyl viologen on NADH availability, flux distribution and fermentation kinetics. *Biotechnol. Bioeng.* **2015**, *112*, 705–715. [CrossRef] [PubMed]
19. Ou, J.; Ma, C.; Xu, N.; Du, Y.; Liu, X.M. High Butanol Production by Regulating Carbon, Redox and Energy in Clostridia. *Front. Chem. Sci. Eng.* **2015**. [CrossRef]

20. Berrios-Rivera, S.J.; Bennett, G.N.; San, K.Y. Metabolic engineering of *Escherichia coli*: Increase of NADH availability by overexpressing an NAD(+)-dependent formate dehydrogenase. *Metab. Eng.* **2002**, *4*, 217–229. [CrossRef] [PubMed]

21. Geertman, J.M.; van Dijken, J.P.; Pronk, J.T. Engineering NADH metabolism in *Saccharomyces cerevisiae*: formate as an electron donor for glycerol production by anaerobic, glucose-limited chemostat cultures. *FEMS Yeast Res.* **2006**, *6*, 1193–1203. [CrossRef] [PubMed]

22. Wang, M.; Hu, L.; Fan, L.; Tan, T. Enhanced 1-Butanol Production in Engineered *Klebsiella pneumoniae* by NADH Regeneration. *Energy Fuels* **2015**, *29*, 1823–1829. [CrossRef]

23. Yu, M.; Zhang, Y.; Tang, I.C.; Yang, S.T. Metabolic engineering of *Clostridium tyrobutyricum* for n-butanol production. *Metab. Eng.* **2011**, *13*, 373–382. [CrossRef] [PubMed]

24. Purdy, D.; O'Keeffe, T.A.; Elmore, M.; Herbert, M.; McLeod, A.; Bokori-Brown, M.; Ostrowski, A.; Minton, N.P. Conjugative transfer of clostridial shuttle vectors from *Escherichia coli* to *Clostridium difficile* through circumvention of the restriction barrier. *Mol. Microbiol.* **2002**, *46*, 439–452. [CrossRef] [PubMed]

25. Dong, H.; Tao, W.; Zhang, Y.; Li, Y. Development of an anhydrotetracycline-inducible gene expression system for solvent-producing *Clostridium acetobutylicum*: A useful tool for strain engineering. *Metab. Eng.* **2012**, *14*, 59–67. [CrossRef] [PubMed]

26. Heap, J.T.; Kuehne, S.A.; Ehsaan, M.; Cartman, S.T.; Cooksley, C.M.; Scott, J.C.; Minton, N.P. The ClosTron: Mutagenesis in Clostridium refined and streamlined. *J. Microbiol. Methods.* **2010**, *80*, 49–55. [CrossRef] [PubMed]

27. Alissandratos, A.; Kim, H.K.; Matthews, H.; Hennessy, J.E.; Philbrook, A.; Easton, C.J. *Clostridium carboxidivorans* strain P7T recombinant formate dehydrogenase catalyzes reduction of CO(2) to formate. *Appl. Environ. Microbiol.* **2013**, *79*, 741–744. [CrossRef] [PubMed]

28. Yamamoto, I.; Saiki, T.; Liu, S.M.; Ljungdahl, L.G. Purification and properties of NADP-dependent formate dehydrogenase from *Clostridium thermoaceticum*, a tungsten-selenium-iron protein. *J. Biol. Chem.* **1983**, *258*, 1826–1832. [PubMed]

29. Heluane, H.; Evans, M.R.; Dagher, S.F.; Bruno-Barcena, J.M. Meta-analysis and functional validation of nutritional requirements of solventogenic Clostridia growing under butanol stress conditions and coutilization of D-glucose and D-xylose. *Appl. Environ. Microbiol.* **2011**, *77*, 4473–4485. [CrossRef] [PubMed]

30. Gheshlaghi, R.; Scharer, J.M.; Moo-Young, M.; Chou, C.P. Metabolic pathways of clostridia for producing butanol. *Biotechnol. Adv.* **2009**, *27*, 764–781. [CrossRef] [PubMed]

31. Amador-Noguez, D.; Brasg, I.A.; Feng, X.J.; Roquet, N.; Rabinowitz, J.D. Metabolome remodeling during the acidogenic-solventogenic transition in *Clostridium acetobutylicum*. *Appl. Environ. Microbiol.* **2011**, *77*, 7984–7997. [CrossRef] [PubMed]

32. Nakayama, S.; Kosaka, T.; Hirakawa, H.; Matsuura, K.; Yoshino, S.; Furukawa, K. Metabolic engineering for solvent productivity by downregulation of the hydrogenase gene cluster hupCBA in *Clostridium saccharoperbutylacetonicum* strain N1-4. *Appl. Microbiol. Biotechnol.* **2008**, *78*, 483–493. [CrossRef] [PubMed]

33. Ma, C.; Ou, J.; McFann, S.; Miller, M.; Liu, X.M. High production of butyric acid by *Clostridium tyrobutyricum* mutant. *Front. Chem. Sci. Eng.* **2015**. [CrossRef]

34. Liao, C.; Seo, S.O.; Celik, V.; Liu, H.; Kong, W.; Wang, Y.; Blaschek., H.; Jin., Y.; Lu, T. Integrated, systems metabolic picture of acetone-butanol-ethanol fermentation by *Clostridium acetobutylicum*. *Proc. Natl. Acad. Sci. USA* **2015**, *112*, 8505–8510. [CrossRef] [PubMed]

5

Biodiesel Production by *Aspergillus niger* Lipase Immobilized on Barium Ferrite Magnetic Nanoparticles

Ahmed I. El-Batal [1,*]**, Ayman A. Farrag** [2]**, Mohamed A. Elsayed** [3] **and Ahmed M. El-Khawaga** [3]

[1]   Drug Radiation Research Department, National Center for Radiation Research and Technology (NCRRT), Atomic Energy Authority, Cairo 11371, Egypt

[2]   Botany and Microbiology Department, Faculty of Science, Al-Azhar University, Cairo 11371, Egypt; dardear2002@yahoo.com

[3]   Chemical Engineering Department, Military Technical College, Cairo 11371, Egypt; aboelfotoh@gmail.com (M.A.E.); ahmedelkhwaga15@gmail.com (A.M.E.-K.)

*   Correspondence: aelbatal2000@gmail.com or Ahmed.Elbatal@eaea.org.eg

Academic Editor: Anthony Guiseppi-Elie

Abstract: In this study, *Aspergillus niger* ADM110 fungi was gamma irradiated to produce lipase enzyme and then immobilized onto magnetic barium ferrite nanoparticles (BFN) for biodiesel production. BFN were prepared by the citrate sol-gel auto-combustion method and characterized by transmission electron microscopy (TEM), X-ray diffraction (XRD), Fourier transform infrared (FTIR) and scanning electron microscopy with energy dispersive analysis of X-ray (SEM/EDAX) analysis. The activities of free and immobilized lipase were measured at various pH and temperature values. The results indicate that BFN–Lipase (5%) can be reused in biodiesel production without any treatment with 17% loss of activity after five cycles and 66% loss in activity in the sixth cycle. The optimum reaction conditions for biodiesel production from waste cooking oil (WCO) using lipase immobilized onto BFN as a catalyst were 45 °C, 4 h and 400 rpm. Acid values of WCO and fatty acid methyl esters (FAMEs) were 1.90 and 0.182 (mg KOH/g oil), respectively. The measured flash point, calorific value and cetane number were 188 °C, 43.1 MJ/Kg and 59.5, respectively. The cloud point (-3 °C), pour point (-9 °C), water content (0.091%) and sulfur content (0.050%), were estimated as well.

Keywords: biodiesel; *Aspergillus niger*; lipase immobilization; barium ferrite magnetic nanoparticles; fatty acid methyl ester

---

1. Introduction

Biodiesel is defined as the fatty acid alkyl monoesters derived from renewable feed stocks such as vegetable oils and animal fats [1]. Biodiesel is an efficient, clean and 100% natural energy alternative to petroleum-based fuels. Biodiesel fuel has many advantages; it is safe for use in all conventional diesel engines, offers the same performance and engine durability as petroleum diesel fuel and is non-flammable and non-toxic. In addition. It reduces tailpipe emissions, visible smoke, noxious fumes, and odors [2]. Biodiesel is better than diesel fuel in terms of sulfur content, flash point, aromatic content and biodegradability [3]. The simple reaction to convert vegetable oil to biodiesel is called transesterification, in which an alcohol and a catalyst are mixed with oil in order to "crack" the oil into esters and glycerol. During this process, the catalyst allows the alcohol to react successively with triglyceride to produce esters and glycerol, as shown in Equation (1). Where, the heavier portion of the reaction medium, glycerol, falls out of the mixture, leaving methyl esters. On the other hand, the methyl esters are purified by distillation. The distilled product, methyl esters of fatty acids, is known as the biodiesel [4]. There are three steps involved in the transesterification of triglyceride (TG) into methyl esters (ME), with the formation of intermediates diglyceride (DG) and monoglyceride (MG), resulting in the production of three moles of ester and one mole of glycerol, as shown in stepwise reactions Equations (2)–(4) [5].

$$\begin{array}{ccccc}
CH_2-COO-R_1 & & & R_1-COO-R' & CH_2-OH \\
| & & & & | \\
CH-COO-R_2 & + & 3R'OH \overset{Catalyst}{\rightleftharpoons} & R_2-COO-R' + & CH-OH \\
| & & & R_3-COO-R' & | \\
CH_2-COO-R_3 & & & & CH_2-OH \\
\text{Triglyceride} & & \text{Alcohol} & \text{Esters} & \text{Glycerol}
\end{array} \tag{1}$$

$$\text{Triglyceride} + \text{ROH} \rightleftharpoons \text{Diglyceride} + \text{RCOOR} \tag{2}$$

$$\text{Diglyceride} + \text{ROH} \rightleftharpoons \text{Monoglyceride} + \text{RCOOR} \tag{3}$$

$$\text{Monoglyceride} + \text{ROH} \rightleftharpoons \text{Glycerol} + \text{RCOOR} \tag{4}$$

A catalyst is usually used to improve the reaction rate and yield. Because the reaction is reversible, excess alcohol is used to shift the equilibrium to the products side. Lipases can catalyze this reaction [6].

Lipases (triacylglycerol acyl hydrolases, EC 3.1.1.3) are largely used as biocatalysts in biotechnology and modern chemistry [7]. These enzymes are also catalyzing several reactions like esterification [8], transesterification, acidolysis, interesterification, alcoholysis, aminolysis, oximolysis, thiotransesterification and ammoniolysis in anhydrous organic solvents [9]. Lipase is widely found in animals, plants and microorganisms [10]. Nowadays, most lipases produced commercially are currently obtained from fungi and yeasts. Fungal lipases have received attention because of their potential use in food processing, pharmaceuticals, cosmetics, oils/fats degradation, detergents and the leather industry [11]. Industrial applications of lipase are limited because it has poor stability, is easily inactivated, and is difficult to separate from the reaction system for reuse. In order to use lipase more economically and efficiently in reaction systems, its activity and operational stability need to be improved by an appropriate choice of immobilization process [12]. Immobilization of enzymes is a well reported technology that allows application of enzymes in many biocatalyzed processes, like in lipase-mediated biodiesel production. This process provides significant advantages, such as reuse of the biocatalyst, easier product recovery, and it frequently enhances the enzyme resistance against inactivation by different denaturants, providing more stable catalysts [13]. Besides, the enzyme immobilization onto magnetic supports such as nanosized magnetite particles allows an additional characteristic compared to other conventional support materials, such as the selective and secure enzyme recovery from the medium under the magnetic force. Hence, there is no need for expensive liquid chromatography systems, centrifuges, filters or other equipment [14]. In this study, the lipase from *Aspergillus nigerADM110* was immobilized onto the magnetic Barium Ferrite Nanoparticles (BFN) under different experimental conditions and characterized. The biodiesel production was carried out and analyzed by GC-MS.

2. Materials and Methods

2.1. Chemicals

All the media components were purchased from Oxoid and Difco. Chemicals and reagents, including olive oil used, were purchased from Sigma-Aldrich (St Louis, MO, USA), were of analytical and HPLC grade, and were used without further purification. *Aspergillus niger ADM110* strain was from Drug Radiation Research Department (NCRRT) [15].

2.2. Lipase Production and Purification

Extracellular lipase was obtained from irradiated *Aspergillus niger ADM110* by submerged production using the optimized medium containing olive oil 5% as the most efficient carbon source and yeast extract 5.0% as the most suitable nitrogen source for the lipase biosynthesis. The fermentation was carried out in 100 mL of sterile production broth seeded with 1.0 mL (4.60×10^7) of spore inoculum and incubated at 25 °C, pH 7, 200 rpm agitation speed and for 72 h. *Aspergillus nigerADM110* was subjected to 0.14 kGy of gamma radiation to improve its lipolytic potential [15]. Crude lipase was obtained by centrifuging the culture broth at 10,000 rpm for 10 min at 4 °C. The lipase purification was carried out by ammonium sulfate precipitation followed by dialysis against 0.05 M sodium phosphate buffer solution (pH 7.2). Extracellular lipase activity was estimated by photometric assay [16].

2.3. Synthesis of Hexagonal Magnetic Barium Ferrite Nanoparticles (BNF)

Barium ferrite nanoparticles were prepared by the citrate sol-gel auto-combustion method [17,18]. Barium nitrate Ba $(NO_3)_2$, ferric nitrate Fe $(NO_3)_3 \cdot 9H_2O$ and citric acid were taken as starting materials. The nitrates were dissolved in the minimum amount of de-ionized water to get a clear solution in the molar ratio 1:10. An aqueous solution containing 12.5 moles of citric acid was mixed with both nitrate solutions (metals: citric acid, 1:3). After complete dissolving of all starting materials, an appropriate amount of ammonia solution was added dropwise with agitation until the pH of the solution reaches 7.0 giving a clear dark-green solution. The resulting solution was then heated in a drying oven at 85 °C for 3 h followed by heating at 110 °C until the transparent solution turned to viscous gel. The viscous gel was then heated at 150 °C until foaming occurred. The foamy gel was maintained at 150 °C until spontaneous ignition occurred. The combustion reaction was completed within a few minutes, and brown ash was formed. The fine powder was calcined (pre-sintered) in the calcination oven at 450 °C for 3 h to remove carbonic materials and then the temperature was increased to 850 °C for 4 h for the final formation of barium ferrite nanoparticles.

2.4. Immobilization of Lipase on BFN Particles

For the immobilization of lipase, 200 mg of BFN particles were added to 0.85% (w/v) sodium chloride solution containing lipase. The mixture was incubated for 4 h using an overhead stirrer. After washing twice with 0.85% (w/v) sodium chloride solution, the immobilized lipase was separated by magnetic decantation of the supernatants and stored at 4 °C. The amount of protein binding to the BFN particles was determined by measuring the protein concentration of the lipase solution and the supernatant by the Lowry method [19].

2.5. Assay of Enzyme Activity

The assay was modified from that described by Pencreac'h et al. [20]. The assay mixture contained 90 μL of 8.25 mM p-nitrophenyl palmitate in isopropanol and 810 μL of 50 mM Tris-HCl, pH 8.0, with 0.5% (w/v) Triton X-100 and 0.12% (w/v) Arabic gum. The mixture was then preheated to 40 °C. To initiate the reaction, 100 μL of lipase solution or suspension of immobilized lipase diluted to the appropriate concentration with 1% (w/v) bovine serum albumin was added. The change in absorbance at 410 nm was monitored for 5 min at 40 °C using a thermostated spectrophotometer (UV-1800, Shimadzu, Kyoto, Japan). The activity was calculated from the difference in absorbance between 2 and 5 min with a standard curve for the hydrolysis product, p-nitrophenol. One activity unit was defined as the production of 1 μmol of p-nitrophenol per min at 40 °C. Measurements were performed in triplicate.

After the reaction, BFN particles were separated magnetically and recycled. The specific activity of the lipase was defined in this study as the enzyme activity (unit) divided by the protein content and expressed as U/mg protein. Relative specific activity was calculated by dividing the specific activity of the immobilized lipase by that of the free lipase [21].

2.6. Biodiesel Production Using the Transestrification Process

Transesterification is the reaction of vegetable oil or animal fat with an alcohol, in most cases methanol, to form esters and glycerol. However, before entering the transesterification reactor, the

FFA (Free Fatty Acids) content and humidity must be removed, to avoid the production of soap. The transesterification reaction is affected by (molar ratio of glycerides to alcohol, the amount of catalyst, reaction temperature, reaction time and shaking speed) [22].

The yield of the transesterification processes was calculated as a sum of the weight of FAME (fatty acid methyl ester) produced to the weight of cooking oil used, multiplied by 100. The formula is given as:

$$\text{Yield of FAME} = \frac{Weight\ of\ fatty\ acid\ methylester}{Weight\ of\ fat\ used} \times 100\%$$

2.6.1. Gas Chromatography Analysis of WCO (Waste Cooking Oil)

GC analysis of waste cooking oil was carried out on a GC system Agilent Technologies 7890A (Santa Clara, USA) equipped with FID, split/splitless injector and Agilent 7693 A automated liquid sampler. Column: HP INNOWAX, 30 m × 0.32 mm ID, 0.25 μm film thickness. Temperature program of the oven: initial temperature 210 °C for 9 min, rate 20 °C/min to 230 °C, 10 min. Detector temperature: 300 °C, injector temperature: 250 °C. Carrier gas: He, column flow 1.5 mL/min, split ratio 1:80. Hydrogen flow 40 mL/min, air flow 400 mL/min, make-up gas (nitrogen) 40 mL/min. Injection volume 1 μL. ChemStation for GC was used for instrument control, data acquisition and data analysis [23].

2.6.2. Fatty Acid Methyl Esters (FAMEs) Analysis (Biodiesel Products)

Five hundred milliliters of the reaction mixture were mixed with 1.0 mL isooctane for two min. Following centrifugal separation, the upper organic layer was collected and washed twice with distilled water and dried over anhydrous Na_2SO_4. The solvent was dried under N_2 steam and dissolved in 0.25 mL of CH_2Cl_2. The previous GC condition of WCO analysis was applied. The prepared FAMEs were then analyzed using particular fatty acid methyl ester standards (methyl palmitate, methyl stearate, methyl oleate, methyl linoleate, and methyl linolenate; Sigma-Aldrich).

2.7. Analytical Methods

The size and morphology of the magnetic nanoparticles were assessed by transmission electron microscopy (TEM) (JEOL, Peabody, MA, USA). The X-ray diffraction (XRD) (Rigaku Corporation, Austin, TX, USA) was performed to check the crystallinity of the magnetic nanoparticles. SEM/EDAX analysis (JEOL, Peabody, MA, USA), and TEM were performed for characterization of the magnetic nanoparticles.

2.8. Statistical Analysis

The analysis of data was carried out according to [24]. Means were compared using the least significant difference (LSD at the 5% level) and Duncan's multiple range tests at the significance $P = 0.05$.

3. Results and Discussion

3.1. Characterization of BFN Particles

The result shows the production of the hexagonal phase with a high overall agreement between the entry database and the sample diffraction pattern. The XRD pattern illustrates the formation of 92.7% weight percent hexagonal barium ferrite and 7.3% some other phases, like $BaFe_2O_4$ (cubic phase), BaO and Fe_2O_3. TEM analysis of the prepared barium ferrite shows the formation of spherical nanoparticles with average particle size diameter ranging from 8 to 25 nm with some unidentified amorphous phases. According to the structure of hexagonal barium ferrite, it has exact element weight ratios of (Ba: Fe: O) equal to (12.3: 60.4: 27.3). By comparing this value with the prepared one, the result was found to be approximately the same by EDAX analysis using SEM.

The prepared $BaFe_{12}O_{19}$ hexaferrite particles were characterized via Fourier transformed Infrared spectra (FTIR) Shimadzu FTIR-8400S. The room temperature infrared spectra of prepared samples was recorded in mid-IR range, 4000 cm^{-1} to 400 cm^{-1}. A few milligrams of $BaFe_{12}O_{19}$ hexaferrite particles were mixed with anhydrous KBr powder and made in the form of a pellet for the measurements.

Figure 1a shows FTIR spectra of $BaFe_{12}O_{19}$ hexaferrite particles. The main absorption bands at 3440.4 cm^{-1} are attributed to the O–H stretching, while the characteristic band at 1631.5 cm^{-1} results from the anti-symmetrical and symmetrical stretching vibration bands of COO^- related to citric acid [25]. The absorption bands at 2925.5 cm^{-1} were due to OH vibrations. The other bands are at 1457, 1430, 1385.6 and 858.17 cm^{-1}, corresponding to the stretching and bending vibrations of C=O, H–C–H, C–H and C–C respectively. The characteristic absorption bands between 586.25 and 450.3 cm^{-1} are assigned to the vibration of the bond between the oxygen atom and the metal ions (F–O), confirming the formation of hexaferrite and corresponding to vibrations of the tetrahedral and octahedral sites for $BaFe_{12}O_{19}$. The Fe–O stretching vibration band of the bulk magnetite is usually at 570 cm^{-1} and the band shifted to high wave numbers because of the finite size of the nanoparticles. The bands located at 1351 cm^{-1} and 858.17 cm^{-1} are associated with the N–O stretching vibration and bending vibration of NO_3^- [26]. The notable bands at about 1385 and 1430 cm^{-1} are attributed to nitrate ion and barium carbonate, respectively.

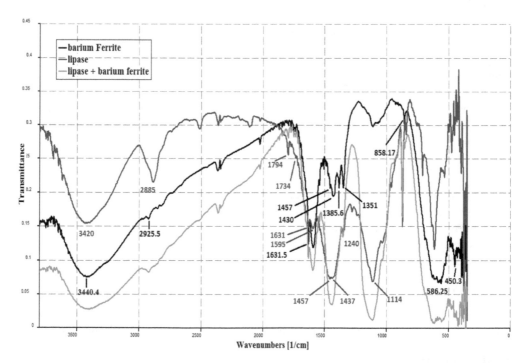

Figure 1. FTIR spectra of: (**a**) free *Aspergillus niger ADM110* lipase enzyme; (**b**) Barium ferrite magnetic nanoparticles; (**c**) Immobilized lipase enzyme on barium ferrite magnetic nanoparticles.

Figure 1b shows the FTIR spectrum of lipase from *Aspergillus nigerADM110*. There are many signals characteristic of different functional groups of the enzyme. They include a typical band at 2885 cm^{-1} assigned to the stretching C–H vibrations of –CH2 and –CH3 groups. A broad and intense band at 3420 cm^{-1} was assigned to the stretching vibrations of –OH group, and a band of the –N–H stretching vibrations, masked by the former one [27]. The characteristic signals coming from the stretching vibrations of carbonyl groups around 1734 cm-1 and stretching vibrations of $\equiv$C–O– appear in the spectrum at 1240 cm^{-1}.

Figure 1c shows the FTIR spectrum of immobilized lipase on barium ferrite magnetic nanoparticles. It was evident that the characteristic bands of protein (*i.e.*, lipase) at 1794 for Amide I and 1457–1437 cm^{-1} for Amide II with characteristic band 1114 cm^{-1} for coil alpha helix. Furthermore, 1631, 1595 cm^{-1} were present in pure lipase and the lipase-bound (barium ferrite nano) Fe_3O_4 nanoparticles, confirming the binding of lipase to $BaFe_{12}O_{19}$ nanoparticles by covalent immobilization. The strong characteristic bands of proteins for the lipase-bound $BaFe_{12}O_{19}$ should be owing to the high enzyme loading [28]. In addition, it is noted that a characteristic band of 1385.6 cm^{-1} was observed in naked Fe_3O_4 nanoparticles. After binding of lipase, this characteristic band disappeared in lipase binding barium ferrite nano. Also, bands at 2885, 2515 and 2111.0 cm^{-1} of lipase disappeared in binding lipase on nanoparticles. Thus, it was suggested that the binding was accomplished via the reaction between the amine group of BaFerrite nanoparticles and the carboxyl group of lipase and confirmed the enzyme immobilization.

3.2. The Activity of the Lipase Immobilized on BFN Particles

It is expected that the binding has been taking place between the hydrophobic particle of barium ferrite and the hydrophobic area of the lipase via interfacial activation.

Various initial lipase concentrations (0.2–2 mg/mL) were tested for immobilization experiments for determining the optimum enzyme loading. As shown in Table 1, the amounts of lipase immobilized increased significantly with increasing the initial lipase concentration, and the activities attained a maximum value at an initial lipase concentration of 1.0 mg/mL. The lipase activity is, however, slowly decreased when the initial lipase concentration was above 1.0 mg/mL. The excessive enzyme loading is known to hinder the substrate conversion due to the increased protein–protein interaction [29]. In the rest of the experiments, the initial lipase concentration was maintained at 1.0 mg/mL.

Table 1. Activity of the lipase immobilized on BFN (barium ferrite nanoparticles) particles.

Initial Lipase Concentration (mg/mL)	Activity of Immobilized Lipase (U/mL)	Protein Immobilized (mg/g Support)	Specific Activity (U/mg Protein)
00.2	4.8a ± 0.56	9.0a ± 0.70	6.15a ± 0.10
00.4	12.5b ± 0.35	15.7b ± 0.49	7.30b ± 0.21
00.6	23.9c ± 0.63	24.8c ± 0.42	8.82c ± 0.57
00.8	27.0d ± 0.56	36.5d ± 1.06	9.44de ± 0.39
01.0	30.5f ± 0.77	49.0e ± 0.91	10.10e ± 0.61
01.4	28.6e ± 0.77	57.8g ± 0.56	8.92cd ± 0.26
01.8	28.4e ± 0.98	52.6f ± 0.91	9.75de ± 0.53
02.0	28.5e ± 0.35	52.5f ± 0.56	9.60de ± 0.35
LSD	01.70	03.25	1.01

Note: Different letters indicate significant differences between treatments (Duncan test, $P \leqslant 0.05$). Means in each column followed by the same letter are not significantly different.

Activities of free and immobilized lipase were measured at various pH and temperatures values. The pH of the free and the immobilized lipase activities was compared and shown in Figure 2. The optimum pH of the immobilized lipase was 8.0, whereas that of the free one was 7.0. Also, the immobilized lipase showed higher activity than the free one, especially at pHs higher than 7.0.

The optimum pH value of free lipase shifted one unit to the alkaline region after binding on the support. It would therefore appear that lipase active sites may well be occluded in some of the composites. Also, the BFN to which the lipase is immobilized retards the denaturing effect of increased pH and/or temperature, which is exhibited by the shift in the optimal pH/temperature for activity. Previous studies [30] have suggested that upon immobilization, the active site becomes more exposed to solvent than that in the folded-dissolved lipase form. Activities of free and immobilized lipase were measured at various temperatures and the results are shown in Figure 3.

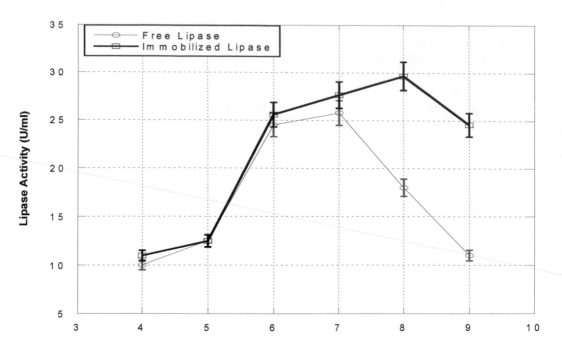

Figure 2. Effect of pH on the activity of free (**red line**) and immobilized (**black line**) lipases.

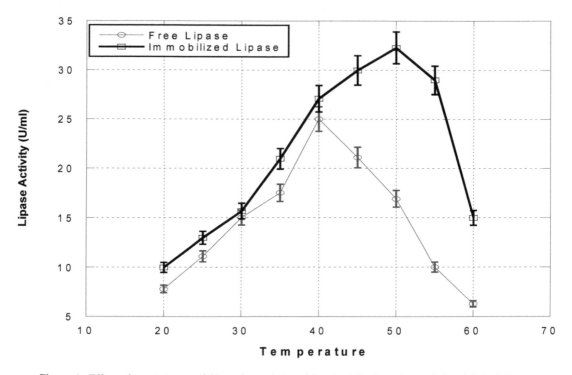

Figure 3. Effect of temperature (°C) on the activity of free (**red line**) and immobilized (**black line**) lipases at pH 8.0.

The optimum temperature of the immobilized lipase was 50 °C, whereas that of the free one was 40 °C. It can be concluded that the immobilization using BFN particles enhanced the thermal stability of lipase, which might be caused by multipoint attachment of the lipase to the support and/or by the hydrophobic interaction [31].

3.3. Recyclability of BFN-Lipase

After the reaction was complete, the immobilized lipase can be easily separated by a magnet. The results indicate that BFN-Lipase (5%) can be reused in biodiesel production without any treatment with 17% loss of activity after 5 cycles and a 66% loss in activity in the sixth cycle, with the activity of 29, 26, 25, 25 and 24 U/mL around the five cycles, respectively. Then the activity was decreased to 10 U/mL after the 6th cycle.

Therefore, no appreciable inhibition due to oil, glycerol or product was detected. Importantly, no aggregation of the particle was observed. The behavior of the prepared sample indicates that BFN particles prevent the undesirable hydrophilic interactions (glycerol and methanol adsorption, protein aggregation, particle aggregation) which are likely to occur in the hydrophobic oils. The results presented in Table 2 show the fatty acid profile of WCO and its biodiesel fatty acid methyl ester.

Table 2. Fatty acid profile of WCO and its biodiesel fatty acid methyl ester.

Fatty Acids	Formula	Common Acronym	Methyl Esters	% Composition by Mass
Oleic acid	$C_{17}H_{33}COOH$	C18: 0	Methyl oleate	46.5d ± 0.91
Palmitic acid	$C_{15}H_{31}COOH$	C16: 0	Methyl palmitate	30.9c ± 0.77
Stearic acid	$C_{17}H_{35}COOH$	C18: 0	Methyl strearate	09.0b ± 0.56
Linoleic acid	$C_{17}H_{31}COOH$	C18: 2	Methyl linoleate	08.5b ± 0.35
Linolenic acid	$C_{17}H_{29}COOH$	C18: 3	Methyl linolenate	05.1a ± 0.28
LSD				02.95

Note: Different letters indicate significant differences between treatments (Duncan test, $P \leqslant 0.05$). Means in each column followed by the same letter are not significantly different.

3.4. Production of Biodiesel Using the Transesterification Process

One liter sample size of waste cooking oil was heated to 60 °C to remove any free water and allowed to settle for 24 h before reacting with methanol.

3.4.1. Effect of Methanol/Waste Cooking Oil Molar Ratio on Biodiesel Production

Stoichiometrically, three moles of methanol are required for each mole of triglyceride, but in practice, a higher molar ratio is needed to drive the reaction towards completion and produce more fatty acid methyl esters (FAMEs) as products.

In this study, biodiesel was obtained under transesterification conditions as follows: reaction temperature, 45 °C; catalyst amount, 5.0 wt %; reaction time, 4 h. The effect of methanol: oil molar ratio increases from 1:1 to 4:1, as listed in Table 3, in which the biodiesel yield increases as the molar ratio increases from 1:1 to 4:1. The maximum biodiesel yield of 90% is obtained when the molar ratio is very close to 4:1. However, beyond the molar ratio of 5:1, the excessively added methanol has no significant effect on the production yield. In contrast, when the amount of methanol is over 4:1, glycerol separation becomes more difficult, thus decreasing the biodiesel yield. Based on this, the optimum molar ratio of methanol to oil is 4:1.

Table 3. Factors affecting biodiesel production using transesterification process.

Factors Affecting Biodiesel Production		Biodiesel Yield (%)
Methanol/oil ratio (mole/mole)	1:1	42a ± 1.41
	2:1	65b ± 2.12
	3:1	87d ± 0.707
	4:1	**90e ± 1.76**
	5:1	76c ± 0.707
LSD		03.75
Enzyme concentration (%)	3	69a ± 0.707
	5	**89b ± 1.06**
	10	91bc ± 1.41
	15	94c ± 0.707
	20	97d ± 1.909
LSD		02.25
Reaction temperature (°C)	15	55a ± 0.494
	25	73b ± 1.41
	35	81c ± 0.636
	45	**91d ± 1.41**
	55	75b ± 0.707
LSD		05.95
Reaction time (h)	2	50a ± 0.707
	4	**88d ± 1.06**
	6	88d ± 1.41
	8	61c ± 0.848
	10	58b ± 0.707
LSD		03.10
Shaking speed (rpm)	100	43a ± 1.41
	200	52b ± 1.48
	300	71c ± 1.41
	400	**88d ± 0.636**
	500	69c ± 1.13
LSD		09.05

Note: Different letters indicate significant differences between treatments (Duncan test, $P \leqslant 0.05$). Means in each column followed by the same letter are not significantly different.

3.4.2. Effect of Enzyme Concentration Percentage on Biodiesel Production

The effect of catalyst dosage was investigated with the mass ratio of immobilized lipase catalyst to waste cooking oil varying within the range of 3.0%–20.0%, under otherwise identical conditions as reaction temperature, 45 °C; methanol/oil, 3:1; reaction time, 4 h. The biodiesel yield was found to increase with increasing catalyst dosage, and the maximum was obtained at the dosage of 20% with the value being 93% at 4 h Table 3. It should be noted that we could not continue increasing catalyst concentration over 20% due to the high cost of immobilized lipase. So, we can deduce that 89% of biodiesel obtained at 5% enzyme concentration is the superior result.

3.4.3. Effect of Reaction Temperature on Biodiesel Production

In this part of the experiments, the reaction temperature was varied within a range of 15–55 °C. Transesterification conditions: methanol/oil, 3:1; catalyst amount, 5.0 wt % (of the feed mass of waste cooking oil, similarly from now on); reaction time (4 h). The results listed in Table 3 indicated that the reaction rate was slow at low temperatures and the biodiesel yield was only 55% at 15 °C after 4 h. The biodiesel yield increased with the increase of the reaction temperature to nearly 91% at 45 °C, but at higher temperatures ($T > 55$ °C), the methanol was vaporized and formed a large number of bubbles in the interface, which inhibited the increase of biodiesel yield, then the yield of biodiesel decreased significantly. Thus, the optimum reaction temperature was 45 °C.

3.4.4. Effect of the Reaction Time on Biodiesel Production

The reaction time tells the fastness of the transesterification process of a particular feedstock. The results in Table 3 indicated that the ester content increased with reaction time from 2 h onward and reached a maximum at a reaction time of 4 h at reaction temperature, 45 °C; methanol/oil, 3:1; reaction time, 4 h; catalyst amount, 5.0 wt % of the feed mass of waste cooking oil, and then remained relatively constant with increasing further the reaction time. An extension of the reaction time from 4 to 6 h had no significant effect on the conversion of triglycerides but led to a reduction in the product yield. Because longer reaction enhanced the hydrolysis of esters (reverse reaction of transesterification), this resulted in the loss of esters.

3.4.5. Effect of Shaking Speed on Biodiesel Production

The agitation rate improves the mixing during the transesterification process. It increases the intact area between oils and immobilized lipase enzyme methanol solution and facilitates the initiation of the reaction.

Transesterification conditions: methanol/oil, 3:1; catalyst amount, 5.0 wt % (of the feed mass of waste cooking oil, similarly hereinafter); reaction time, 4 h and at 45 °C. In methanolysis conducted with different rates of stirring, such as 100, 200, 300, 400 and 500 revolutions per minute (rpm), it was observed that the reaction of methanolysis was practically incomplete at 100 rpm and only exhibited a yield of 43%, as in Table 3. The yield was increased when the mixing intensity was accelerated up to 400 rpm.

3.5. Physicochemical Properties of the Used Cooking Oil and Its Corresponding Production of Biodiesel

3.5.1. Kinematic Viscosity

Viscosity refers to a fluid's resistance to flow at a given temperature. Fuel that is too viscous can hinder the operation of an engine. Kinematic viscosity measures the ease with which a fluid will flow under force. It is different from absolute viscosity, also called dynamic viscosity. Kinematic viscosity is obtained by dividing the dynamic viscosity by the density of the fluid [32]. Kinematic viscosity allows comparison between the engine performances of different fuels, independent of the density of the fuels. Two fuels with the same kinematic viscosity should have the same hydraulic fuel properties, even though one fuel may be denser than the other.

The kinematic viscosity of the used waste cooking oil was 60.0 cSt and its corresponding biodiesel was 5.83 cSt (Table 4).

Table 4. Physicochemical properties of the produced biodiesel.

No.	Characteristics	Result	Unit	Test Method
1	Kinematic viscosity at 40 °C	5.83	$mm^2 \cdot s^{-1}$	ASTM D445
2	Density at 15.5 °C	0.850	$g \cdot cm^{-3}$	ASTM D1298
3	Calorific value	43.1	MJ/Kg	ASTM D-224
4	Total sulfur content	0.050	mass%	ASTM D4294
5	Flash point	188	°C	ASTM D92
6	Pour point	−9	°C	ASTM D97
7	Cloud point	−3	°C	ASTM D2500
8	Cetane number	59.5	—	ASTM D613
9	Water content	0.091	vol%	ASTM D6304
10	Acid number	0.182	mg KOH g^{-1}	ASTM D664
11	Distillation temperature (DT)	95% Recovery at 340	°C	ASTM D86
12	Iodine number	102	mg I_2/100 g oil	ASTM D4737
13	Saponification value	206	mg KOH/g oil	ASTM D 5558

3.5.2. Density at 15 °C

The density of biodiesel is used to judge the homogeneity of biodiesel. This property is important, mainly in airless combustion systems because it influences the efficiency of atomization of the fuel [33]. The result obtained (Table 4) showed that the methyl ester produced in this study had a density of 0.850 g/mL, which falls in the range 0.82–0.87 g/mL, specified according to Egyptian diesel oil.

3.5.3. Distillation at Atmospheric Pressure

Distillation is used to determine the boiling range of the biodiesel product quantitatively. Even though distillation at the atmospheric pressure provides some clues or information about the properties of methyl esters such as the boiling range of the biodiesel product, distillation at the reduced pressure (vacuum distillation) is required because the biodiesel will thermally decompose using atmospheric distillation. The plot pattern of the volume of distillate recovered *versus* temperature in Figure 4 indicates this fact. The decrease in temperature during the distillation process shows the decomposition of methyl ester which results in the formation of low boiling molecular substances.

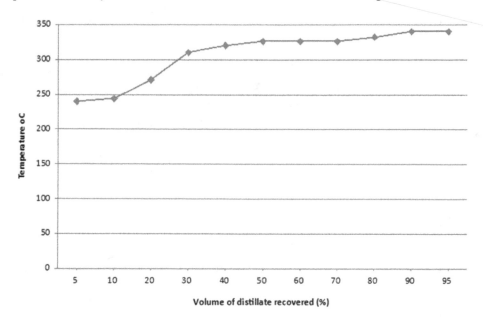

Figure 4. Relationship between volume of distillate recovered and temperature °C.

3.5.4. Flash Point (FP)

The flash point temperature is the measure of the fuel to form a flammable mixture with air. It is only one of many properties which must be considered in assessing flammability hazard of material. For biodiesel, a flash point of below 130 °C is found to be out of specification (ASTM D 6751, 2009). The measured flash point of this biodiesel was 188 °C (Table 4), indicating very small or negligible methanol levels in the biodiesel. This high flash point can prevent auto ignition and fire hazard at a high temperatures during transportation and storage periods.

3.5.5. Heat of Combustion/Calorific Value

Gross heat of combustion (HG) is a property proving the suitability of using fatty compounds as diesel fuel. The calorific value of the obtained biodiesel; 43.1 MJ/Kg is an acceptable value (Table 4). Although it is lower than the Egyptian standards for petrodiesel (44.3 MJ/Kg), it is higher than that of the European standard for biodiesel (32.9 MJ/Kg).

3.5.6. The Cetane Number

The cetane number (CN) is a dimensionless descriptor of the ignition quality of a diesel fuel (DF) [34]. It is a prime indicator of DF quality. It is a measure of how easily the fuel will ignite in the engine. The CN is related to the time that passes between injection of the fuel into the cylinder and the start of ignition. Because of the higher oxygen content, biodiesel has a higher cetane number compared to petroleum diesel. The cetane index of waste cooking oil from the experiment was found to be 59.5 (Table 4). The cetane number of methyl esters of rapeseed oil, soybean oil, palm oil, lard and beef tallow were found to be 58, 53, 65, 65 and 75, respectively [35]. Among these biodiesel feed stocks, beef tallow has the highest cetane number. The higher cetane number indicates, the higher engine performance of beef tallow compared to other fuels, resulting in the lower emission of all pollutants other than oxides of nitrogen (NOx). As beef tallow has the higher amount of saturated fatty acids, the increase in the saturated fatty acids content positively enhanced the cetane number of biodiesel. The obtained data revealed the higher cetane number, 59.5 compared to Egyptian petrodiesel standards of 55, but within the recommended universal biodiesel standards (min 47).

3.5.7. Cloud Point (CP) and Pour Point

The cloud and pour points are also important properties of biodiesel fuel. The cloud point is the temperature at which crystals first appear in the fuel when cooled [36]. At temperatures below CP, larger crystals fuse together and form large agglomerates that can restrict or cut off flow through fuel lines and filters and cause start-up and performance problems the next morning [37]. The temperature at which crystal agglomeration is extensive enough to prevent free pouring of fluid is determined by measurement of its pour point [38]. These properties are related to the use of biodiesel in cold temperatures. The cloud and pour points are (-3 and -9 °C), respectively (Table 4). So the produced biodiesel would be more suitable to cold conditions as compared to petrodiesel.

3.5.8. Water and Sulfur Content

Biodiesel is sulfur free, but this test is an indicator of contamination of protein material and/or carryover catalyst material or neutralization material from the production process (ASTM D4294). Sulfur reduces the function of catalysts and causes SOx emissions as in diesel engines and has an adverse effect with the formation of black smoke. Hence, a low level of sulfur, or no sulfur, is beneficial to the performance of the diesel engine due to the lower emissions and the decrease in the levels of corrosive sulfuric acid accumulating in the engine crankcase oil. Sulfur content of biodiesel is from 0.031 to 0.05. The synthesized kind is found to be in this range.

Water can cause corrosion of tanks and equipment, and if detergent is present, the water can cause emulsions or a hazy appearance. Water supports microbiological growth at the fuel/water interfaces in fuel systems. Biodiesel can contain as much as 1500 parts per million of dissolved water [39]. The produced biodiesel is characterized by low water and sulfur content, 0.091% and 0.050% respectively (Table 4), compared with the Egyptian petrodiesel standards, <0.15% and 1.2%, respectively.

3.5.9. Iodine Number

The iodine value is an important measure for the determination of the unsaturation in fatty acids, which is only dependent on the vegetable oil. This property greatly influences fuel oxidation and the deposits formed in the injector of diesel engines [40]. With the increasing iodine value of unsaturated fatty acids, the effect of polymerization is stronger. In this study, the iodine value of the WCO used was 125 mg I$_2$/100 g oil, and 102 mg I$_2$/100 g oil for the corresponding biodiesel (Table 4). According to EN 14214, the biodiesel must have an iodine value <120 mg I$_2$/100 g oil in the sample.

3.5.10. Acid Number

The acid number is the quantity of base, expressed as milligrams of KOH per gram of sample, required to titrate a sample to a specified end point. Acid number determination is an important test to assess the quality of a particular biodiesel. It can indicate the degree of hydrolysis of the methyl ester, a particularly important aspect when considering storage and transportation, as large quantities of free fatty acids can cause corrosion in tanks [41]. The acid value of feedstock and produced methyl esters was 1.90 and 0.182 (mg KOH/g oil), respectively (Table 4). The average percentage decrease of acid value from WCO to the produced biodiesel was about 90.4%, indicating a good transesterification process. This result meets the EN 14214 and D-6751 standards, but it was higher than that of Egyptian petrodiesel standards.

3.5.11. Saponification Value (mg KOH/g Oil)

Saponification value is the quantity of base, expressed as milligrams of KOH per gram of sample, required to saponify one gram of fat or oil. The obtained result (Table 4) indicated that the methyl esters produced had higher saponification value 206 (mg KOH/g oil) than the corresponding oils 201 (mg KOH/g oil).

4. Conclusions

Biodiesel production technology is competitive in terms of it being a low-cost and alternative source of energy, which should be not only sustainable but also environmentally friendly. Catalyst immobilization addresses several issues: repetitive and continuous use, localization of the interaction, prevention of product contamination, reduction of effluent problems and material handling, and effective control of the reaction parameters. In this study, we developed novel hydrophobic magnetic nanoparticles (barium ferrite nanoparticles) for lipase immobilization. This allows additional characteristics compared to other conventional support materials, such as selective and easy enzyme recovery from the medium under the magnetic force. The second advantage is that we can reuse the BFN-Lipase without any rinsing treatment for five cycles, with an activity of 29, 26, 25, 25 and 24 U/mL during the five cycles. The activities of free and immobilized lipase were measured at various pH and temperatures values. The optimum pH of the immobilized lipase was 8.0, whereas that of the free one was 7.0, and the optimum temperature of the immobilized lipase was 50 °C, whereas that of the free one was 40 °C. The fatty acid profile of collected waste cooking oil as feedstock for lipase production was determined using GC/MS and showed the presence of oleic, palmitic, stearic, linoleic and linolenic acids. The enzymatic conversion of waste cooking oil to biodiesel (transesterification) was carried out using methyl alcohol. Different factors affecting transesterification reaction, such as methanol/oil molar ratio, reaction temperature, shaking, reaction time and enzyme concentration, were studied. The previous results showed that the use of immobilized lipase enzyme on barium ferrite nanoparticles had high efficiency as a catalyst in the process of biodiesel production from waste cooking oil. Tests also demonstrated that the properties of biodiesel attained were fit as a substitute for diesel in addition to being a friend of the environment.

Acknowledgments: The authors would like to thank the Nanotechnology Research Unit (Principal Investigator. Ahmed El-Batal), Drug Radiation Research Department, National Center for Radiation Research and Technology (NCRRT), Egypt, for financing and supporting this study under the project Nutraceuticals and Functional Foods Production by using Nano/Biotechnological and Irradiation Processes.

Author Contributions: Ahmed I. El-Batal and Ayman A. Farrag conceived and designed the experiments; Ahmed M. El-Khawaga performed the experiments; Ahmed I. El-Batal and Mohamed A. Elsayed analyzed the data; and contributed reagents/materials/analysis tools; Ahmed I. El-Batal and Ahmed M. El-Khawaga wrote the paper. All authors read and approved the final manuscript.

Conflicts of Interest: The authors declare that there is no conflict of interests regarding the publication of this article.

References

1. Yu, C.Y.; Huang, L.Y.; Kuan, I.; Lee, S.L. Optimized Production of Biodiesel from Waste Cooking Oil by Lipase Immobilized on Magnetic Nanoparticles. *Int. J. Mol. Sci.* **2013**, *14*, 24074–24086. [CrossRef] [PubMed]

2. Demirbas, A. Biodiesel production via non-catalytic SCF method and biodiesel fuel characteristics. *Energy Convers. Manag.* **2006**, *47*, 2271–2282. [CrossRef]

3. Bala, B.K. Studies on biodiesels from transformation of vegetable oils for diesel engines. *Energy Educ. Sci. Technol.* **2005**, *15*, 1–45.

4. Pighinelli, A.L.; Ferrari, R.A.; Miguel, A.M.; Park, K.J. High oleic sunflower biodiesel: Quality control and different purification methods. *Grasas Aceites* **2011**, *62*, 171–180.

5. Pogaku, R.; Raman, J.K.; Ravikumar, G. Evaluation of Activation Energy and Thermodynamic Properties of Enzyme-Catalysed Transesterification Reactions. *Adv. Chem. Eng. Sci.* **2012**, *2*, 150–154. [CrossRef]

6. Ahmia, A.; Danane, F.; Bessah, R.; Boumesbah, I. Raw material for biodiesel production. Valorization of used edible oil. *Rev. Energies Renouv.* **2014**, *17*, 335–343.

7. Gupta, R.; Kumari, A.; Syal, P.; Singh, Y. Molecular and functional diversity of yeast and fungal lipases: Their role in biotechnology and cellular physiology. *Prog. Lipid Res.* **2015**, *57*, 40–54. [CrossRef] [PubMed]

8. Babaei, M.; Karimi, A.; Hejazi, M.A. Use of mesoporous MnO_2 as a support for immobilization of lipase from *Candida rugosa*. *Chem. Ind. Chem. Eng. Q.* **2014**, *20*, 371–378. [CrossRef]

9. Villeneuve, P.; Muderhwa, J.M.; Graille, J.; Haas, M.J. Customizing lipases for biocatalysis: A survey of chemical, physical and molecular biological approaches. *J. Mol. Catal. B Enzym.* **2000**, *9*, 113–148. [CrossRef]

10. Metin, A.Ü. Immobilization studies and biochemical properties of free and immobilized *Candida rugosa* lipase onto hydrophobic group carrying polymeric support. *Macromol. Res.* **2013**, *21*, 176–183. [CrossRef]

11. Mendes, A.A.; Castro, H.F.D.; Pereira, E.B.; Furigo Júnio, F. Application of lipases for wastewater treatment containing high levels of lipids. *Quím. Nova* **2005**, *28*, 296–305. [CrossRef]

12. Turinayo, Y.K.; Kalanzi, F.; Mudoma, J.M.; Kiwuso, P.; Asiimwe, G.M.; Esegu, J.P.O.; Balitta, P.; Mwanja, C. Physicochemical Characterization of Jatropha curcas Linn Oil for Biodiesel Production in Nebbi and Mokono Districts in Uganda. *J. Sustain. Bioenergy Syst.* **2015**, *5*, 104–113. [CrossRef]

13. Cesarini, S.; Infanzón, B.; Pastor, F.I.; Diza, P. Fast and economic immobilization methods described for non-commercial Pseudomonas lipases. *BMC Biotechnol.* **2014**, *14*, 1071–1075. [CrossRef] [PubMed]

14. El-Hadi, A.A.; Saleh, H.I.; Moustafa, S.A.; Ahmed, H.M. Immobilization of Mucor racemosus NRRL 3631 lipase and characterization of silica-coated magnetite (Fe3O4) nanoparticles. *Egypt. Pharm. J.* **2013**, *12*, 28–35.

15. El-Batal, A.I.; Farrag Ahmed, A.; Elsayed, M.A.; Soltan, A.M.; El-Khawaga, A.M. Enhancement of Lipase Biosynthesis by *Aspergillus nigerADM110* using Gamma Radiation. *Egypt. J. Med. Microbiol.* **2015**, *24*, 87–94.

16. Xu, J.; Sun, J.; Wang, Y.; Sheng, J.; Wang, F.; Sun, M. Application of iron magnetic nanoparticles in protein immobilization. *Molecules* **2014**, *19*, 11465–11486. [CrossRef] [PubMed]

17. Bahadur, D.; Rajakumar, S.; Kumar, A. Influence of fuel ratios on auto combustion synthesis of barium ferrite nano particles. *J. Chem. Sci.* **2006**, *118*, 15–21. [CrossRef]

18. Abedini Khorrami, S.; Islampour, R.; Bakhtiari, H.; Sadr, M.N.Q. The effect of molar ratio on structural and magnetic properties of $BaFe_{12}O_{19}$ nanoparticles prepared by sol-gel auto-combustion method. *Int. J. Nano Dimens.* **2013**, *3*, 191–197.

19. Lowry, O.H.; Rosebrough, N.J.; Farr, A.L.; Randall, R.J. Protein measurement with the Folin phenol reagent. *J. Boil. Chem.* **1951**, *193*, 265–275.

20. Pencreac'h, G.; Baratti, J.C. Hydrolysis of *p*-nitrophenyl palmitate in *n*-heptane by the *Pseudomonas cepacia* lipase: A simple test for the determination of lipase activity in organic media. *Enzym. Microb. Technol.* **1996**, *18*, 417–422. [CrossRef]

21. Liu, X.; Guan, Y.; Shen, R.; Liu, H. Immobilization of lipase onto micron-size magnetic beads. *J. Chromatogr. B* **2005**, *822*, 91–97. [CrossRef] [PubMed]

22. Xu, Q.-Q.; Li, Q.; Yin, J.-Z.; Guo, D.; Qiao, B.-Q. Continuous production of biodiesel from soybean flakes by extraction coupling with transesterification under supercritical conditions. *Fuel Process. Technol.* **2016**, *144*, 37–41. [CrossRef]

23. Milina, R.; Mustafa, Z. Gas chromatographic investigations of compositional profiles of biodiesel from different origin. *Pet. Coal* **2013**, *55*, 12–19.

24. Snedecor, G.; Cochran, W. *Statistical Methods*, 8th ed.; Iowa State University Press: Ames, IA, USA, 1989.

25. Kaur, T.; Srivastava, A. Effect of pH on Magnetic Properties of Doped Barium Hexaferrite. *Int. J. Res. Mech. Eng. Technol.* **2013**, *3*, 171–173.

26. Chauhan, C.; Jotania, R.; Jotania, K. Structural properties of cobalt substituted barium hexaferrite nanoparticles prepared by a thermal treatment method. *J. Nanosyst. Phys. Chem. Math.* **2013**, *4*, 363–369.

27. Wang, J.; Meng, G.; Tao, K.; Feng, M.; Zhao, X.; Li, Z.; Xi, H.; Xia, D.; Lu, J.R. Immobilization of lipases on alkyl silane modified magnetic nanoparticles: Effect of alkyl chain length on enzyme activity. *PLoS ONE* **2012**, *7*, e43478. [CrossRef] [PubMed]

28. Rani, K.; Goyal, S.; Chauhan, C. Novel approach of alkaline protease mediated biodegradation analysis of mustard oil driven emulsified bovine serum albumin nanospheres for controlled release of entrapped Pennisetum glaucum (Pearl Millet) amylase. *Am. J. Adv. Drug Deliv.* **2015**, *3*, 135–148.

29. Lee, D.-G.; Ponvel, K.M.; Kim, M.; Hwang, S.; Ahn, I.S.; Lee, C.H. Immobilization of lipase on hydrophobic nano-sized magnetite particles. *J. Mol. Catal. B Enzym.* **2009**, *57*, 62–66. [CrossRef]

30. Ting, W.-J.; Tung, K.-Y.; Giridhar, R.; Wu, W.T. Application of binary immobilized *Candida rugosa* lipase for hydrolysis of soybean oil. *J. Mol. Catal. B Enzym.* **2006**, *42*, 32–38. [CrossRef]

31. Ferrario, V.; Veny, H.; De Angelis, E.; Navarini, L.; Ebert, C.; Gardossi, L. Lipases immobilization for effective synthesis of biodiesel starting from coffee waste oils. *Biomolecules* **2013**, *3*, 514–534. [CrossRef] [PubMed]

32. Koria, L.; Nithya, G. Analysis of *Datura stramonium* Linn. biodiesel by gas chromatography-mass spectrometry (gc-ms) and influence of fatty acid composition on the fuel related characteristics. *J. Phytol.* **2012**, *4*, 6–9.

33. Felizardo, P.; Correia, M.J.N.; Raposo, I.; Mendes, J.F.; Rui, B.; Bordado, J.M. Production of biodiesel from waste frying oils. *Waste Manag.* **2006**, *26*, 487–494. [CrossRef] [PubMed]

34. Chhetri, A.B.; Watts, K.C.; Islam, M.R. Waste cooking oil as an alternate feedstock for biodiesel production. *Energies* **2008**, *1*, 3–18. [CrossRef]

35. Hilber, T.; Ahn, E.; Mittelbach, M.; Schmitd, E. Animal fats perform well in biodiesel. *Render Magazine*, February 2006; pp. 16–18.

36. Shrestha, D.; Van Gerpen, J.; Thompson, J. Effectiveness of cold flow additives on various biodiesels, diesel, and their blends. *Trans. ASABE* **2008**, *51*, 1365–1370. [CrossRef]

37. Dunn, R.O. Cold flow properties of biodiesel: A guide to getting an accurate analysis. *Biofuels* **2015**, *6*, 115–128. [CrossRef]

38. Ibiari, N.N.; El-Enin, S.A.A.; Attia, N.K.; El-Diwani, G. Ultrasonic comparative assessment for biodiesel production from rapeseed. *J. Am. Sci.* **2010**, *6*, 937–943.

39. Brown, E.K.; Palmquist, M.; Luning Prak, D.J.; Mueller, S.S.; Bowen, L.L.; Sweely, K.; Ruiz, O.N.; Trulove, P.C. Interaction of Selected Fuels with Water: Impact on Physical Properties and Microbial Growth. *J. Pet. Environ. Biotechnol.* **2015**, *6*. [CrossRef]

40. Cancela, Á.; Maceiras, R.; Alfonsín, V.; Sánchez, Á. Transesterification of waste frying oil under ultrasonic irradiation. *Eur. J. Sustain. Dev.* **2015**, *4*, 401–406. [CrossRef]

41. Akanawa, T.T.; Moges, H.G.; Babu, R.; Bisrat, D.; Akanawa, T.T.; Babu, R.; Bisrat, D. Castor Seed from Melkasa Agricultural Research Centre, East Showa, Ethiopia and it's biodiesel performance in Four Stroke Diesel Engine. *Int. J. Renew. Energy Dev.* **2014**, *3*, 99–105.

6

Stable Gene Regulatory Network Modeling From Steady-State Data

Joy Edward Larvie [1], Mohammad Gorji Sefidmazgi [1,‡], Abdollah Homaifar [1,*], Scott H. Harrison [2], Ali Karimoddini [1] and Anthony Guiseppi-Elie [3]

[1] Department of Electrical and Computer Engineering, North Carolina A&T State University, 1601 E. Market Street, Greensboro, NC 27411, USA; jelarvie@aggies.ncat.edu (J.E.L.); mgorjise@email.arizona.edu (M.G.S.); akarimod@ncat.edu (A.K.)

[2] Department of Biology, North Carolina A&T State University, 1601 E. Market Street, Greensboro, NC 27411, USA; scotth@ncat.edu

[3] Department of Biomedical Engineering, Texas A&M University, 5045 ETB, College Station, TX 77843, USA; guiseppi@tamu.edu

* Correspondence: homaifar@ncat.edu

† This paper is an extended version of our paper published in 41st Annual Northeast Biomedical Engineering Conference (NEBEC).

‡ Current affiliation: School of Information, The University of Arizona, 1103 E. 2nd St, Tucson, AZ 85721, USA

Academic Editor: Aldo R. Boccaccini

Abstract: Gene regulatory networks represent an abstract mapping of gene regulations in living cells. They aim to capture dependencies among molecular entities such as transcription factors, proteins and metabolites. In most applications, the regulatory network structure is unknown, and has to be reverse engineered from experimental data consisting of expression levels of the genes usually measured as messenger RNA concentrations in microarray experiments. Steady-state gene expression data are obtained from measurements of the variations in expression activity following the application of small perturbations to equilibrium states in genetic perturbation experiments. In this paper, the least absolute shrinkage and selection operator-vector autoregressive (LASSO-VAR) originally proposed for the analysis of economic time series data is adapted to include a stability constraint for the recovery of a sparse and stable regulatory network that describes data obtained from noisy perturbation experiments. The approach is applied to real experimental data obtained for the SOS pathway in *Escherichia coli* and the cell cycle pathway for yeast *Saccharomyces cerevisiae*. Significant features of this method are the ability to recover networks without inputting prior knowledge of the network topology, and the ability to be efficiently applied to large scale networks due to the convex nature of the method.

Keywords: gene regulatory network; reverse engineering; sparse network; stable network; convexity

1. Introduction

A number of technological advances, such as oligonucleotide arrays, serial analysis of gene expression (SAGE) and cDNA microarrays [1], have enabled biomedical researchers to expeditiously and simultaneously collect large amounts of metabolomic, transcriptomic, proteomic data [2] in a single experiment, providing a wealth of information for elucidating gene regulation, functions and interactions [3,4]. Over time, repositories such as the Gene Expression Omnibus (GEO) [5] and the Biological General Repository for Interaction Datasets (BioGRID) [6] are mapping functional information and ontologies to expression data sets [2,7]. Gene expression measurement data acquired from microarray experiments typically occur in two contexts: steady-state data which provides information on interaction directions, and temporal data that allows for the investigation of temporal patterns in biological networks [8,9].

Owing to their inherent ability to encapsulate the high dimensional data of biological processes and pathways, networks have become an important tool in functional genomics [10,11]. Researchers refer to any such network that provides a system level interaction among genes as a gene regulatory network (GRN) [12,13]. GRNs are usually represented by directed graphs with nodes as genes, and edges depicting either an inhibition (negative regulation) or an activation (positive regulation) imposed by a gene over another through the production of a protein [14,15].

The process of identifying genetic interactions from measured gene expression data is referred to as reverse engineering or network inference or recovery [7]. Inferring the topology of GRNs and isolating functional subnetworks are computationally challenging tasks in contemporary functional genomics, and these efforts are valuable for advancing scientific insight and for capitalizing on the time and costs associated with experimental data [16–19]. GRNs typically contain information about the pathway to which a gene belongs and the genes it interacts with [16], and this helps to reveal potential pathway initiators and drug targets [8]. Further analysis, to map interactions among phenotypic and genotypic characteristics, can provide a framework for the identification of biomarkers for medical diagnosis and prognosis [20,21].

A plethora of modeling approaches such as co-expression clustering [22], Boolean network [23,24], Bayesian network [25] and ordinary differential equation (ODE) [8] models have been proposed for recovering genetic networks. Cluster analysis and the sequential search for patterns of gene expression related with some pathological state of interest usually provide only indirect information about the structure of the network [7]. Alternatively, grouping of co-expressed genes may be achieved using information-theoretic methods. Both approaches, however, lack causality [9]. Causality may be recovered through Bayesian networks which can handle directed graphs [9,26]. However, Bayesian networks typically do not accommodate cycles, and, hence, are unable to handle feedback motifs that are common in gene regulatory networks [26]. Causality and feedback motifs, however, are no longer a problem when the network is modeled as a set of differential equations [26]. Excellent as they are at modeling causality and feedback motifs, differential equations are only suitable for small-scale networks [9].

These existing techniques, however, rely heavily on temporal expression data which can be very difficult to acquire, and also require high computational effort [8,26]. Major considerations of sparsity, stability and causality must be captured in the biological network recovery process [2]. In this paper, the least absolute shrinkage and selection operator-vector autoregressive (LASSO-VAR) model, originally proposed for the analysis of economic time series data in [27], is adapted to include a stability constraint defined and used by [26] for the recovery of sparse and stable regulatory networks that describe steady-state data obtained from noisy perturbation experiments. The fact that LASSO-VAR is a vector autoregressive process implies that Granger causality can be inferred. The technique only requires one tuning parameter, which works to penalize non-sparse networks. The selection of this parameter is based on its mean square forecast error. The identification algorithm proposed is applicable for the identification of regulatory roles of individual genes and control genes in the network. It is also applicable for identifying genes that directly impact the bioactivity of a compound in the cell. The approach is applied to real experimental data obtained for the SOS pathway in *Escherichia coli* and the cell cycle pathway for yeast *Saccharomyces cerevisiae*. The significant features of this method are the ability to recover networks without *a priori* knowledge of the network topology, and to be efficiently applied to large scale networks due to the convex nature of the method.

2. Methodology

This section introduces the stable LASSO-VAR, the identification technique being adapted for reverse engineering gene regulatory networks from steady-state data [28]. In its original form, the LASSO-VAR technique described in [27] finds applications in the analysis and prediction of economic and financial time series. It is an extension of the VAR model to include a selection and shrinkage operator known as the LASSO. The inherent advantages of the LASSO-VAR are the ability to perform dimension reduction and variable selection, as well as being able to test Granger causality. For this reason, this paper adapts the LASSO-VAR concept and incorporates a stability constraint as a convex constraint to allow for the inference of a stable genetic network from steady-state data.

2.1. Network Identification Approach

The vector autoregressive (VAR) model is known to be one of the most flexible and easy to use models for analyzing multivariate time series [27]. It has found applications in neurosciences for the estimation of functional connectivity between several brain areas [29], and most recently in system biology for the reconstruction of gene regulatory networks [2].

In the general case, an N-dimensional multiple time series gene expression data $y_1, ..., y_T$ with $y_t = (y_{1t}, ..., y_{Nt})'$ can be assumed to be generated by a stationary, stable VAR(p) process as [30]:

$$y_t = v + A_1 y_{t-1} + ... + A_p y_{t-p} + u_t, \tag{1}$$

where p denotes the order of the vector autoregressive process (*i.e.*, the vector autoregressive lag length), y_t is an $(N \times 1)$ random vector, A_i is a fixed $(N \times N)$ coefficient matrix, $v = (v_1, ..., v_N)'$ is a fixed $(N \times 1)$ vector of intercept terms allowing for the possibility of a nonzero mean $E(y_t)$, $u_t = (u_{1t}, ..., u_{Nt})'$ is a N-dimensional white noise, thus, $E(u_t) = 0$, $E(u_t u_t') = \Sigma_u$ and $E(u_t u_s') = 0$ for $s \neq t$. The covariance matrix Σ_u is assumed to be nonsingular. The framework of the general VAR(p) allows for the testing of Granger causality [31]. The concept of Granger causality is founded on the idea that a cause must precede an effect. This concept was originally proposed by Granger in [32].

In the present context, however, the number of genes considered in most microarray experiments generally runs from several thousands to millions, thereby making it impossible to accommodate the most general form of the Granger causality test [31]. Thus, a VAR(1) (VAR of order one) model is usually employed to allow for a pairwise comparison study as seen in [31] and [29]. Stated simply, in the case of a VAR(1), if gene b at time t is affected by a gene a at time $(t-1)$, the latter should help to predict the target gene expression [29].

A first order VAR model is defined as [30]:

$$y_t = v + A y_{t-1} + u_t. \tag{2}$$

For convenience, (2) is usually expressed in compact matrix notation as [30]:

$$\mathbf{Y} = \mathbf{v} + \mathbf{A}\mathbf{Z} + \mathbf{U}, \tag{3}$$

where $\mathbf{Y} = (y_1, y_2, ..., y_T)$ is an $N \times T$ data or response matrix, $\mathbf{A}$ is an $N \times N$ unknown coefficient matrix, $\mathbf{Z} = (Z_0, ..., Z_{T-1})$ is an $N \times T$ covariate matrix and $\mathbf{U} = (u_1, ..., u_T)$ is an $N \times T$ matrix.

The solution to (3) is given as follows [30]:

$$A = ((ZZ^T)^{-1}Z \otimes I_N)Y, \tag{4}$$

where I_N is an $N \times N$ identity matrix, and $\otimes$ is the Kronecker product or direct product.

In high dimensional space, the VAR processes become computationally intractable [27]. As such, the model usually contains unwanted parameters which leads to less efficient parameter estimates [33]. The LASSO-VAR model addresses the intractability issue by zeroing some elements of the coefficient matrix, which removes unnecessary variables [27,33]. The requirement that the coefficient matrix, A, be sparse is due to the loose connectivity that biological networks generally exhibit [8,26].

This requirement is addressed by applying an L_1 penalty to the convex least squares objective function, resulting in [27]:

$$\frac{1}{2}\|Y - v - AZ\|_F^2 + \lambda\|A\|_1,\tag{5}$$

where $\|X\|_F^2 = \sum_{i=1}^m \sum_{j=1}^n |x_{ij}|^2$ is the square of the Frobenius norm of X (*i.e.*, the sum of the absolute squares of its elements), $\|X\|_1 = \Sigma_{jk}|X_{jk}|$ is the sum of the absolute values of X, and $\lambda \geq 0$ is a penalty parameter.

According to [30], if all eigenvalues of the coefficient matrix A of the VAR(1) process have absolute values less than 1, the sequence $A^i, i = 0, 1, ...$ is absolutely summable; as such, the infinite sum $\sum_{i=1}^{\infty}$ exists in mean square. Hence, in general, a VAR(1) is said to be stable *iff* all eigenvalues of A have absolute value less than one. It is mathematically equivalent to [30]:

$$\mathbf{det}(I_N - A\tau) \neq 0 \text{ for } |\tau| \leq 1.\tag{6}$$

The original formulation of the LASSO-VAR technique by [27] lacks the ability to infer a stable network. This setback means that stability cannot be inferred from gene perturbation experiments. To solve this inadequacy, a stability constraint that relies on the theorem by Geršgorin as discussed in [34] is incorporated into the LASSO-VAR objective function (5).

Geršgorin's theorem states that all the eigenvalues of an $n \times n$ matrix $A = (a_{ij})$ are in the union of the discs whose boundaries are circles $C_1, C_2, ..., C_n$ with centers at the points $a_{11}, a_{22}, ..., a_{nn}$ and the radii are: $r_i = \sum_{j=1, j\neq i}^n |a_{ij}|$. Stated more compactly, every eigenvalue τ must be contained in at least one of the circles characterized by the rows of A for an $n \times n$ matrix A. In essence, the eigenvalues of a square matrix can not be too far from its diagonal entries. It follows that [34]:

$$|\tau - a_{ii}| \leq \sum_{j=1, j\neq i}^n |a_{ij}|, \; i = 1, 2, ..., n.\tag{7}$$

As such, the real part of each eigenvalue must satisfy one of the conditions [34]

$$\text{Re}[\tau] \leq a_{ii} + \sum_{j=1, j\neq i}^n |a_{ij}|, \quad i = 1, 2, ..., n.\tag{8}$$

Since $V^{-1}AV$ and A have the same eigenvalues for all invertible matrix V, it is possible to apply Geršgorin's theorem to $V^{-1}AV$. For a good choice of V, one can find some tighter bounds for the eigenvalues. [26]. A particularly convenient choice is $V \triangleq \text{diag}(v_1, ..., v_n)$, with $v_i > 0$ for all $i = 1, ..., n$. Then, $V^{-1}AV = (v_j a_{ij}/v_i)$. It follows therefore that, $\forall V \in \mathbf{V}$, the real part of an eigenvalue of A must satisfy [26]:

$$\text{Re}[\tau] \leq a_{ii} + \sum_{j=1, j\neq i}^n \frac{v_j}{v_i}|a_{ij}|, \quad i = 1, 2, ..., n.\tag{9}$$

The stability requirement of the algorithm stems from the steady-state nature of the gene expression data adopted. Stability of the network simply refers to the robustness of the network to topology and parameter changes, as well as instrumental and biological noise [35]. The inherent stability description of the VAR model therefore allows the incorporation of a stability constraint that helps to address the specification of a stable gene regulatory network from steady-state data.

Incorporating this concept as a constraint retains the convex nature of the objective function; hence, it has the associated properties of scalability and global optimality. The resulting overall optimization problem is given as:

$$\textbf{minimize } \frac{1}{2}\|Y - AZ\|_F^2$$

$$\textbf{subject to } \|A\|_1 < \lambda, \, 0 \leq \lambda \leq 1 \qquad (10)$$

$$a_{ii} \leq - \sum_{j=1, j \neq i}^{n} \frac{v_j}{v_i} |a_{ij}|, \, i = 1, ..., n, \, v_i, \, v_j > 0$$

The choice of v_i is dependent on the stability requirements defined in the problem formulation. Zavlanos *et al.* [26] provide a convenient way to choose the weights v_i.

Define the deleted absolute sum for row i as $R_i(A) \triangleq \sum_{j \neq i} |a_{ij}|$. Then, for $\beta \triangleq \frac{1}{n} \sum_{i=1}^{n} (|a_{ii}| - R_i(A))$ the weights v_i are chosen given Figure 1 as follows [26]:

$$v_i \triangleq \begin{cases} 1 + \dfrac{|a_{ii}| - R_i(A) - \beta}{\delta + (|a_{ii}| - R_i(A) - \beta)}, & \text{if } |a_{ii}| - R_i(A) > \beta \\[2ex] \dfrac{\delta}{\delta - (|a_{ii}| - R_i(A) - \beta)}, & \text{if } |a_{ii}| - R_i(A) \leq \beta \end{cases} \qquad (11)$$

Solving the constrained optimization problem in (10) iteratively yields a sparse, stable coefficient matrix that models the causal interactions among the genes under observation as desired.

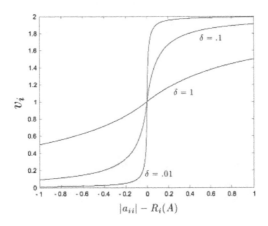

Figure 1. Plot of v_i as a function of the entries $|a_{ii}| - R_i(A)$, for average $\beta = 0$ and different values of the parameter $0 < \delta \leq 1$. Taken from [26].

In shrinkage problems, a formalized approach for the selection of an optimal penalty parameter value is achieved by employing either a k-fold or a leave-one-out cross validation [27]. Due to time-dependence, however, traditional cross-validation techniques are not well-suited for the problem formulation. The optimal penalty parameter is selected by minimizing the one-step ahead mean square forecast error (MSFE) [27]. This process starts by dividing the data into three periods: one for initialization, one for training, and one for forecast evaluation. Two time indices are defined as: $T_1 = \lfloor \frac{T}{3} \rfloor$, and $T_2 = \lfloor \frac{2T}{3} \rfloor$, where $\lfloor x \rfloor$ is the largest integer less than or equal to x. The period $T_1 + 1$ through T_2 is used for training and $T_2 + 1$ through T for the evaluation of forecast accuracy in a rolling manner. In addition, for the one-step ahead forecast based on all observations from $1, ..., t$ is defined as $\hat{y}^{\lambda}_{t+1}$ [27]. The objective therefore is to minimize [27]:

$$MSFE(\lambda) = \frac{1}{T_2 - T_1 - 1} \sum_{t=T_1}^{T_2-1} \|\hat{y}^{\lambda}_{t+1} - y_{t+1}\|^2_F. \tag{12}$$

MSFE represents the most appropriate criterion given the use of the least squares objective function. Instead of parallelizing the cross-validation procedure, this approach uses the result from the previous period as an initialization, substantially reducing computational time [27].

The algorithm was implemented in MATLAB® R2014a using the cvx 2.1 toolbox for convex optimization problems [36].

2.2. Performance Evaluation

In order to evaluate the performance of the proposed network identification algorithm for reconstructing stable gene regulatory networks from datasets, statistical measures are employed. For predictive analysis, the confusion matrix (Table 1), represents a table with two rows and two columns that report the number of True Positives (TPs), False Positives (FPs), True Negatives (TNs) and False Negatives (FNs). Since the parameter, λ, regulates the weight imposed on sparsity, the terms "Positives" and "Negatives" here refer to non-zero and zero interactions between genes, respectively.

Table 1. Confusion matrix.

	True Network		Total
Inferred Network	True Positive	False Postive	**P′**
	False Negative	True Negative	**N′**
Total	**P**	**N**	

TP represents the interaction that exists in both the true network and inferred network, FP denotes the interaction that does not exist in the true network but was falsely inferred, TN is the interaction that does not exist in either the true network or the inferred network, while FN represents the interaction that does exist in the actual network but is not recovered by the network identification method.

Three other criteria *Sensitivity* (sen), *Specificity* (spc) and *Precision* (not commonly used) are also employed as evaluation methods. *Sensitivity* (sen), is the fraction of the number of recovered true regulations to all regulations in the model. *Specificity* (spc), is the ratio of correctly found no-interactions to all no-interactions in the model. *Precision* (pre), measures the fraction of the number of correctly found regulations to all found regulations in the inferred network. These three performance criteria are defined as follows [37]:

$$Sensitivity = \frac{TP}{TP + FN}, \tag{13a}$$

$$Specificity = \frac{TN}{TN + FP}, \tag{13b}$$

$$Precision = \frac{TP}{TP + FP}. \tag{13c}$$

For the purposes of comparison with other network identification techniques, the performance evaluation graphs are restricted to sensitivity and specificity.

Overall, the following specific steps are performed for modeling the GRN using LASSO-VAR. The time series of gene expressions are converted to a matrix where the rows are expression of various genes and the columns are observations at different time points. Assuming the model of Equation (2), the optimization problem as Equation (10) can be generated. Solving this optimization using the iterative approach yields the matrix A with sparse and stable structure. The optimal values of hyperparameters (λ and v_i) should also be found. The structure of gene regulatory network is found using the matrix A. The results of our method are then compared with target GRNs that are accepted or inferred by other means.

Data for GRN recovery of the *Escherichia coli* SOS network was from a perturbation experiment for relative RNA expression changes from Table S6 of [8]. Data for GRN recovery of the cell cycle pathway in yeast *Saccharomyces cerevisiae* was based on the alpha time series of [38], including 18 time points at 7 min interval over 119 min. Data came from the yeast cell cycle analysis database [39] with the analysis conducted on a set of 14 genes. Standard (and systematic) gene names for these genes are: FUS3 (YBL016W), SIC1 (YLR079W), FAR1 (YJL157C), CDC6 (YJL194W), CDC20 (YGL116W), CDC28 (YBR160W), CLN1 (YMR199W), CLN2 (YPL256C), CLN3 (YAL040C), CLB5 (YPR120C), CLB6 (YGR109C), SWI4 (YER111C), SWI6 (YLR182W) and MBP1 (YDL056W).

3. Results and Discussion

In this section, the efficiency of the proposed network identification is analyzed by studying networks for which the experimental data as well as the ground truth is available. The studied datasets consist of real experimental dataset in a known subnetwork of the SOS pathway in *Escherichia coli*, provided in [26], and the cell cycle pathway in yeast *Saccharomyces cerevisiae*.

With datasets that have a known network, it is possible to evaluate the performance of the algorithm, allowing for the measurement of the false positives, false negatives, *etc*. The effect of different values of the penalty parameter λ on the performance of the algorithm is also investigated.

3.1. SOS Pathway in Escherichia coli

The proposed identification algorithms is first applied to a sub-network of the SOS pathway in *Escherichia coli*, using the gene perturbation experimental data set provided in [8]. The SOS pathway is known to control the survival and repair of cells after DNA damage [8]. This pathway typically involves the genes *recA* and *lexA* directly regulating over 30 genes, and indirectly controlling over 100 genes [8].

The sub-network considered, shown in Figure 2a, consists of nine genes and several transcription factors and metabolites [8] whose expression levels are measured over nine different perturbations. According to Gardner *et al.* [8], the nine transcripts in the test network (Figure 2a) were chosen to enable evaluation of the performance of their proposed algorithm. These nine transcripts include the key mediators of the SOS response (*lexA* and *recA*) and sets of genes with known regulatory roles (*ssb*, *recF*, *dinI*, and *umuDC*) and unknown regulatory roles (*rpoD*, *rpoH*, and *rpoS*). The presence of genes with regulatory roles that are already known allows this network to be used to validate an inference algorithm [8].

3.1.1. Network Recovery

In order to evaluate the performance of the proposed algorithm, those links that are correctly recovered in the model are determined based on knowledge of the true network. An inferred connection is regarded to be accurate if there exists a known RNA, protein, or metabolite pathway between the two transcripts, and if the sign of the net effect of regulatory interaction (*i.e.*, inhibition or activation) is correct, as determined by the known network in Figure 2a. In general, since RNA concentrations (*i.e.*, expressions) were measured and not metabolite or protein species, the recovered regulatory network model does not necessarily depict physical connections; rather, the links show effective functional associations between transcripts.

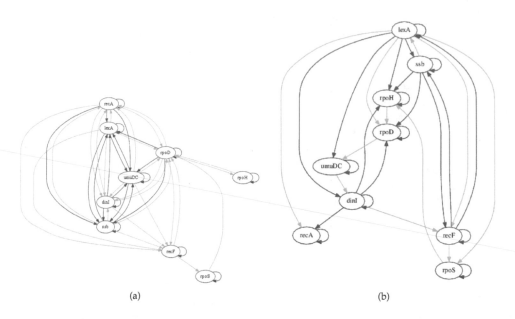

(a)   (b)

Figure 2. Known and recovered GRN for SOS pathway in *E. coli*. (**a**) Diagram of interactions in the SOS network. DNA lesions caused by mitomycin C (MMC) (**blue hexagon**) are converted to single-stranded DNA during chromosomal replication. Upon binding to ssDNA, the RecA protein is activated (RecA*) and serves as a coprotease for the LexA protein. The LexA protein is cleaved, thereby diminishing the repression of genes that mediate multiple protective responses. Green arrows denote positive regulation, while red arrows denote negative regulation. *Adapted from* [8]. (**b**) Diagram of the recovered gene regulatory network of the SOS pathway in *Escherichia coli*. Green arrows denote positive regulation, while red arrows denote negative regulation.

Figure 2b shows the inferred gene regulatory network from the steady-state data. The identification algorithm accurately identified the key regulatory associations in the network. For instance, the model correctly shows that *lexA* activates *recA* while negatively regulating its own transcription, whereas *recA* negatively regulates its own transcription. In addition, the model identified *lexA* as having the greatest regulatory influence on the other genes in the network. Due to the differences in network topology (*e.g.*, *recA*, *lexA* and *CDC20*), inaccuracies are expected from either the current published model, the LASSO-VAR GRN recovery, or both. Some of these potential differences may alternatively be dependent on the dynamic state of the system as inferred from the temporal context.

The plots in Figure 3 show the variations in algorithm performance as the penalty parameter λ varies between 0 and 1. In Figure 3a, the total number of false identifications (*i.e.*, false activations, inhibitions and no-interactions) and the net connectivity of the network are measured against the different λ values. The net connectivity provides a measure of the total number of interactions inferred by the algorithm. As such since sparsity is required, lower values in the net connectivity is desired. Figure 3b shows the variations in the performance metrics, sensitivity and specificity, as λ changes.

The choice of the "best" penalty parameter for the given application represents the λ value that produces the best trade-off between the number of false identification, net connectivity, sensitivity and specificity. In this regard, values of 0.3 and 0.4 produce 46% false identification, 37% net connectivity, sensitivity of 60% and specificity of 31%. Eventually, $\lambda = 0.3$ is selected based on its superior MSFE value of 5.20 as opposed to 5.22 for $\lambda = 0.4$ and satisfies the desired constraints of stability, sparsity and causality.

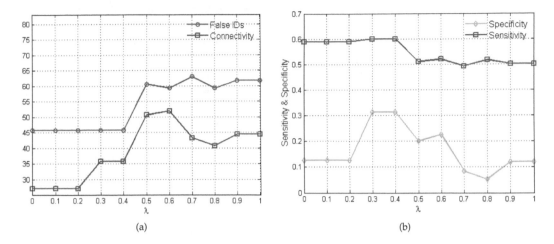

(a)

(b)

Figure 3. Variations in λ and the corresponding algorithm performance. (**a**) plot of λ *versus* total number of false identifications and net connectivity in percentages. (**b**) plot of λ *versus* sensitivity and specificity.

Table 2 shows how the proposed network identification algorithm without *a priori* knowledge of the network structure compares with that proposed by Zavlanos *et al.* [26] with 30% *a priori* knowledge of the network.

Table 2. Comparisons of the inferred network for the SOS pathway in *E. coli* using LASSO-VAR and Zavlanos' method.

	TP	FP	TN	FN	Sensitivity	Specificity	Precision
LASSO-VAR	39	11	5	26	60%	31%	78%
ZAVLANOS [26]	40	10	15	16	71%	60%	80%

Table 2 shows how the proposed algorithm compares to that proposed by Zavlanos *et al.* In all, the recovered network has four false activations, eight false inhibitions, 25 false no-interactions, and 37 false identifications in total, while satisfying the desired constraints of stability, sparsity and causality. The penalty parameter, λ selected is 0.4 based to its MSFE value.

3.2. Yeast Saccharomyces Cerevisiae Cell Cycle

From 1998, when Spellman *et al.* published the yeast *Saccharomyces cerevisiae* (*i.e.*, budding yeast) cell cycle microarray expression levels [38], many computational methods have been applied to these data. To demonstrate the applicability of the proposed algorithm in this study, a subset from yeast *Saccharomyces cerevisiae* microarray time series dataset including 14 genes, FUS3, SIC1, FAR1, CDC6, CDC20, CDC28, CLN1, CLN2, CLN3, CLB5, CLB6, SWI4, SWI6 and MBP1, is perturbed and used. The details of *S. cerevisiae* cell cycle control are well known, as shown in Figure 4a.

The 14 genes are known to be involved in the early cell cycle of the yeast *Saccharomyces cerevisiae*. The cell cycle describes the series of events that precedes its division and duplication [40]. The mitotic cell cycle in yeast is accomplished through a reproducible sequence of events: DNA replication (S phase) and mitosis (M phase) separated temporally by gaps, G1 and G2 phases. At the G1 phase, CDC28 associates with CLN1, CLN2 and CLN3, while CLB5 and CLB6 controls CDC during S, G2, and M phases [41]. Cell cycle progression begins upon the activity of CLN3/CDC28. When the levels of CLN3/CDC28 accumulate more than a certain threshold, SWI4/SWI6 and MBF1/SWI6 are activated, promoting transcription of CLN1 and CLN2 [41]. CLN1/CDC28 and CLN2/CDC28 promote activation of other associated kinase, which drives DNA replication. SIC1 and FAR1 are the substrates and inhibitors of CDC28. CDC6 and CDC20 affect the cell division control proteins. Mitogen-activated protein kinase affect this progression through FUS3 [41]. The dataset generated by Spellman *et al.* [38] contains three time series measured using different cell synchronization methods: α factor-based arrest (referred to as alpha, includes 18 time points at 7 min interval over 119 min), size-based (*elu*, 14 time points at 30 min interval over 390 min), and arrest of a cdc15 temperature-sensitive mutant (*cdc15*, 24 time points, the first four and last three of which are at 20 min interval and the rest are at 10 min interval over 290 min). The *alpha* dataset is used and then studied in more detail as has been explored in literature [42].

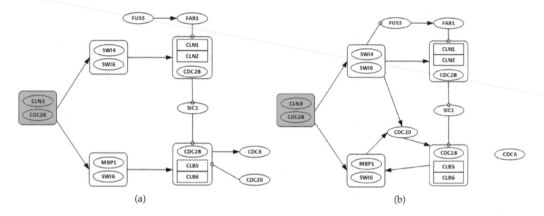

(a)   (b)

Figure 4. Known and recovered GRN for cell cycle pathway in yeast *Saccharomyces cerevisiae*. (**a**) target pathways of the 14 genes. CDC28 associates with cyclin CLN3 at the start of mitosis to cause the activation of SBF (SWI4/SWI6) and MBF (MBP1/SWI6), promoting the transcription of CLN1, CLN2. At G1 phase CDC28 associates with G1-cyclins CLN1 to CLN3, while B-type cyclins CLB1 to CLB6 regulate CDC28 during S, G2, and M phases. CLN1 and CLN2 interacting with CDC28 promote activation of B-type cyclin associated Cyclin-dependent kinase (CDK), which drives DNA replication and entry into mitosis. *Adapted from* [41]. (**b**) recovered yeast cell cycle pathway. The arrows show the direction of regulation. Some key regulations like activation (positive regulation) of the SBF (SWI4/SWI6) and MBF (MBP1/SWI6) complexes by the starter complex (CDC28/CLN3) are recovered.

Data for the expected topology of this network were extracted from the Kyoto Encyclopedia of Genes and Genomes (KEGG) [41], which is a major collection of knowledge for molecular and genetic pathways and includes information on experimental observations in organisms. The KEGG regulatory pathway represents current knowledge on the protein and gene interaction networks.

3.2.1. Network Recovery

As discussed in the preceding section, the KEGG pathway is considered as the target network for comparison. Complexes including one or several genes are considered as a '*gene*' in the network. There are 10 complexes, including CLN3/CDC28, SWI4/SWI6, MBP1/SWI6, CLN1/CLN2/CDC18, and CLB5/CLB6/CDC28. Other nodes that are made of one single gene only, CDC20, CDC6, SIC1, FAR1, and FUS. The following assumptions are made:

1. Genes CLN3 and CDC28 are only considered as possible regulators, as they are starters of the cell cycle network.
2. All discovered links from any gene in one complex to any other genes in a different complex are considered as a single regulation.
3. All regulations among genes in the same complex are ignored.

Figure 4b is the recovered network from the perturbed *alpha* gene expression data for the yeast *Saccharomyces cerevisiae* cell cycle pathway.

The recovered network has seven true positives and five false positives. The algorithm recovers key regulations. For instance, the activation (positive regulation) of the SWI4/SWI6 and MBP1/SWI6 complexes by the starter complex are identified. The graphs in Figure 5 show the changes in algorithm performance for varying λ values. In Figure 5a, the relationship between λ and total false identification as well as the net connectivity of the network. The plot in Figure 5b show the variations in λ and the corresponding effects on the performance of the algorithm. Again, the terms false identifications and net connectivity have the same meanings as discussed in Section 3.1.1. The results in this application are quite uniform due to the assumptions made in grouping some genes into complexes as required. $\lambda \in [0, 0.2]$ provides the best trade-off between the number of false identification, net connectivity, sensitivity and specificity. They produce sensitivity and specificity values of 65% and 94%, respectively. Based on its lower MSFE value, $\lambda = 0.2$ is chosen.

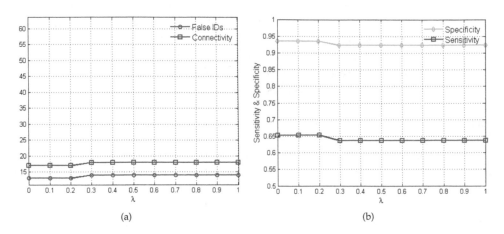

(a)   (b)

Figure 5. Variations in λ and the corresponding algorithm performance. (a) plot of λ *versus* total number of false identifications and net connectivity in percentages; (b) plot of λ *versus* sensitivity and specificity.

4. Conclusion and Discussion

In this paper, the least absolute shrinkage and selection operator—vector autoregressive (LASSO-VAR) model–has been adapted in solving the problem of identifying a minimal model that best explains genetic perturbation data obtained at a network's equilibrium state. The fact that the network identification algorithm is an autoregressive technique means it has the inherent ability to model Granger causality, making it possible to identify regulatory roles among the genes under consideration. Additionally, the technique handles the sparsity constraint imposed by the loose-connectivity restriction of biological networks through the application of the L_1 penalty term. Due to the steady-state nature of the expression data, a stability constraint is imposed on the original LASSO-VAR objective function, which allows robustness of the inferred networks to slight variations in the input.

To evaluate the reliability and efficiency of LASSO-VAR for recovering stable, sparse and causal regulatory interactions from steady-state gene expression data, data from the SOS pathway in *E. coli* was first used. The performance of the algorithm was measured and compared with results obtained in literature. This comparison was based on two statistical evaluation criteria, sensitivity and specificity, which allowed the accuracies of the inferred network structure using the LASSO-VAR technique to be quantified. LASSO-VAR performed without prior knowledge at a roughly comparable level to the alternative Zavanos method that requires some prior knowledge. The efficiency of the stable LASSO-VAR for learning the network structure was then evaluated using the perturbed gene expression data of 14 genes in yeast *Saccharomyces cerevisiae* cell cycle reported in [41]. The network inferred from the yeast data by LASSO-VAR is compared with the known network from the cell cycle pathway of the yeast *Saccharomyces cerevisiae* using the evaluation criteria: sensitivity and specificity. Results showed the ability of the identification algorithm to infer the regulatory network for the cell cycle pathway.

The surge of large biological data sets [43,44] and the aggressive efforts at methods for ontology-driven annotation and data modeling [45,46] have helped to provide opportunities for a more objective and reproducible basis for analysis. The formulation of our pathway recovery analysis model fulfills those criteria necessary for it to be highly scalable with data for biological networks by: (1) avoiding the context-limiting aspects of *a priori* knowledge and presumptions of frequentist-type statistics which are difficult to implement based on the inherent sparsity of biological networks; and (2) being strictly data-driven in matrix-based numerical forms with one controlling parameter—*i.e.*, being outside of subjective standards of knowledge and curation. The criterion for stability as we investigate it as a necessary condition for steady-state data, although seemingly trivial, is nonetheless fundamental for how structural and functional robust aspects for pathways are identified. In general, scoring outcomes from our methodology identified key mechanisms of pathways through a scoring gradient. We expect there to be significant dividends from this approach that go beyond our initial usage of a reference database for tallying of true *versus* false results. The quantitative support underlying false positive and false negative results for the target model may aid in the development and testing of new hypotheses, or quality control measures, agendas that are important in large-scale investments into empirical data collection such as for perturbation studies.

Canonical pathways provide well-vetted models that have a deep legacy in empirical published studies [46,47]. It is still the case, however, that uncharted dynamics and natural variation go beyond the limited range of studied organisms and would furthermore impact predictive modeling even for the two chosen models of this study for which knowledge remains limited [48,49]. For instance, synchronization of expression across multiple genes is likely lost following its initial measurement at a starting point of an experimental assay. The gain or loss of known interactions as input data would furthermore be expected to impact the predictive power of LASSO-VAR. There is, therefore, a need to more systematically study LASSO-VAR across these frontier contexts involving a larger number of simulated and empirical data sets. This would in particular aid the empirical study of computational performance, beyond theoretical expectations for restricting the parameter space through the penalty parameter [27]. This would also provide a robust capacity for selecting which genes to base a model upon and for constructing the analysis in a way that separates initialization and training from forecast evaluation, to guide contrasting modes of usage for how LASSO-VAR could be used in supervised *versus* unsupervised modes of analytical evaluations. As multiple gene expression data resources and annotations may be harnessed for this effort [50–53], it remains an essential next step to identify computational and theoretical limits for objectively inferring GRN configurations based on input data complexity and sizes. The overall outcome for such an effort would help resolve the lack of overlap between pathway databases and approaches to analytical treatment, both with respect to content [50,54] and foundational criteria such as how to define start and stop points for individual pathways [50]. Future usage of this approach could identify pathways for assembly and disassembly of differentially constructed multicomponent cellular objects recovered at the same point of time. Such usage of cellular component data to infer subunit associations would add to the explanatory potential of this algorithm, and help to guard against the *post hoc, ergo propter hoc* fallacy for how changes in cellular composition would otherwise be inferred from analyses conducted only upon time series.

Acknowledgments: The authors wish to acknowledge Michael Zavlanos for making the source codes to his genetic network identification algorithm available. His source package has been modified and used in this work. This work is partially supported by the National Science Foundation (NSF) under Cooperative Agreement No. CCF-1029731. In addition, the third and the fifth authors would like to acknowledge the support from the Air Force Research Laboratory and Office of the Secretary of Defense (OSD) for sponsoring this research under agreement number FA8750-15-2-0116. The views and conclusions contained herein are those of the authors and should not be interpreted as necessarily representing the official policies or endorsements, either expressed or implied, of the Air Force Research Laboratory, OSD, NSF or the U.S. Government.

Author Contributions: Joy Edward Larvie has made substantial contributions to the conception and design of the study, analysis and interpretation of biological data. Mohammad Gorji Sefidmazgi has made substantial contributions to the analysis and interpretation of mathematical results. Joy Edward Larvie and Mohammad Gorji Sefidmazgi have been involved in drafting the manuscript. Abdollah Homaifar and Ali Karimoddini have critically read and revised the manuscript for important mathematical content. Scott Harrison and Anthony Guiseppi-Elie have provided critical biological interpretation to the results and have critically read and revised the manuscript for analytical utility. All authors read and approved the final manuscript.

Conflicts of Interest: The authors declare no conflict of interest.

References

1. Quackenbush, J. Computational analysis of microarray data. *Nat. Rev. Genet.* **2001**, *2*, 418–427.

2. Michailidis, G.; d'Alché Buc, F. Autoregressive models for gene regulatory network inference: Sparsity, stability and causality issues. *Math. Biosci.* **2013**, *246*, 326–334.

3. Hasty, J.; McMillen, D.; Isaacs, F.; Collins, J.J. Computational studies of gene regulatory networks: In numero molecular biology. *Nat. Rev. Genet.* **2001**, *2*, 268–279.

4. Margolin, A.A.; Nemenman, I.; Basso, K.; Wiggins, C.; Stolovitzky, G.; Favera, R.D.; Califano, A. ARACNE: an algorithm for the reconstruction of gene regulatory networks in a mammalian cellular context. *BMC Bioinform.* **2006**, *7*, S7.

5. Stark, C.; Breitkreutz, B.J.; Reguly, T.; Boucher, L.; Breitkreutz, A.; Tyers, M. BioGRID: A general repository for interaction datasets. *Nucleic Acids Res.* **2006**, *34*, D535–D539.

6. Edgar, R.; Domrachev, M.; Lash, A.E. Gene Expression Omnibus: NCBI gene expression and hybridization array data repository. *Nucleic Acids Res.* **2002**, *30*, 207–210.

7. Tegnér, J.; Yeung, M.K.S.; Hasty, J.; Collins, J.J. Reverse engineering gene networks: Integrating genetic perturbations with dynamical modeling. *PNAS* **2003**, *100*, 5944–5949.

8. Gardner, T.S.; Di Bernardo, D.; Lorenz, D.; Collins, J.J. Inferring genetic networks and identifying compound mode of action via expression profiling. *Science* **2003**, *301*, 102–105.

9. Chai, L.E.; Loh, S.K.; Low, S.T.; Mohamad, M.S.; Deris, S.; Zakaria, Z. A review on the computational approaches for gene regulatory network construction. *Comput. Biol. Med.* **2014**, *48*, 55–65.

10. Lopes, F.M.; de Oliveira, E.A.; Cesar, R.M. Inference of gene regulatory networks from time series by Tsallis entropy. *BMC Syst. Biol.* **2011**, *5*, 61.

11. Wang, Y.X.R.; Huang, H. Review on statistical methods for gene network reconstruction using expression data. *J. Theor. Biol.* **2014**, *362*, 53–61.

12. Polynikis, A.; Hogan, S.J.; di Bernardo, M. Comparing different ODE modelling approaches for gene regulatory networks. *J. Theor. Biol.* **2009**, *261*, 511–530.

13. De Jong, H. Modeling and simulation of genetic regulatory systems: a literature review. *J. Comput. Biol.* **2002**, *9*, 67–103.

14. Hecker, M.; Lambeck, S.; Toepfer, S.; van Someren, E.; Guthke, R. Gene regulatory network inference: Data integration in dynamic models—A review. *Biosystems* **2009**, *96*, 86–103.

15. Ahmad, J.; Bourdon, J.; Eveillard, D.; Fromentin, J.; Roux, O.; Sinoquet, C. Temporal constraints of a gene regulatory network: Refining a qualitative simulation. *Biosystems* **2009**, *98*, 149–159.

16. Someren, E.V.; Wessels, L.; Backer, E.; Reinders, M. Genetic network modeling. *Pharmacogenomics* **2002**, *3*, 507–525.

17. Hartemink, A.J. Reverse engineering gene regulatory networks. *Nat. Biotechnol.* **2005**, *23*, 554–555.

18. Yeung, M.S.; Tegnér, J.; Collins, J.J. Reverse engineering gene networks using singular value decomposition and robust regression. *Proc. Natl. Acad. Sci. USA* **2002**, *99*, 6163–6168.

19. Wang, Y.; Joshi, T.; Zhang, X.S.; Xu, D.; Chen, L. Inferring gene regulatory networks from multiple microarray datasets. *Bioinformatics* **2006**, *22*, 2413–2420.

20. Karlebach, G.; Shamir, R. Modelling and analysis of gene regulatory networks. *Nat. Rev. Mol. Cell Biol.* **2008**, *9*, 770–780.

21. Kordmahalleh, M.M.; Sefidmazgi, M.G.; Homaifar, A.; Karimoddini, A.; Guiseppi-Elie, A.; Graves, J.L. Delayed and Hidden Variables Interactions in Gene Regulatory Networks. In Proceedings of the 2014 IEEE International Conference on Bioinformatics and Bioengineering (BIBE), Boca Raton, FL, USA, 10–12 November 2014; pp. 23–29.

22. D'haeseleer, P.; Liang, S.; Somogyi, R. Genetic network inference: From co-expression clustering to reverse engineering. *Bioinformatics* **2000**, *16*, 707–726.

23. Kauffman, S. Metabolic stability and epigenesis in randomly constructed genetic nets. *J. Theor. Biol.* **1969**, *22*, 437–467.

24. Bornholdt, S. Less is more in modeling large genetic networks. *Science* **2005**, *310*, 449.

25. Friedman, N.; Linial, M.; Nachman, I.; Pe'er, D. Using Bayesian networks to analyze expression data. *J. Comput. Biol.* **2000**, *7*, 601–620.

26. Zavlanos, M.M.; Julius, A.A.; Boyd, S.P.; Pappas, G.J. Inferring stable genetic networks from steady-state data. *Automatica* **2011**, *47*, 1113–1122.

27. Nicholson, W.; Matteson, D.; Bien, J. *Structured Regularization for Large Vector Autoregression*; Technical report; Cornell University: Ithaca, NY, USA, 2014.

28. Larvie, J.E.; Gorji, M.S.; Homaifar, A. Inferring stable gene regulatory networks from steady-state data. In Proceedings of the 2015 41st Annual Northeast Biomedical Engineering Conference (NEBEC), Troy, NY, USA, 17–19 April 2015; pp. 1–2.

29. Fujita, A.; Sato, J.R.; Garay-Malpartida, H.M.; Yamaguchi, R.; Miyano, S.; Sogayar, M.C.; Ferreira, C.E. Modeling gene expression regulatory networks with the sparse vector autoregressive model. *BMC Syst. Biol.* **2007**, *1*, 39.

30. Lütkepohl, H. *New Introduction to Multiple Time Series Analysis*; Springer: Berlin, Germany, 2007.

31. Mukhopadhyay, N.D.; Chatterjee, S. Causality and pathway search in microarray time series experiment. *Bioinformatics* **2007**, *23*, 442–449.

32. Granger, C.W.J. Investigating Causal Relations by Econometric Models and Cross-Spectral Methods. *Econometrica* **1969**, *37*, 424–438.

33. Hsu, N.J.; Hung, H.L.; Chang, Y.M. Subset selection for vector autoregressive processes using Lasso. *Comput. Stat. Data Anal.* **2008**, *52*, 3645 – 3657.

34. Chen, J. Sufficient conditions on stability of interval matrices: connections and new results. *IEEE Trans. Autom. Control* **1992**, *37*, 541–544.

35. Rajapakse, J.C.; Mundra, P.A. Stability of building gene regulatory networks with sparse autoregressive models. *BMC Bioinform.* **2011**, *12*, S17.

36. CVX Research Inc. CVX: Matlab Software for Disciplined Convex Programming, Version 2.0, 2012. Available online: http://cvxr.com/cvx (accessed on 15 October 2015).

37. De Muth, J. *Basic Statistics and Pharmaceutical Statistical Applications*, 2nd ed.; Pharmacy Education Series, Taylor & Francis: Boca Raton, FL, USA, 2006.

38. Spellman, P.T.; Sherlock, G.; Zhang, M.Q.; Iyer, V.R.; Anders, K.; Eisen, M.B.; Brown, P.O.; Botstein, D.; Futcher, B. Comprehensive Identification of Cell Cycle-regulated Genes of the Yeast Saccharomyces cerevisiae by Microarray Hybridization. *Mol. Biol. Cell* **1998**, *9*, 3273–3297.

39. The Yeast Cell Cycle Analysis Database. Available online: http://genome-www.stanford.edu/cellcycle/data/rawdata/combined.txt (accessed on 15 November 2015).

40. Lodish, H. *Molecular Cell Biology*, 5th ed.; W. H. Freeman: New York, NY, USA, 2003.

41. KEGG PATHWAY: map04111. Available online: http://www.genome.jp/dbget-bin/www_bget?map04111 (accessed on 27 April 2015).

42. Kim, S.Y.; Imoto, S.; Miyano, S. Inferring gene networks from time series microarray data using dynamic Bayesian networks. *Brief. Bioinform.* **2003**, *4*, 228–235.

43. Schatz, M.; Langmead, B. The DNA data deluge. *IEEE Spectr.* **2013**, *50*, 28–33.

44. Barrett, T. Gene Expression Omnibus (GEO). Available online: http://www.ncbi.nlm.nih.gov/geo/ (accessed on 15 October 2015).

45. Gene Ontology Consortium. Gene Ontology Consortium: going forward. *Nucleic Acids Res.* **2015**, *43*, D1049–D1056.

46. Fehrmann, R.S.N.; Karjalainen, J.M.; Krajewska, M.; Westra, H.J.; Maloney, D.; Simeonov, A.; Pers, T.H.; Hirschhorn, J.N.; Jansen, R.C.; Schultes, E.A.; *et al.* Gene expression analysis identifies global gene dosage sensitivity in cancer. *Nat. Genet.* **2015**, *47*, 115–125.

47. Zhou, H.; Jin, J.; Zhang, H.; Yi, B.; Wozniak, M.; Wong, L. IntPath–an integrated pathway gene relationship database for model organisms and important pathogens. *BMC Syst. Biol.* **2012**, *6* (Suppl. 2), S2.

48. Michel, B. After 30 Years of Study, the Bacterial SOS Response Still Surprises Us. *PLoS Biol.* **2005**, *3*, e255.

49. Alberghina, L.; Mavelli, G.; Drovandi, G.; Palumbo, P.; Pessina, S.; Tripodi, F.; Coccetti, P.; Vanoni, M. Cell growth and cell cycle in Saccharomyces cerevisiae: Basic regulatory design and protein-protein interaction network. *Biotechnol. Adv.* **2012**, *30*, 52–72.

50. Caspi, R.; Dreher, K.; Karp, P.D. The challenge of constructing, classifying, and representing metabolic pathways. *FEMS Microbiol. Lett.* **2013**, *345*, 85–93.

51. Krämer, A.; Green, J.; Pollard, J.; Tugendreich, S. Causal analysis approaches in Ingenuity Pathway Analysis. *Bioinformatics* **2014**, *30*, 523–530.

52. Croft, D.; Mundo, A.F.; Haw, R.; Milacic, M.; Weiser, J.; Wu, G.; Caudy, M.; Garapati, P.; Gillespie, M.; Kamdar, M.R.; *et al.* The Reactome pathway knowledgebase. *Nucleic Acids Res.* **2014**, *42*, D472–D477.

53. Morgat, A.; Coissac, E.; Coudert, E.; Axelsen, K.B.; Keller, G.; Bairoch, A.; Bridge, A.; Bougueleret, L.; Xenarios, I.; Viari, A. UniPathway: a resource for the exploration and annotation of metabolic pathways. *Nucleic Acids Res.* **2012**, *40*, D761–D769.

54. Karp, P.D. Pathway databases: a case study in computational symbolic theories. *Science* **2001**, *293*, 2040–2044.

7

Nano-Modeling and Computation in Bio and Brain Dynamics

Paolo Di Sia [1,2,*,†] and Ignazio Licata [2,3,†]

[1] Department of Philosophy, Education and Psychology, University of Verona, Lungadige Porta Vittoria 17, Verona 37129, Italy

[2] ISEM, Institute for Scientific Methodology, Palermo 90146, Italy; Ignazio.licata@ejtp.info

[3] School of Advanced International Studies on Applied Theoretical and Non Linear Methodologies in Physics, Bari 70121, Italy

* Correspondence: paolo.disia@gmail.com

† These authors contributed equally to this work.

Academic Editor: Gou-Jen Wang

Abstract: The study of brain dynamics currently utilizes the new features of nanobiotechnology and bioengineering. New geometric and analytical approaches appear very promising in all scientific areas, particularly in the study of brain processes. Efforts to engage in deep comprehension lead to a change in the inner brain parameters, in order to mimic the external transformation by the proper use of sensors and effectors. This paper highlights some crossing research areas of natural computing, nanotechnology, and brain modeling and considers two interesting theoretical approaches related to brain dynamics: (a) the memory in neural network, not as a passive element for storing information, but integrated in the neural parameters as synaptic conductances; and (b) a new transport model based on analytical expressions of the most important transport parameters, which works from sub-pico-level to macro-level, able both to understand existing data and to give new predictions. Complex biological systems are highly dependent on the context, which suggests a "more nature-oriented" computational philosophy.

Keywords: neuro-nanoscience; cognitive science; bioengineering; brain; carrier transport; theoretical modeling; neural geometry; memristor; electrical circuits

1. Introduction

The great potential of nanobiotechnology is based on the ability to deal with complex hierarchically structured systems from the macroscale to the nanoscale [1]; it requires novel theoretical approaches and the competence of creating models, able to explain the dynamics at such a scale [2–5]. Geometric and analytical approaches seem to be very promising in all scientific areas, including the study of brain processes.

Deep comprehension and adaptiveness cause a change in the inner brain parameters (conductance of synapses), in order to mimic the outer transformation by the appropriate use of sensors and effectors. The basic mathematical aspects can be illustrated with the use of a toy model related to "Network Resistors with Adaptive Memory" (memristors). Designed by Chua in 1971 [6], only in recent years it has been possible to develop effective realizations [7,8]. The memristor is an electrical circuit with "analogic" properties, able to vary the resistance after a variation in the current and to preserve the last state at the interruption of the energy flow. In a toy model of the brain, this introduces an element of memory that takes into account the enormous non-linear complexity of the homeo-cognitive equilibrium states. This promotes the utility of going back to models of natural computation and therefore of looking at the Turing computation as a "coarse grain" of processes, which are best described by geometric manifolds [9–13].

Technology advancement provides a finer modeling, new solutions, and a capability of active interaction between the environment, machines, and humans and the possibility of not necessarily scaling, as per Moore's law [14–17].

Dynamics in the brain are based on transport models; their improvement is a mandatory step in deep comprehension of brain functioning. Advances in analytical modeling can adequately study the nano-dynamics in the brain and lead to interesting ideas for future developments.

Learning experiences produce a "chain action" of signaling among neurons in some areas of the brain, with the modification of neuron connections in particular brain areas, resulting in reorganization. Research on brain plasticity and circuitry also indicates that the brain is always learning, in both formal and informal contexts.

Natural computing refers to observed computational processes and "human-designed/ inspired-by-nature" computing. Analyzing complex natural phenomena in terms of computational processes, we reinforce the understanding of both nature and essence of computation. Peculiar to this kind of approach is the use of concepts, principles, and mechanisms underlying natural systems.

This paper aims to highlight some areas of interest for research, combining natural computing, nanotechnology, and brain modelization. It is structured as follows: after an overview of the nanoscience in the brain (Section 2), we consider the technical details of a recently appeared analytical transport model (Section 3). In Section 4, examples of application concerning geometrical images in neural spaces and nano-diffusion together with results are considered, which is followed by conclusions (Section 5).

2. Nanoscience in the Brain

Chemical communication and key bio-molecular interactions in the brain occur at the nanoscale; therefore, the idea of taking advantages of nanoscience for advances in the study of brain structure and function is becoming increasingly popular. In the human brain, there are 85 billion neurons and an estimated 100 trillion synapses [18]; as experimental "nano-brain" techniques, we remember:

(1) Snapshots of connections in the brain by making thin physical slices and stacking electron microscopy images (this technique does not provide dynamic and chemical information);
(2) A dynamic voltage map of the brain, dealing with the brain as a close relative of a computer [19,20], with the aim to get to the emergent properties underlying the use and storage of information by a mapping network activity *vs.* single or small numbers of multiple unit recordings;
(3) The attempt to obtain functional chemical maps of the neurotransmitter systems in the brain, for investigating the genetic and chemical brain heterogeneity and the interactions among neurotransmitter systems.

In all these cases, the length scale ranges from the "centimeter" scale (cm) (in mapping brain regions and networks), to the "micrometer" scale (μm) (cells and local connectivity), to the "nanometer" scale (nm) (synapses), to the single-molecule scale [4]. The current ability in performing neurochemical and electrophysiological measurements needs to be miniaturized, sped up, and multiplexed. Electrical measurements at time scales of milliseconds are not complicated, but getting to the nanometer scale and simultaneously making tens of thousands *in vivo* measurements is very difficult. Obtaining dynamic chemical maps at this scale is a bigger challenge, as there are problems in analysis, interpretation, and visualization of data.

3. Transport Processes at Nano-Level: Technical Details

Research at the theoretical level help science in all sectors. Recently, a new analytical model that generalizes the Drude-Lorentz and Smith models for transport processes in solid-state physics and soft condensed matter has been used [21–23]. It provides analytical time-dependent expressions of the three most important parameters related to transport processes:

(a) The velocities correlation function of a system $< \vec{v}(t) \cdot \vec{v}(0) >_T$ at the temperature T, from which it is possible to obtain the velocity of a carrier at generic time t;

(b) The mean squared deviation of position $R^2(t)$, defined as $R^2(t) = \left\langle [\vec{R}(t) - \vec{R}(0)]^2 \right\rangle$, from which the position of a carrier in time is obtainable;

(c) The diffusion coefficient D, defined as $D(t) = (1/2)(dR^2(t)/dt)$, which gives important information about the temporal propagation of carriers inside a nanostructure [22–24].

With this model, it is possible both to fit experimental data, confirming known results, and to discover new features and details. The presence of a gauge factor inside the model allows its use from sub-pico-level to macro-level [25,26].

Starting by the time-dependent perturbation theory, analytical calculation leads to relations for the velocities correlation function, the mean square deviation of position, and the diffusion coefficient of carriers moving in a nanostructure. The general calculation is performed via contour integration in the complex plane. Analytical expressions of the velocities correlation function $< \vec{v}(t) \cdot \vec{v}(0) >_T$, the mean square deviation of position:

$$R^2(t) = 2 \int_0^t dt' \, (t - t') \left\langle \vec{v}(t') \cdot \vec{v}(0) \right\rangle \tag{1}$$

and the diffusion coefficient:

$$D(t) = \int_0^t dt' \left\langle \vec{v}(t') \cdot \vec{v}(0) \right\rangle \tag{2}$$

allow a complete dynamical study of carriers.

The classical analytical expressions of the diffusion coefficient D are as follows:

$$D(t) = \left(\frac{k_B T}{m^*} \right) \left(\frac{\tau}{\alpha_I} \right) \cdot \left[\exp\left(-\frac{(1 - \alpha_I)}{2} \frac{t}{\tau} \right) - \exp\left(-\frac{(1 + \alpha_I)}{2} \frac{t}{\tau} \right) \right] \tag{3}$$

$$D = 2 \left(\frac{K_B T}{m^*} \right) \left[\frac{\tau}{\alpha_R} \sin\left(\frac{\alpha_R}{2} \frac{t}{\tau} \right) \exp\left(-\frac{t}{2\tau} \right) \right] \tag{4}$$

The parameters of the model are two real numbers defined in this way:

$$\alpha_I = \sqrt{1 - 4\tau^2 \omega_0^2} \tag{5}$$

$$\alpha_R = \sqrt{4\tau^2 \omega_0^2 - 1} \tag{6}$$

with τ and ω_0 relaxation time and center frequency, respectively, and m^* effective mass of the carrier. Generalizations of the model take quantum [23] and relativistic effects [26] into account.

4. Examples of Application and Results

(a) An inertial system is used to describe the ordinary differential equation (ODE) types based on standard physical terminology, which is well defined. An example of non-trivial inertial system is the geodesic (kinetic energy) for a mechanical rotatory system with the inertial moment Ii,j. In this case, the geodesic is non-inertial, with an invariant given by the respective computable geodesic.

(b) To take into account a minimum of biological plausibility, it is necessary to introduce a simplified membrane model. In a toy membrane just having, for example, three potassium channels (twelve gates), nine open gates can be configured into a variety of topological states, with the possible results that channels are not open, that one is open, or that two are open. Given a vector q in the channel state, we can compute the probability for the given configuration q of states in the channels. Associating a probability to any configuration, there are configurations with very low probability and configurations with high probability. Given the join probability P, we compute the variation of the probability with respect to the state q_j. Given the current i_j, we can compute the flux of states Φ_j for the current as a random variable. Assuming the invariant form $\Phi_j + \lambda D_j P = 0$, the flux is controlled by the probability in an inverse way, and it is zero when the probability is a constant value. Therefore, we have three different powers: W_1, the ordinary power for the ionic current without noise, $W_{1,2}$, the flux of power from current to the noise current, and W_2 as power in the noise currents.

(c)   About the Fisher information in neurodynamics, computing the average of power as the cost function with a minimum value, we can consider a parametrized family of probability distributions $S = \{P(x, t, q_1, q_2, ..., q_n)\}$, with random variables x and t, and q_i real vector parameters specifying the distribution. The family is regarded as an n-dimensional manifold having q as a coordinate system; it is a Riemannian manifold, and $G(q)$ plays the role of a Fisher information matrix. The underlying idea of information geometry is to think the family of probability distributions as a space, each distribution being a "point," while the parameters q plays the role of coordinates. There is a natural unique way for measuring the extent to which neighboring "points" can be distinguished from each other; it has all wished properties for imposing upon a measure of distance, thus keeping "distinguishable" the distance. Considering the well known Kullback-Leibler divergence, with noise equal to zero, the Fisher information assumes the maximum value and the geodesic becomes the classical geodesic. In the case of noise, the information approaches zero, and the cost function is reduced. Resistor networks provide the natural generalization of the lattice models for which percolation thresholds and percolation probabilities can be considered. The geodesic results composed by two parts: the "synchronic and crisp geodesic" and the "noise change of the crisp and synchronic geodesic" (Figure 1).

Figure 1. The geodesic as solution of the ODE. With noise, the geodesic is transformed in a more complex structure related to the Fisher information. The total effect is the percolation random geodesic.

Therefore, the electric neural activity can be represented in the manifold state space, where the minimum path (geodesic) between two points in the multi-space of currents is a function of the neural parameters as resistors, with or without memory. Simple cases given by electric activity of a little part of the membrane of axons, dendrites, or soma, thus ignoring the presence of the voltage-gated channels in the membrane, can be done. The power is comparable to the Lagrangian in mechanics (Hamilton principle) or the Fermat principle in optics (minimum time). In the context of Freeman's neurodynamics, we hypothesize that the minimum condition in any neural network gives the meaning of "intentionality". A neural network changes the reference and the neurodynamics in a trajectory with minimum dissipation of power or geodesic. Therefore, any neural network, or the equivalent electric circuit, generates a deformation of the current space and geodesic trajectories. For every part of the neural network, it is possible to give a similar electric circuit in this way.

(d)   In bio-molecules, it is easy to understand that the natural computation is more suitable than the Turing computation. The proposed nanobio approach takes back the attention toward geometric patterns and attractors. As examples of application, we consider a nanomaterial of great interest in nanomedicine, the fullerene in tubular form, *i.e.* carbon nanotubes (CNTs), and we study the behavior of their diffusion using the new proposed analytical model. For the used values, the temperature $T = 310$ K, three values of the parameter α_I (0.1, 0.5, 0.9), an average relaxation time $\tau_{av} = 10^{-13}$ s (the relaxation time in soft condensed matter takes values of order of 10^{-12}–10^{-14} s), and two values of m^* in relation to the variation of the chiral vector (n,m) have been fixed [27]:

$$\text{(a)} \quad (n, m) = (3, 1) \rightarrow m_{\text{eff}} = 0.507 m_e$$

$$\text{(b)} \quad (n, m) = (7, 3) \rightarrow m_{\text{eff}} = 0.116 m_e$$

$$(m_e = \text{mass of the electron} = 9.109 \times 10^{-31} \text{ kg})$$

Figures 2 and 3 show the variation of the diffusion $vs.$ time for cases (a) and (b) respectively. It is important to emphasize that the variation of the parameter α_I (or α_R) also implies a variation of the shape of diffusion because of the appearance of α_I (or α_R) in the arguments of exponentials of Equations (3) and (4).

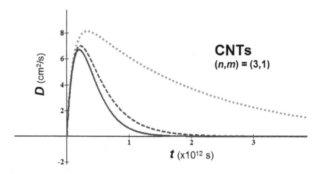

Figure 2. D $vs.$ t for CNTs with $(n,m) = (3,1)$. $\alpha_I = 0.1 =$ blue solid line; $\alpha_I = 0.5 =$ red dashed line; $\alpha_I = 0.9 =$ green dot line.

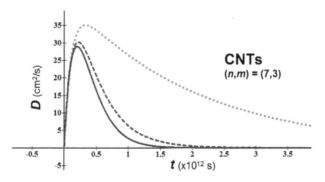

Figure 3. D $vs.$ t for CNTs with $(n,m) = (7,3)$. $\alpha_I = 0.1 =$ blue solid line; $\alpha_I = 0.5 =$ red dashed line; $\alpha_I = 0.9 =$ green dot line.

For the peak values of diffusion (in cm^2/s) obtainable by Figures 2 and 3 we have:
CTNs (a): 6.74 – 7.05 – 8.15;
CTNs (b): 28.89 – 30.29 – 35.15
The considered examples clearly show:

(a) The usefulness of a analogic and continuous computation, to which Turing returned with his work on morphogenesis. This kind of computation takes into account the fact that, in these systems, the boundary conditions and the environment are often important; therefore, not everything can be algorithmized;

(b)   The usefulness of nano-modeling at this level, which is able to locally provide interesting help in the study of brain dynamics and brain processes.

5. Conclusions

In this work we have considered two interesting theoretical approaches related to brain dynamics:

(a)   The memory in the neural network, as a non-passive element for storing information: Memory is integrated in the neural parameters as synaptic conductances, which give the geodesic trajectories in the non-orthogonal space of the free states. The optimal non-linear dynamics is a geodesic inside the deformed space that directs the neural computation. This approach provides the ability to mathematically set up the Freeman hypothesis on the intentionality as "optimal emergency" in the "system/environment" relations [28].

(b)   A new transport model, based on analytical expressions of the three most important parameters related to transport processes: It holds both for the motion of carriers inside a nanostructure, as considered in this paper, and for the motion of nanoparticles inside the human body, because of an inner gauge factor, allowing its use from sub-pico-level to macro-level. The model can be used to understand and manage existing data, but also to give predictions concerning, for example, the best nanomaterial in a particular situation with peculiar characteristics.

Other possibilities in the direction of a required variation of diffusion concern variations in temperature, variations of the effective mass through doping and chiral vector, the variation of the parameters α_I and α_R which are functions of the frequency and the relaxation time. The diffusivity of a nano-substance traveling in the human body is an important parameter for a fast diagnosis of possible diseases, potentially leading to a rapid treatment.

We emphasize that the proposed nanobio approach to the brain directs attention to geometric patterns and attractors, which is a general return to analogic-geometric models, made possible by the fine advances in nano-modeling. Despite the simulations using Turing computation, it is clear that the complex biological systems are highly dependent on the context, which suggests a computationally philosophy, more oriented to natural computation [29,30].

Author Contributions: These authors contributed equally to this work.

Conflicts of Interest: The authors declare no conflict of interest.

References

1.   MacLennan, B.J. Molecular coordination of hierarchical self-assembly. *Nano Commun. Netw.* **2012**, *3*, 116–128. [CrossRef]
2.   Andrews, A.M.; Weiss, P.S. Nano in the Brain: Nano-Neuroscience. *ACS Nano* **2012**, *6*, 8463–8464. [CrossRef] [PubMed]
3.   Claridge, S.A.; Schwartz, J.J.; Weiss, P.S. Electrons, Photons, and Force: Quantitative Single-Molecule Measurements from Physics to Biology. *ACS Nano* **2011**, *5*, 693–729. [CrossRef] [PubMed]
4.   Di Sia, P. Analytical Nano-Modeling for Neuroscience and Cognitive Science. *J. Bioinform. Intell. Control* **2014**, *3*, 268–272. [CrossRef]
5.   Raichle, M.E. A brief history of human brain mapping. *Trends Neurosci.* **2009**, *32*, 118–126. [CrossRef] [PubMed]
6.   Chua, L.O. Memristor, Hodgkin-Huxley, and Edge of Chaos. *Nanotechnology* **2013**, *24*, 383001. [CrossRef] [PubMed]
7.   Tetzlaff, R. *Memristors and Memristive Systems*; Springer: Berlin, Germany, 2013.
8.   Thomas, A. Memristor-based neural networks. *J. Phys. D Appl. Phys.* **2013**, *46*, 093001. [CrossRef]
9.   Nugent, M.A.; Molter, T.W. A HaH Computing-Metastable Switches to Attractors to Machine Learning. *PLoS ONE* **2014**, *9*, 0085175. [CrossRef] [PubMed]
10.   Resconi, G.; Licata, I. Computation by Intention and Electronic Image of the Brain. *J. Appl. Comput. Math.* **2015**, *4*, 232–244.

11. Licata, I. Effective Physical Processes and Active Information in Quantum Computing. *Quantum Biosyst.* **2007**, *1*, 51–65.

12. Licata, I. Beyond Turing: Hypercomputation and Quantum Morphogenesis. *Asia Pac. Math. Newslett.* **2012**, *2*, 20–24.

13. MacLennan, B. Natural computation and non-Turing models of computation. *Theor. Comput. Sci.* **2004**, *317*, 115–145. [CrossRef]

14. Alivisatos, A.P.; Andrews, A.M.; Boyden, E.S.; Chun, M.; Church, G.M.; Deisseroth, K.; Donoghue, J.P.; Fraser, S.E.; Lippincott-Schwartz, J.; Looger, L.L.; *et al.* Nanotools for neuroscience and brain activity mapping. *ACS Nano* **2013**, *7*, 1850–1866. [CrossRef] [PubMed]

15. Wang, Z.L. Toward self-powered sensor networks. *Nano Today* **2010**, *5*, 512–514. [CrossRef]

16. Hao, Y.; Foster, R. Wireless body sensor networks for health monitoring applications. *Physiol. Meas.* **2008**, *29*, R27–R56. [CrossRef] [PubMed]

17. Kim, T.H.; Jang, E.Y.; Lee, N.J.; Choi, D.J.; Lee, K.J.; Jang, J.T.; Choi, J.S.; Moon, S.H.; Cheon, J. Nanoparticle Assemblies as Memristors. *Nano Lett.* **2009**, *9*, 2229–2233. [CrossRef] [PubMed]

18. Mishra, A.K. *Nanomedicine for Drug Delivery and Therapeutics*; Wiley: New York, NY, USA, 2013.

19. Azevedo, F.A.C.; Carvalho, L.R.B.; Grinberg, L.T.; Farfel, J.M.; Ferretti, R.E.L.; Leite, R.E.P.; Filho, W.J.; Lent, R.; Herculano-Houzel, S. Equal Numbers of Neuronal and Nonneuronal Cells Make the Human Brain an Isometrically Scaled-Up Primate Brain. *J. Comp. Neurol.* **2009**, *513*, 532–541. [CrossRef] [PubMed]

20. Alivisatos, A.P.; Chun, M.; Church, G.M.; Greenspan, R.J.; Roukes, M.L.; Chun Yuste, R. The Brain Activity Map Project and the Challenge of Functional Connectomics. *Neuron* **2012**, *74*, 970–974. [CrossRef] [PubMed]

21. Jones, R.A.L. *Soft Condensed Matter: Oxford Master Series in Condensed Matter Physics*; Oxford University Press: Oxford, UK, 2002; Volume 6.

22. Di Sia, P. An Analytical Transport Model for Nanomaterials. *J. Comput. Theor. Nanosci.* **2011**, *8*, 84–89. [CrossRef]

23. Di Sia, P. An Analytical Transport Model for Nanomaterials: The Quantum Version. *J. Comput. Theor. Nanosci.* **2012**, *9*, 31–34. [CrossRef]

24. Di Sia, P. Nanotechnology between Classical and Quantum Scale: Applications of a new interesting analytical Model. *Adv. Sci. Lett.* **2012**, *17*, 82–86. [CrossRef]

25. Di Sia, P. Interesting Details about Diffusion of Nanoparticles for Diagnosis and Treatment in Medicine by a new analytical theoretical Model. *J. Nanotechnol. Diagn. Treat.* **2014**, *2*, 6–10.

26. Di Sia, P. Relativistic nano-transport and artificial neural networks: Details by a new analytical model. *Int. J. Artif. Intell. Mechatron.* **2014**, *3*, 96–100.

27. Marulanda, J.M.; Srivastava, A. Carrier Density and Effective Mass Calculation for carbon Nanotubes. *Phys. Status Solidi (b)* **2008**, *245*, 2558–2562. [CrossRef]

28. Freeman, W. Nonlinear Dynamics of Intentionality. *J. Mind Behav.* **1997**, *18*, 291–304.

29. Resconi, G.; Licata, I. Beyond an Input/Output Paradigm for Systems: Design Systems by Intrinsic Geometry. *Systems* **2014**, *2*, 661–686. [CrossRef]

30. Rozenberg, G.; Back, T.; Kok, J. *Handbook of Natural Computing*; Springer: Berlin, Germany, 2012.

8

Construction and Experimental Validation of a Quantitative Kinetic Model of Nitric Oxide Stress in Enterohemorrhagic *Escherichia coli* O157:H7

Jonathan L. Robinson and Mark P. Brynildsen *

Department of Chemical and Biological Engineering, Princeton University, Princeton, NJ 08544, USA;
jlrtwo@princeton.edu
* Correspondence: mbrynild@princeton.edu

Academic Editors: Mark Blenner and Michael D. Lynch

Abstract: Enterohemorrhagic *Escherichia coli* (EHEC) are responsible for large outbreaks of hemorrhagic colitis, which can progress to life-threatening hemolytic uremic syndrome (HUS) due to the release of Shiga-like toxins (Stx). The presence of a functional nitric oxide (NO·) reductase (NorV), which protects EHEC from NO· produced by immune cells, was previously found to correlate with high HUS incidence, and it was shown that NorV activity enabled prolonged EHEC survival and increased Stx production within macrophages. To enable quantitative study of EHEC NO· defenses and facilitate the development of NO·-potentiating therapeutics, we translated an existing kinetic model of the *E. coli* K-12 NO· response to an EHEC O157:H7 strain. To do this, we trained uncertain model parameters on measurements of [NO·] and [O_2] in EHEC cultures, assessed parametric and prediction uncertainty with the use of a Markov chain Monte Carlo approach, and confirmed the predictive accuracy of the model with experimental data from genetic mutants lacking NorV or Hmp (NO· dioxygenase). Collectively, these results establish a methodology for the translation of quantitative models of NO· stress in model organisms to pathogenic sub-species, which is a critical step toward the application of these models for the study of infectious disease.

Keywords: nitric oxide; enterohemorrhagic *E. coli*; kinetic model; ensemble modeling; Hmp; NorV; microaerobic; anaerobic

1. Introduction

Pathogenic *Escherichia coli* are responsible for a broad range of infections within humans depending on their pathotype, and these can generally be classified as enteric (diarrheagenic) or extraintestinal pathogenic *E. coli* (ExPEC) [1–3]. ExPEC can cause infections at most locations within the body (e.g., meningitis, pneumonia, sepsis, abdominal infection) [1], and the most common pathotype is uropathogenic *E. coli* (UPEC) [4], which are responsible for ~65%–75% of urinary tract infections (UTIs) [5]. Diarrheagenic *E. coli* are generally divided into six pathotypes; enterotoxigenic (ETEC), enteropathogenic (EPEC), enteroaggregative (EAEC), enteroinvasive (EIEC), diffusely adherent (DAEC), and enterohemorrhagic (EHEC). EHEC are potentially deadly, food-borne pathogens responsible for large outbreaks of hemorrhagic colitis (bloody diarrhea). These outbreaks often receive international attention, such as the 1996 outbreak in Japan that sickened over 8000 people [6], the 2006 spinach contamination in California affecting 205 individuals [7,8], and more recently the 2011 outbreak in Germany where ~4000 cases led to the highest incidence of hemolytic uremic syndrome (HUS) on record [9,10].

HUS can occur in up to 20%–25% of patients, and is the leading cause of death from EHEC [3,11,12]. The condition is caused by EHEC Shiga-like toxins (Stx) that are harbored on prophage and released during the phage lytic cycle in response to DNA damage [13]. When released, these toxins proceed to attack renal tissue where they inhibit translation through disruption of the 60S ribosomal subunit, leading to kidney failure [3,7,14–16]. Unfortunately, antibiotics are not recommended for treatment of EHEC infections for several reasons: (1) fluoroquinolones can stimulate prophage induction and Stx release; (2) β-lactams lyse EHEC to increase Stx exposure; and (3) destruction of the gut microbiome can enhance Stx absorption [3,11,17,18]. Furthermore, effective vaccines have yet to be developed, due to the challenges associated with a lack of human disease symptoms in EHEC-infected murine models, and Stx neutralizers have yet to show improved patient outcomes [3,11,19]. For these reasons, treatment of EHEC is largely restricted to supportive care [3], and the frequent outbreaks serve as important reminders of how helpless we are in our fight against these pathogens [11].

Kulasekara and colleagues conducted a genomic analysis of 100 different EHEC isolates, and found that the presence of a functional NO· reductase (NorV) correlated with an increased incidence of HUS [7]. EHEC strains possessing a functional NorV enzyme were associated with HUS incidences of up to 25%, whereas those with the inactive form of the enzyme (possessing a 204 nt in-frame deletion in the *norV* gene) exhibited fewer cases (up to a 10-fold reduction) of HUS [7,10]. Recent outbreaks have supported this association, such as the German outbreak in 2011, where the causative O104:H4 strain possessed a functional NorV [10]. The greater severity of infections caused by functional NorV-bearing EHEC isolates has been attributed to the protection NorV provides from the nitrosative stress exerted by immune cells, such as macrophages [20]. NorV is the main anaerobic NO· detoxification enzyme in *E. coli* [21–24], and NorV-proficient EHEC have been shown to exhibit reduced NO· levels, prolonged survival, and increased Stx production within macrophages compared to their NorV-deficient counterparts [20].

Under aerobic conditions, the major NO· detoxification system in *E. coli* is NO· dioxygenase (Hmp) [25–27]. Hmp has been identified as a virulence factor in many pathogens [28], including UPEC [29], *Salmonella enterica* serovar Typhimurium [30,31], *Yersinia pestis* [32], *Staphylococcus aureus* [33], and *Vibrio cholerae* [34]. Interestingly, a study by Vareille and colleagues found that NO· inhibits Stx production and release from EHEC under oxygenated conditions [35]. Likewise, Branchu and colleagues demonstrated that NO· inhibits the expression of many genes of the EHEC enterocyte effacement (LEE) pathogenicity island under aerobic conditions [36]. Since the vast majority of NO· in *E. coli* cultures is detoxified by Hmp under such conditions [23,25,27], it is tempting to postulate that EHEC can use Hmp to promote expression of its primary virulence factors under NO· stress in the presence of O_2.

Since inhibitors of NorV or Hmp would have the potential to reduce kidney failure and death from EHEC, and humans do not possess a homologue to either protein (Section 2.10), they represent attractive targets for the development of therapies for EHEC infections. Furthermore, inhibitors specific to NO· detoxification enzymes should minimally impact the normal microbiome because NO· defenses are not essential functions for bacterial growth under normal conditions. Unfortunately, known chemical inhibitors of NO· reductases or dioxygenases (e.g., carbon monoxide or cyanide [37,38]) are toxic to humans, and have a greatly weakened effect on the flavodiiron active site used by the *E. coli* NorV [22,23]. Alternative targets in the NO· defense network of EHEC must therefore be identified; however, the broad reactivity of NO· and its reaction products give rise to a large, complex, and interconnected reaction network that includes biological effects ranging from iron-sulfur cluster destruction to inhibition of respiration and DNA damage [27,28,39–41]. The biological outcome of NO· exposure is dictated by a complex kinetic competition within this network, which necessitates the use of computational models for accurate interpretation, understanding, and analysis [27,28]. Such models can quantify the impact of different perturbations on the NO· response, and aid in identifying the underlying mechanisms [27,28,42,43]. Because previous models of NO· stress were developed for a non-pathogenic model organism (*E. coli* K-12), we sought to translate this approach and demonstrate its performance in the clinically-relevant pathogen, EHEC.

Here, we have constructed a quantitative kinetic model of the NO· stress response network in EHEC, which is composed of a system of differential mass balances that was translated from

E. coli K-12 through a process involving literature and database examination, BLAST comparison of protein sequences, and relaxation of uncertain parameters. The model was trained on experimental measurements of $[O_2]$ and $[NO\cdot]$ from EHEC cultures and an MCMC algorithm was employed to populate an ensemble of viable models that could be used to assess parametric and prediction uncertainty. Given the low O_2 tension typically associated with EHEC infection sites, experiments were performed under microaerobic (50 μM O_2) and anaerobic (0 μM O_2) conditions. The trained ensemble of models was able to quantitatively capture $[NO\cdot]$ dynamics in EHEC cultures treated with an $NO\cdot$-releasing chemical, DPTA NONOate, under both O_2 conditions. Furthermore, the ensemble was used to make forward predictions of $[NO\cdot]$ dynamics in cultures of EHEC mutants lacking either of the two major $NO\cdot$ detoxification enzymes, NorV and Hmp. The corresponding experiments were performed, and measurements exhibited excellent agreement with predictions. These results demonstrate that quantitative kinetic modeling of $NO\cdot$ stress can be extended to clinically-relevant strains under physiological O_2 environments, which will facilitate deeper understanding of EHEC $NO\cdot$ defenses and could foster the development of alternative therapeutics for EHEC infections.

2. Materials and Methods

2.1. Chemicals and Growth Media

Cells were grown in Luria-Bertani (LB) Broth (BD Difco), or MOPS minimal media (Teknova) with 10 mM glucose. $NO\cdot$ was delivered to the cultures using DPTA NONOate ((Z)-1-[N-(3-aminopropyl)-N-(3-ammoniopropyl)amino]diazen-1-ium-1,2-diolate) (Cayman Chemical, Ann Arbor, MI, USA), which spontaneously dissociates with a half-life of ~2.5 h at 37 °C and pH 7.4 to release 2 equivalents of $NO\cdot$ per parent compound. For plasmid retention, all growth media contained 30 μg/mL kanamycin (Fisher Scientific) under oxygenated conditions, or 100 μg/mL kanamycin for anaerobic conditions (due to the reduced activity of kanamycin in the absence of respiration).

2.2. Bacterial Strains

Strains used in this study (Table 1) were derived from enterohemorrhagic *E. coli* O157:H7 TUV93-0 [44]. This strain natively possesses a 204 nt in-frame deletion within the *norV* gene that renders the protein inactive. To introduce a functional *norV*, TUV93-0 was transformed with a plasmid harboring the intact gene and its promoter (Section 2.3), yielding the TUV93-0 $hmp^+/norV^+$ strain. The *norV*-null strain used here, TUV $hmp^+/norV^-$, refers to the native TUV93-0 transformed with an empty vector (pUA66 [45]), which was included to impart a similar metabolic burden to that in the $norV^+$ strain. The TUV $hmp^-/norV^+$ strain was generated by replacing *hmp* on the TUV93-0 chromosome with a kan^R cassette using the lambda Red recombinase system [46], where the kan^R cassette DNA was amplified from the purified genomic DNA of the Δ*hmp*::kan^R mutant in the Keio Collection [47] using primers 5′-TGAGATACATCAATTAAGATGCAAAA-3′ (forward) and 5′-AAGGGTTGCCGGATGTTT-3′ (reverse). The mutant was cured of the kan^R marker using FLP recombinase encoded on the pCP20 plasmid [46], and confirmed via cPCR with two sets of primers: 5′-CCGAATCATTGTGCGATAACA-3′ with 5′-GCAAAATCGGTGACGGTAAA-3′ to check for the scar sequence, and 5′-TCCCTTTACTGGTGGAAACG-3′ with 5′-CACGCCCAGATCCACTAACT-3′ to confirm absence of *hmp* in the genome. The cured TUV Δ*hmp* mutant was then transformed with the *norV* complementation plasmid to restore NorV functionality, yielding the TUV $hmp^-/norV^+$ strain.

Table 1. *E. coli* strains used in the present study.

Strain	Genotype	Reference
TUV93-0	EHEC O157:H7 EDL933 Stx$^-$	Leong, J.M. [44]
TUV $hmp^+/norV^+$	TUV93-0 + pUA66-P$_{norV}$-*norV*	This work
TUV $hmp^+/norV^-$	TUV93-0 + pUA66	This work
TUV $hmp^-/norV^+$	TUV93-0 Δ*hmp* + pUA66-P$_{norV}$-*norV*	This work

2.3. Plasmid Construction

The *norV* complementation plasmid (pUA66-P$_{norV}$-*norV*) consisted of an SC101 origin of replication (low copy), a kan^R resistance marker, and the *norV* gene with 188 nt of its 5' UTR (P$_{norV}$), which was cloned from purified *E. coli* K-12 MG1655 genomic DNA. Only 188 nt of the 5' UTR of *norV* was incorporated to avoid inclusion of the *norR* start codon, which begins 189 nt upstream of the *norV* coding sequence (on the complement DNA strand), and this 188 nt sequence is identical in *E. coli* K-12 and EHEC except for two single-nucleotide mutations. Briefly, the genomic DNA of an overnight culture of WT MG1655 grown in LB media was purified using the DNeasy Blood & Tissue Kit (Qiagen), following the manufacturer's instructions for Gram-negative bacteria. The P$_{norV}$-*norV* DNA fragment was amplified from the purified genomic DNA using Phusion® High-Fidelity DNA Polymerase (New England Biolabs, Ipswich, MA, USA; NEB) with primers 5'-GCGCATCTCGAGTACGATCTTTGCCTCACTGTCAATTT-3' (forward) and 5'-GCGCGG TCTAGATCATTTTGCCTCCGATG-3' (reverse), which possessed XhoI and XbaI restriction enzyme (RE) cut sites, respectively. The P$_{norV}$-*norV* amplicon was gel-purified with the QIAquick Gel Extraction Kit (Qiagen), RE digested with XhoI and XbaI (NEB), and PCR-purified (Qiagen). Meanwhile, purified pUA66 plasmid (possessing the SC101 origin and kan^R cassette) [45] was RE digested with XhoI and XbaI (NEB), PCR-purified, and treated with Antarctic Phosphatase (NEB) to prevent self-ligation of the linearized plasmid backbone. After Antarctic Phosphatase treatment, the plasmid backbone was PCR-purified. The P$_{norV}$-*norV* DNA fragment and linearized plasmid backbone were ligated using the Quick Ligation™ Kit (NEB), and transformed into XL1-Blue competent cells (Zymo Research, Irvine, CA, USA), which were immediately plated onto LB-agar plates containing 50 μg/mL kanamycin. Colonies were selected and grown overnight in LB with 30 μg/mL kanamycin, and the plasmid was purified (QIAprep Spin Miniprep Kit, Qiagen). The plasmid sequence was confirmed with PCR and sequencing of the P$_{norV}$-*norV* region of the plasmid (Genewiz).

2.4. Glovebox Setup and Operation

In order to perform measurements under microaerobic and anaerobic conditions, experiments were conducted in a Coy hypoxic chamber with an anaerobic upgrade. For microaerobic conditions, the atmosphere was maintained at 5.0% O_2 v/v (corresponding to 50 μM dissolved O_2 in MOPS minimal media at 37 °C), with 0.2% v/v CO_2, and the balance N_2. To achieve anaerobic conditions, the atmosphere contained 2% H_2 to scavenge trace O_2 via reduction to H_2O on a palladium catalyst, as well as 0.2% CO_2, and the balance N_2.

2.5. Bioreactor Apparatus

A bioreactor was operated in the glovebox to facilitate NO· measurements in a culture exposed to environments with controlled O_2 concentrations. The reactor consisted of a 50 mL conical tube with 10 mL of MOPS + 10 mM glucose media, open to the glovebox environment, and stirred constantly with a 0.5 inch magnetic stir bar. The conical tube was suspended in a magnetically stirred beaker of water maintained at 37 °C with a stirring hotplate.

2.6. NO· Treatment Assay

One mL of LB media with 10 mM glucose was inoculated with a small scrape of *E. coli* cells from a −80 °C frozen stock, and grown at 37 °C and 250 r.p.m. under aerobic conditions for 4 h. A new test tube of 1 mL fresh LB + 10 mM glucose was inoculated with 10 μL of the pregrowth, and placed in the glovebox to grow overnight (16 h) at 37 °C and 200 r.p.m. under O_2 environments of 50 μM (microaerobic) or 0 μM (anaerobic). A 250 mL baffled shake flask containing 20 mL of fresh MOPS minimal media with 10 mM glucose, which had been equilibrated with the glovebox atmosphere overnight, was inoculated with the overnight culture to an OD$_{600}$ of 0.01, and grown at 37 °C and 200 r.p.m. Upon reaching an OD$_{600}$ of 0.2, the flask culture was used to inoculate the bioreactor (containing 10 mL of MOPS + 10 mM glucose) to an OD_{600} of 0.05, which was immediately treated with 50 μM DPTA NONOate. The concentration of NO· was monitored continuously for 1 h following treatment using an ISO-NOP electrode (World Precision Instruments).

2.7. Respiration (O_2 Consumption) Assay

To train the respiratory module of the model, TUV $hmp^+/norV^+$ was grown identically as described for the NO· treatment assays (Section 2.6), except it was not treated with DPTA NONOate following inoculation into the bioreactor. Instead, dissolved $[O_2]$ in the culture was continuously monitored (1 read/sec) for up to 10 min post-inoculation, using a fiber-optic O_2 sensor (FireStingO2 robust miniprobe; PyroScience).

2.8. Model Simulation

The kinetic model was constructed as described previously [27], and adapted for EHEC physiology using the process described in Section 3.1. Briefly, the model is composed of a set of ordinary differential equations (ODEs) describing the change in biochemical species concentrations over time, as a function of the associated reaction rates. Expressed in matrix form, the governing set of equations is written:

$$\frac{dC}{dt} = S \cdot r \tag{1}$$

where C is a vector of biochemical species concentrations, S is the reaction stoichiometry matrix, and r is a vector of reaction rates. Reaction rates are a function of associated species concentrations and kinetic parameter values. The reactions and species were partitioned into intracellular and extracellular compartments to enable experimental parameterization and validation of model predictions [27,48], where NO· and O_2 were assumed to diffuse rapidly across the membrane [49,50]. For further details on model compartmentalization, see [48]. Simulations were run in MATLAB (The MathWorks, Inc.) using the *ode15s* function to numerically integrate the system of ODEs, and solve for biochemical species concentrations as a function of time. The model is available for download on our research group website (https://www.princeton.edu/cbe/people/faculty/brynildsen/group/software/).

2.9. Parameter Optimization

Model parameters were optimized using a two-stage process. The first stage employed the MATLAB *lsqcurvefit* function, whereby parameter values were optimized such that the variance ($σ^2$)-normalized sum of the squared residuals (SSR) between the measured (y_{meas}) and simulated (y_{sim}) data (e.g., NO· or O_2 concentrations) was minimized:

$$SSR = \sum_{i=1}^{n} \frac{\left(y_{i,meas} - y_{i,sim}\right)^2}{\sigma^2_{i,meas}} \tag{2}$$

The non-convex nature of the optimization problem yields local minima, which are dependent on the initial parameter values. To improve coverage of the solution space, least-squares minimizations were repeated 1000 times, each initiated with random parameter values drawn from a uniform distribution that spanned the permitted bounds.

The second stage of parameter optimization involved a Markov chain Monte Carlo (MCMC) method [51], whereby a random walk through parameter space was performed, starting from the best-fit (lowest SSR) parameter set obtained from the previous nonlinear least squares minimization. Relative quality of fit was quantified by the evidence ratio (ER), which was determined from the Akaike Information Criteria (AIC) corrected for small sample sizes [52,53], as described previously [48]. Briefly, the AIC was calculated for each parameter set using the following formula [54]:

$$\text{AIC}_i = n\ln\left(\frac{\text{SSR}_i}{n}\right) + 2k + \frac{2k\,(k+1)}{n-k-1} \tag{3}$$

where n represents the number of data points and k is the number of fit parameters plus one (to account for the SSR estimation) [54]. The weight of evidence (w) of each parameter set in a collection of P parameter sets was calculated from the AIC values [54]:

$$w_i = \frac{\exp(-\Delta_i/2)}{\displaystyle\sum_{i=1}^{P}\exp(-\Delta_i/2)}, \; \Delta_i = \text{AIC}_i - \min(\text{AIC}) \tag{4}$$

From the Akaike weights, the evidence ratios (ER) of each parameter set, which represents the likelihood of each parameter set relative to the best-fit parameter set, were calculated [55]:

$$\text{ER}_i = \frac{w_{\text{best}}}{w_i} \tag{5}$$

where w_{best} is the Akaike weight of the parameter set with the lowest AIC value (best fit). Parameter sets exhibiting an ER > 10 (*i.e.*, less than 10% as likely as the best-fit parameter set) were discarded. If the MCMC process improved the AIC such that the ER of the initial parameter set was greater than 10, the walk was repeated, using the new best-fit parameter set as the new initial set.

In addition to identifying more optimal parameter sets, the MCMC method revealed information on the flexibility, or confidence of parameter values. Confidence intervals (CIs) for parameters were calculated as the range of parameter values among all parameter sets with an ER < 10. Parameters with relatively narrow CIs were more informed by the optimization than those with a relatively wide CI.

2.10. BLAST Comparison of Proteins

Similarities of amino acid sequences between *E. coli* K-12 MG1655 and *E. coli* O157:H7 EDL933 proteins were determined via BLAST on the BioCyc Database [56], where MG1655 amino acid sequences were queried in the EDL933 genome with an expectation value threshold of 10, no filtering of query sequence for low-complexity regions, and the default BLOSUM62 substitution matrix.

For the searches of a Hmp or NorV homologue in humans, a BLAST analysis was conducted on the NCBI Database, querying for the *E. coli* EDL933 Hmp or *E. coli* MG1655 NorV amino acid sequence (MG1655 was used for NorV since the sequence is mutated in EDL933) in the *Homo sapiens* genome (Annotation Release 107) using default algorithm parameters (expectation threshold of 10, BLOSUM62 matrix).

3. Results

3.1. Model Translation from E. coli K-12 to Enterohemorrhagic E. coli O157:H7

To facilitate quantitative investigations of the EHEC NO· defense network, we constructed a kinetic model comprised of the relevant reactions and biomolecules involved in NO· stress. Beginning with a previously constructed model of the NO· response network of non-pathogenic *E. coli* (K-12 MG1655), the model was translated to represent EHEC physiology, accounting for potential and known differences in elements such as transcriptional regulation, metabolite concentrations, and enzyme kinetics. A schematic summarizing the model translation procedure is presented in Figure 1.

Sparse data existed in the literature on EHEC enzyme kinetics at the time of the present study, preventing the use of literature data to inform model parameters. We therefore performed a BLAST analysis to compare protein sequences between EDL933 (the parent of TUV93-0) and MG1655 for all enzymes in the model, with the assumption that enzymes with sufficiently similar amino acid sequences ($\geq$99% match) would exhibit similar kinetics. Of the 20 enzymes (42 subunit proteins) present in the model, only 4 (5 subunits) exhibited amino acid sequences with less than 99% similarity between EDL933 and MG1655 (Table S1). These consisted of enzymes involved in DNA base excision repair (XthA and AlkA, 98.1% and 97.2% similar to that of MG1655, respectively), a subunit of NADH dehydrogenase I (NuoN, 87.2% similar), and NO· reductase NorVW (85.4% and 97.9% similarity for NorV and NorW, respectively). The difference in NorV for EDL933 has been noted previously, where this and some other EHEC strains possess a *norV* sequence with a 204 nt in-frame deletion that renders the protein inactive (commonly referred to as *norVs*) [7,20]. To enable analysis of both Hmp and NorV activity in the present study, a functional NorV was introduced into EHEC on a plasmid (Section 2.2 and Section 2.3).

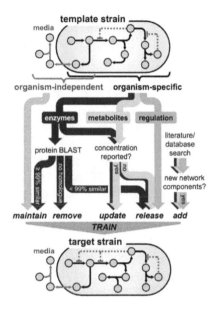

Figure 1. Schematic of the process of translating the NO· stress model from one bacterial strain ("template strain") to another ("target strain"). Beginning with the original model constructed for the template strain, model components were categorized as organism-independent (e.g., NO· exchange with the gas phase) or organism-dependent (e.g., NO· reductase activity). Although the extracellular (media) compartment contains a high frequency of organism-independent components, many are also present intracellularly (e.g., NO· autoxidation). Parameters governing organism-independent processes were maintained at the same value, whereas those dependent on the species were handled differently depending on whether they governed enzyme activity, metabolite/biomolecule concentration, or transcriptional regulation. A BLAST analysis was conducted for each enzyme in the model, where parameters associated with enzymes exhibiting ≥99% amino acid similarity between the template and target strain were maintained at their original value, <99% similarity were released (allowed to vary) for subsequent training, and enzymes with no homologue in the target strain were removed entirely. If metabolite or enzyme concentrations were reported in the literature for the target strain, their values were updated in the model; however, if the concentrations had not been measured previously, they were released to be estimated during the training process. Given the relative complexity and large number of factors influencing the dynamics of transcriptional regulation, parameters governing these processes were released during model training. Finally, any additional network components (enzymes, metabolites, regulatory interactions) found in the literature to be involved in the NO· stress network of the target organism that were not present in the original construction were added to the model. Upon finalizing the model structure, the unknown/released parameters were trained (optimized) on experimental measurements with the target organism (e.g., NO· detoxification and respiratory O_2 consumption).

Given the potential differences in metabolite concentrations and transcriptional regulation between the two *E. coli* strains, we relaxed all regulatory parameters and species concentrations, allowing them to vary within one order of magnitude of the MG1655 value when training on experimental measurements.

3.2. Model Parameter Training and Sensitivity Analysis

Extracellular parameters specific to the experimental apparatus (rate of DPTA NONOate dissociation, $k_{NONOate}$, autoxidation rate of NO·, $k_{NO·-O2}$, and the NO· volumetric mass transfer coefficient, $k_L a_{NO}$.) were trained on [NO·] measured in cell-free MOPS minimal media containing 10 mM glucose following treatment with 50 μM DPTA NONOate in an environment of 50 μM O_2 (Figure S1 and Table S2). DPTA NONOate is an NO·-releasing chemical that dissociates with a half-life of approximately 2.5 h (at 37 °C, pH 7.4) to release two moles of NO· per mole of parent compound. Given the modest decrease in pH of anaerobically grown EHEC cultures (pH = 7.2 instead of 7.4), the DPTA release rate ($k_{NONOate}$) was trained on cell-free MOPS media treated with 50 μM DPTA under anaerobic conditions, with the pH adjusted to 7.2 with HCl, while the other two extracellular parameters ($k_{NO·-O2}$ and $k_L a_{NO}$.) were held constant. The volumetric mass transfer coefficient governing the exchange of dissolved O_2 with the gas phase ($k_L a_{O2}$) was determined by measuring dissolved [O_2] in cell-free media following degassing with N_2. The data were plotted as $\ln([O_2]_{sat}—[O_2])$ vs. time, and a line was fit to the points, where the negative of the slope corresponded to the $k_L a_{O2}$ ($1.25 \times 10^{-3}·s^{-1}$). After obtaining the extracellular parameter values, parameters governing the respiratory module (cytochrome ubiquinol oxidases and NADH reductases) were trained on O_2 measurements in a culture of EHEC inoculated into the bioreactor to an OD_{600} of 0.05 (Figure S2 and Table S3).

Uncertain model parameters (68), defined as those not present in the literature or not sufficiently similar to MG1655, were trained on [NO·] measurements in TUV $hmp^+/norV^+$ cultures treated with 50 μM DPTA NONOate under microaerobic (50 μM O_2) and anaerobic (0 μM O_2) conditions (Table S4). The NO· concentration was monitored continuously (>1 read/sec) following DPTA treatment (Figure 2).

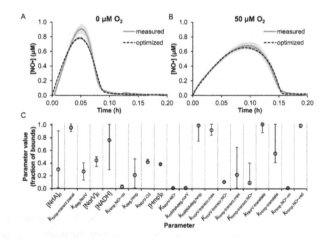

Figure 2. Model training on [NO·] measurements in enterohemorrhagic *Escherichia coli* (EHEC) cultures at 0 and 50 μM O_2. TUV $hmp^+/norV^+$ were treated with 50 μM DPTA NONOate at an OD_{600} of 0.05 under conditions of (**A**) 0 μM O_2 or (**B**) 50 μM O_2, and the resulting [NO·] was measured (solid yellow line; mean of 3 independent experiments, with light yellow shading representing the SEM). Simulated [NO·] curves for each condition are shown (dashed black lines), and were obtained using the best-fit parameter set from the model training process, where gray shading represents the range of simulated [NO·] curves from the ensemble of viable parameter sets (ER < 10). (**C**) Parameter values (expressed as a fraction of the allowed bounds) obtained from the MCMC analysis are shown, where open circles are the best-fit parameter set (ER = 1), and error bars represent the viable range (min and max) that maintained an ER < 10. Descriptions of parameter names can be found in Table S4.

A nonlinear least-squares minimization was performed to determine parameter values yielding the minimum sum of the squared residuals (SSR) between the simulated and measured [NO·] at 0 and 50 μM O_2 (see Section 2). To perform an efficient MCMC search of the parameter space [51], a follow-up sensitivity analysis was executed to assess which parameters had an appreciable impact on [NO·] dynamics, whereby each of the parameters were individually varied among 100 evenly spaced values spanning their allowed range. Parameters whose variation resulted in more than a 5% increase in the SSR (Figure S3) were explored further using the MCMC approach (Figure 2C and Table S4) (see Section 2).

The parameters found to exert the greatest impact on the agreement between simulated and measured [NO·] curves (SSR) were primarily associated with Hmp and NorV, which was expected given that these enzymes are known to be the dominant E. coli NO· detoxification systems under oxygenated and anaerobic conditions, respectively [21–24,27,57–59]. Concentrations of NrfA and NADH were also found to impact the SSR, though to a lesser extent relative to Hmp- and NorV-related parameters, and the MCMC analysis revealed that their values were only mildly constrained.

3.3. Prediction of EHEC NO· Detoxification Dynamics in the Absence of Hmp or NorV

The predictive accuracy of the trained EHEC model was assessed with genetic removal of the two major NO· detoxification enzymes, Hmp and NorV. The genes were deleted synthetically (in the model) by setting their concentration and transcription rate to zero. Using the ensemble of parameter sets that sufficiently captured the TUV $hmp^+/norV^+$ [NO·] curves (ER < 10), NO· treatment (50 μM DPTA NONOate) was simulated for cultures of TUV $hmp^-/norV^+$ and $hmp^+/norV^-$ in environments of 0 and 50 μM O_2 (Figure 3).

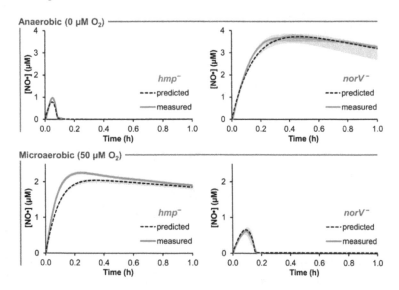

Figure 3. Comparison of predicted and measured [NO·] at 0 and 50 μM O_2 for hmp^- and $norV^-$ EHEC. 50 μM DPTA NONOate treatment was simulated for conditions of 0 and 50 μM O_2 in cultures of TUV $hmp^-/norV^+$ and $hmp^+/norV^-$ at an OD_{600} of 0.05, where the black dashed line was obtained using the best-fit parameter set, and gray shading represents prediction uncertainty (range of viable parameter sets with ER < 10). The corresponding experiments for each condition and mutant were performed, and are shown as solid green ($hmp^-/norV^+$) or solid blue ($hmp^+/norV^-$) curves (mean of 3 independent experiments, with light shading of the same color representing the SEM).

Under anaerobic (0 μM O_2) conditions, NO· consumption by the $norV^-$ EHEC was predicted to be largely negligible, while the simulated hmp^- [NO·] was virtually identical to that of the $hmp^+/norV^+$ strain. Prediction uncertainty was relatively low for hmp^- at 0 μM O_2 (<0.05 μM variation in simulated

[NO·]), but was slightly heightened for $norV^-$ (maximum variation of ~0.6 μM NO·). For the 50 μM O_2 conditions, the predicted behavior for the two mutants was exchanged, where $norV^-$ simulations now exhibited NO· clearance similar to that of TUV $hmp^+/norV^+$, while NO· detoxification was severely impaired in hmp^- cultures. Simulations exhibited little uncertainty in [NO·] for either mutant at 50 μM O_2, with ≤0.1 μM variation in predicted [NO·].

Corresponding experimental NO· treatment assays were performed with TUV $hmp^-/norV^+$ and $hmp^+/norV^-$ cultures at 0 and 50 μM O_2. The resulting measured [NO·] curves were in excellent agreement with model predictions for both mutants under both O_2 conditions (Figure 3). Experimental confirmation of the quantitative accuracy in predicted NO· dynamics for mutants lacking either of the two major NO· detoxification systems in EHEC demonstrated a successful translation of the NO· kinetic modeling approach to this medically-relevant pathogen.

3.4. Predicted Distribution of NO· Consumption

Upon confirming that model simulations could quantitatively capture EHEC NO· detoxification dynamics under two different O_2 regimes, and accurately predict the behavior of genetic mutants lacking either of the major NO· defense systems (TUV $hmp^-/norV^+$ and $hmp^+/norV^-$), we sought to use the model to quantify the distribution of NO· flux through the reaction network in TUV $hmp^+/norV^+$ cultures treated with 50 μM DPTA NONOate. DPTA treatment was simulated for conditions of 0 and 50 μM O_2, and the resulting cumulative NO· consumption by each of the available pathways was quantified (Figure 4).

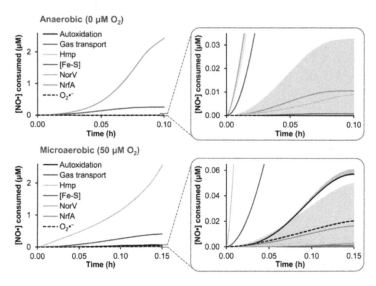

Figure 4. Simulated distribution of NO· consumption in an EHEC culture. The model was used to simulate treatment of a TUV $hmp^+/norV^+$ culture at an OD_{600} of 0.05 with 50 μM DPTA NONOate under conditions of 0 or 50 μM O_2, and the resulting distribution of cumulative NO· consumption among the available pathways is shown, up to the approximate time when NO· was cleared (<0.1 μM NO·) from the culture. Lines are predictions obtained using the best-fit parameter set, while lighter shading of a similar color represent prediction uncertainty (range of viable parameter sets with ER < 10). "Autoxidation" is NO· autoxidation, "Gas transport" is loss of NO· to the gas phase, "Hmp" is Hmp-mediated NO· detoxification, "[Fe-S]" is nitrosylation of iron-sulfur clusters, "NorV" is NorV-mediated NO· detoxification, "NrfA" is NrfA-mediated detoxification, and "$O_2·^-$" is reaction of NO· with superoxide.

Consistent with NO· detoxification measured in TUV hmp^- and $norV^-$ cultures, the dominant NO· consumption pathway at 0 and 50 μM O_2 was predicted to be NorV and Hmp, respectively, accounting for over 80% of NO· detoxification up to the time NO· was cleared (<0.1 μM NO·). The majority of the remaining NO· flux for both conditions was loss to the gas phase (~10%–15% of NO· consumption), whereas all other pathways (autoxidation, iron-sulfur cluster ([Fe-S]) nitrosylation, NrfA detoxification, and reaction with superoxide ($O_2 \cdot^-$)) did not exceed 2% of the cumulative NO· consumption. Although NO· autoxidation and reaction with $O_2 \cdot^-$ was absent under anaerobic conditions due to the lack of O_2, their combined contribution was predicted to remain modest (<2.5% of the NO· flux) even at 50 μM O_2. Nitrosylation of [Fe-S] had a largely negligible impact on NO· under either O_2 condition, accounting for less than 0.1% of the NO· consumed.

From the simulation results, it was clear that the greatest level of relative uncertainty was regarding the participation of NrfA. Since one of the NrfA parameters (initial concentration) was identified as having an impact on the [NO·] curve during the model training procedure (Figure S3), and the MCMC analysis revealed a mild constraint of its value (Figure 2C), it was not surprising that its predicted contribution to NO· consumption would be highly variable among the ensemble of viable models. Further model training on $nrfA^-$ mutant data could resolve the uncertainty; however, given its minimal contribution to NO· detoxification under either 0 or 50 μM O_2 conditions (Figure 3), increased resolution of the contribution of NrfA under the conditions used here was not necessary.

4. Discussion

EHEC is a potentially life-threatening pathogen with an extremely low infectious dose (<50 bacteria [13,60,61]). Antibiotic treatments are not recommended due to their exacerbation of Stx-related damage, which can increase the risk for HUS development [3,11,17,18,62]. Although alternative treatment approaches are being explored, such as the development of vaccines or Stx-targeting strategies, they are hindered by a lack of animal models capable of accurately mimicking a human EHEC infection, and the possibility of toxins other than Stx contributing to HUS, respectively [63]. Furthermore, efforts to neutralize the toxins after they have been produced may not be quick enough to prevent damage, whereas inhibition of toxin production would halt damage at the source [11]. Given that EHEC harboring an active NorV exhibit increased Stx production within macrophages [20], therapeutics designed to target this NO· defense system are an attractive solution. Recent studies have shown that NO· stress can decrease Stx production [35], and EHEC lacking NO· detoxification machinery exhibit attenuated survival within macrophages [20]. Here, we have developed and experimentally validated a computational tool to enable quantitative investigation of the broad and complex NO· response network in EHEC.

We previously constructed a kinetic model of the bacterial NO· response of E. coli K-12 MG1655, which accounted for the complex biochemical reaction network associated with NO· stress, including iron-sulfur cluster damage, thiol nitrosation, DNA deamination, enzymatic detoxification, and reversible respiratory inhibition [27]. Model predictions were validated with experimental measurements of biochemical species concentrations (NO·, O_2, NO_2^-, and NO_3^-), and novel network dynamics were revealed, such as the reduced utility of Hmp with increasing NO· delivery rate [27]. In addition, the model has been used to aid in characterizing the relationship between NO· payload, release rate, and cytotoxicity [42], as well as elucidating the mechanism underlying the enhanced NO· sensitivity of an E. coli mutant lacking the ClpP protease [43]. Furthermore, this approach was demonstrated to be an effective tool for studying the dynamics of a different broadly reactive metabolite, hydrogen peroxide (H_2O_2) [48].

Here, the NO· kinetic model was adapted to an E. coli O157:H7 strain, and trained on experimental data. The adapted model exhibited excellent agreement with [NO·] dynamics measured in EHEC cultures following treatment with the NO·-releasing compound, DPTA NONOate, under both microaerobic and anaerobic conditions (50 and 0 μM O_2, respectively). Forward predictions of [NO·] in mutant EHEC cultures lacking the major aerobic (Hmp) or anaerobic (NorV) NO· detoxification enzyme at both 0 and 50 μM O_2 were in excellent agreement with the corresponding experimental measurements, which demonstrated the accuracy and versatility of the translated model.

These results, which demonstrated that Hmp and NorV dominate NO· detoxification in EHEC in the presence and absence of O_2, respectively, are consistent with previous studies of *E. coli* NO· detoxification [20–22,25,57,64]. The NO· dioxygenase function of Hmp requires O_2 as a substrate, and is therefore inactive under anaerobic conditions [23,57]. Hmp has been demonstrated to harbor an additional, O_2-indepedent NO· reductase activity, but the rate of this reaction is orders of magnitude slower than the dioxygenation reaction or NorV-mediated NO· reduction [64]. NorV, which is capable of reducing NO· under anaerobic conditions, possesses an O_2-sensitive flavodiiron catalytic site that is rapidly inactivated upon exposure to oxygenated environments [23]. Although both *hmp* and *norV* genes are expressed in response to NO· exposure in the presence or absence of O_2 [43,65,66], the substrate requirements and O_2 sensitivity of their catalytic activity limits the function of these enzymes in each other's respective O_2 environments. However, the intermediate regime between 0 and 50 μM O_2 is far less studied, and the quantitative contribution of Hmp and NorV under these conditions remains ill-defined.

We envision that the computational tool developed here will facilitate future quantitative investigations of the complex NO· stress network in EHEC. For example, mutations found to enhance the NO· sensitivity of EHEC could be interrogated for their underlying mechanism using a model-guided approach, as demonstrated for Δ*clpP* in *E. coli* K-12 [43]. Given the breadth and complexity of the NO· biochemical network, an NO··-sensitive phenotype could arise through an immense number of possible mechanisms. By using the model as a framework to interpret the perturbed dynamics of a pathogen's NO· response, it could elucidate the specific network components and/or functions involved in the altered behavior, or at least reduce the amount of feasible mechanisms to an experimentally-tractable number. In addition, such a model-guided approach could be translated to the mechanistic investigation of chemical perturbations to the NO· stress network. High-throughput chemical screens could be performed to identify compounds that selectively impair EHEC NO· defenses, and the model could provide a quantitative framework to guide investigations of the compounds' mechanisms of action. Further, the model could be expanded to incorporate processes governing Stx production, and reveal important relationships between NO· stress and EHEC's primary virulence factors. Indeed, it has already been shown that NO· interferes with Stx production in EHEC [35], and therefore, a deeper understanding of how these two systems interact could reveal novel treatment strategies.

5. Conclusions

Since antibiotics are not recommended for treatment of EHEC infections due to their enhancement of Stx-related damage, alternative treatment approaches, such as those that target virulence factors, are needed. NO· detoxification is a newly identified virulence system for EHEC [7,20], and the computational tool developed in this work will enable quantitative understanding of NO· stress in EHEC to be gained. Such knowledge could lead to novel anti-infective modalities for the treatment of EHEC, which are sorely needed for this dangerous pathogen. Indeed, continued outbreaks of EHEC offer frequent reminders that our current therapeutic options are insufficient, and novel approaches, such as targeting its NO· defenses, are required to combat this potentially deadly pathogen.

Supplementary Materials
Figure S1: Training of extracellular model parameters. 50 μM DPTA NONOate was delivered to cell-free media (MOPS minimal media with 10 mM glucose) at 50 and 0 μM O_2, and [NO·] was measured (solid red lines; mean of 3 independent experiments, with light red shading representing the SEM). For the 0 μM O_2 condition, the media pH was adjusted to 7.2 (from 7.4) with HCl, to mimic the slightly acidified conditions measured in the bioreactor culture for anaerobic assays with EHEC. Model parameters specific to the experimental apparatus and media conditions ($k_{NO·−O2}$, $k_La_{NO·}$, and $k_{NONOate}$) were optimized on the [NO·] curve measured at 50 μM O_2 (dashed black line using the best-fit parameter set, with gray shading representing the range of viable parameter sets with ER < 10). Since a decrease in pH increases the rate of NONOate dissociation, the $k_{NONOate}$ parameter was released and trained on [NO·] measured in the pH-adjusted media at 0 μM O_2. The simulation result using the trained $k_{NONOate}$ parameter is shown (dashed black line, obtained using the best-fit parameter value, with gray shading representing viable parameter values with ER < 10), Figure S2: Training of parameters associated with

the respiratory module. TUV $hmp^+/norV^+$ in mid-exponential phase were delivered to the bioreactor to an OD_{600} of 0.05, and the O_2 consumption was measured (solid purple line; mean of 3 independent experiments with light purple shading representing the SEM). Model parameters associated with the aerobic respiratory module (NADH dehydrogenases and cytochrome ubiquinol oxidases) were trained on the measured $[O_2]$. The simulated $[O_2]$ generated by the trained model is shown (dashed black line was obtained using the best-fit parameter set, with gray shading representing the range of viable parameter sets with ER < 10), Figure S3: Impact of individual parameter variation on [NO·] distribution. A sensitivity analysis was performed on the 68 parameters trained on the [NO·] curves measured in TUV $hmp^+/norV^+$ cultures following treatment with 50 μM DPTA NONOate at 0 and 50 μM O_2. Each parameter was individually varied within its permitted bounds, and the resulting increase in SSR was quantified (fold increase = SSR/SSR_0). Shown is the maximum fold increase in SSR achieved for each parameter, showing only the 20 parameters exhibiting a maximum increase of ⩾5% (1.05-fold), Table S1: BLAST analysis. Proteins with <99% amino acid (AA) similarity are in bold, Table S2: Training of extracellular model parameters. Parameter values were released and allowed to vary within the "Minimum" and "Maximum" bounds. Optimal (best-fit, yielding minimum SSR between measured and simulated [NO·]) parameter values are reported, along with their confidence interval (CI), defined as the range of the parameter among viable parameter sets with ER < 10, Table S3: Training of model parameters associated with the respiratory module. Parameter values were released and allowed to vary within the "Minimum" and "Maximum" bounds. Optimal (best-fit, yielding minimum SSR between measured and simulated $[O_2]$) parameter values are reported, along with their confidence interval (CI), defined as the range of the parameter among viable parameter sets with ER < 10, Table S4: Training of organism-specific model parameters. Parameter values were released and allowed to vary within the "Minimum" and "Maximum" bounds. Optimal (best-fit, yielding minimum SSR between measured and simulated [NO·]) parameter values are reported, along with their confidence interval (CI), defined as the range of the parameter among viable parameter sets with ER < 10. Only the 20 parameters identified as having a substantial impact on the SSR (>5% increase) are shown.

Acknowledgments: We thank James B. Kaper for helpful discussions, and acknowledge the National BioResource Project (NIG, Japan) for distribution of the Keio collection. We thank John M. Leong for kindly providing the EHEC TUV93-0 strain, and Shayan Rakhit for his early contributions to the project. Computationally-expensive optimizations and simulations were performed using the Terascale Infrastructure for Groundbreaking Research in Science and Engineering (TIGRESS) high performance computer center at Princeton University, which is jointly supported by the Princeton Institute for Computational Science and Engineering (PICSciE) and the Princeton University Office of Information Technology's Research Computing department. This work was supported by the National Science Foundation (CBET-1453325), and Princeton University (Forese Family Fund for Innovation, start-up funds).

Author Contributions: Jonathan L. Robinson and Mark P. Brynildsen conceived and designed the experiments, analyzed the data, and wrote the paper. Jonathan L. Robinson performed the experiments.

Conflicts of Interest: The authors declare no conflict of interest.

References

1. Russo, T.A.; Johnson, J.R. Proposal for a new inclusive designation for extraintestinal pathogenic isolates of *Escherichia coli*: ExPEC. *J. Infect. Dis.* **2000**, *181*, 1753–1754. [CrossRef] [PubMed]

2. Wiles, T.J.; Kulesus, R.R.; Mulvey, M.A. Origins and virulence mechanisms of uropathogenic *Escherichia coli*. *Exp. Mol. Pathol.* **2008**, *85*, 11–19. [CrossRef] [PubMed]

3. Nataro, J.P.; Kaper, J.B. Diarrheagenic *Escherichia coli*. *Clin. Microbiol. Rev.* **1998**, *11*, 142–201. [PubMed]

4. Kaper, J.B.; Nataro, J.P.; Mobley, H.L. Pathogenic *Escherichia coli*. *Nat. Rev. Microbiol.* **2004**, *2*, 123–140. [CrossRef] [PubMed]

5. Flores-Mireles, A.L.; Walker, J.N.; Caparon, M.; Hultgren, S.J. Urinary tract infections: Epidemiology, mechanisms of infection and treatment options. *Nat. Rev. Microbiol.* **2015**, *13*, 269–284. [CrossRef] [PubMed]

6. Michino, H.; Araki, K.; Minami, S.; Takaya, S.; Sakai, N.; Miyazaki, M.; Ono, A.; Yanagawa, H. Massive outbreak of *Escherichia coli* O157:H7 infection in schoolchildren in Sakai City, Japan, associated with consumption of white radish sprouts. *Am. J. Epidemiol.* **1999**, *150*, 787–796. [CrossRef] [PubMed]

7. Kulasekara, B.R.; Jacobs, M.; Zhou, Y.; Wu, Z.N.; Sims, E.; Saenphimmachak, C.; Rohmer, L.; Ritchie, J.M.; Radey, M.; McKevitt, M.; *et al.* Analysis of the genome of the *Escherichia coli* O157:H7 2006 spinach-associated outbreak isolate indicates candidate genes that may enhance virulence. *Infect. Immun.* **2009**, *77*, 3713–3721. [CrossRef] [PubMed]

8. Centers for Disease Control and Prevention. Ongoing multistate outbreak of *Escherichia coli* serotype O157:H7 infections associated with consumption of fresh spinach—United States, September 2006. *MMWR Morb. Mortal. Wkly. Rep.* **2006**, *55*, 1045–1046.

9.   Frank, C.; Werber, D.; Cramer, J.P.; Askar, M.; Faber, M.; an der Heiden, M.; Bernard, H.; Fruth, A.; Prager, R.;
     Spode, A.; *et al.* Epidemic profile of Shiga-toxin-producing *Escherichia coli* O104:H4 outbreak in Germany.
     N. Engl. J. Med. **2011**, *365*, 1771–1780. [CrossRef] [PubMed]

10. Mellmann, A.; Harmsen, D.; Cummings, C.A.; Zentz, E.B.; Leopold, S.R.; Rico, A.; Prior, K.;
     Szczepanowski, R.; Ji, Y.M.; Zhang, W.L.; *et al.* Prospective genomic characterization of the German
     enterohemorrhagic *Escherichia coli* O104:H4 outbreak by rapid next generation sequencing technology.
     PLoS ONE **2011**, *6*, e22751. [CrossRef] [PubMed]

11. Goldwater, P.N.; Bettelheim, K.A. Treatment of enterohemorrhagic *Escherichia coli* (EHEC) infection and
     hemolytic uremic syndrome (HUS). *BMC Med.* **2012**, *10*, 12. [CrossRef] [PubMed]

12. Torres, A.G.; Zhou, X.; Kaper, J.B. Adherence of diarrheagenic *Escherichia coli* strains to epithelial cells.
     Infect. Immun. **2005**, *73*, 18–29. [CrossRef] [PubMed]

13. Kaper, J.B.; Karmali, M.A. The continuing evolution of a bacterial pathogen. *Proc. Natl. Acad. Sci. USA*
     2008, *105*, 4535–4536. [CrossRef] [PubMed]

14. Tarr, P.I.; Gordon, C.A.; Chandler, W.L. Shiga-toxin-producing *Escherichia coli* and haemolytic uraemic
     syndrome. *Lancet* **2005**, *365*, 1073–1086. [CrossRef]

15. Obrig, T.G.; Moran, T.P.; Brown, J.E. The mode of action of Shiga toxin on peptide elongation of eukaryotic
     protein synthesis. *Biochem. J.* **1987**, *244*, 287–294. [CrossRef] [PubMed]

16. Schuller, S.; Heuschkel, R.; Torrente, F.; Kaper, J.B.; Phillips, A.D. Shiga toxin binding in normal and inflamed
     human intestinal mucosa. *Microbes Infect.* **2007**, *9*, 35–39. [CrossRef] [PubMed]

17. Zhang, X.P.; McDaniel, A.D.; Wolf, L.E.; Keusch, G.T.; Waldor, M.K.; Acheson, D.W.K. Quinolone antibiotics
     induce shiga toxin-encoding bacteriophages, toxin production, and death in mice. *J. Infect. Dis.* **2000**, *181*,
     664–670. [CrossRef] [PubMed]

18. Bielaszewska, M.; Idelevich, E.A.; Zhang, W.; Bauwens, A.; Schaumburg, F.; Mellmann, A.; Peters, G.;
     Karch, H. Effects of antibiotics on Shiga toxin 2 production and bacteriophage induction by epidemic
     Escherichia coli O104:H4 strain. *Antimicrob. Agents Chemother.* **2012**, *56*, 3277–3282. [CrossRef] [PubMed]

19. Garcia-Angulo, V.A.; Kalita, A.; Torres, A.G. Advances in the development of enterohemorrhagic
     Escherichia coli vaccines using murine models of infection. *Vaccine* **2013**, *31*, 3229–3235. [CrossRef] [PubMed]

20. Shimizu, T.; Tsutsuki, H.; Matsumoto, A.; Nakaya, H.; Noda, M. The nitric oxide reductase of
     enterohaemorrhagic *Escherichia coli* plays an important role for the survival within macrophages.
     Mol. Microbiol. **2012**, *85*, 492–512. [CrossRef] [PubMed]

21. Gomes, C.M.; Giuffre, A.; Forte, E.; Vicente, J.B.; Saraiva, L.M.; Brunori, M.; Teixeira, M. A novel type of
     nitric-oxide reductase. *Escherichia coli* flavorubredoxin. *J. Biol. Chem.* **2002**, *277*, 25273–25276. [PubMed]

22. Gardner, A.M.; Helmick, R.A.; Gardner, P.R. Flavorubredoxin, an inducible catalyst for nitric oxide reduction
     and detoxification in *Escherichia coli*. *J. Biol. Chem.* **2002**, *277*, 8172–8177. [CrossRef] [PubMed]

23. Gardner, A.M.; Gardner, P.R. Flavohemoglobin detoxifies nitric oxide in aerobic, but not anaerobic, *Escherichia
     coli*—Evidence for a novel inducible anaerobic nitric oxide-scavenging activity. *J. Biol. Chem.* **2002**, *277*,
     8166–8171. [CrossRef] [PubMed]

24. Vine, C.E.; Cole, J.A. Nitrosative stress in *Escherichia coli*: Reduction of nitric oxide. *Biochem. Soc. Trans.*
     2011, *39*, 213–215. [CrossRef] [PubMed]

25. Gardner, P.R.; Gardner, A.M.; Martin, L.A.; Salzman, A.L. Nitric oxide dioxygenase: An enzymic function for
     flavohemoglobin. *Proc. Natl. Acad. Sci. USA* **1998**, *95*, 10378–10383. [CrossRef] [PubMed]

26. Hausladen, A.; Gow, A.J.; Stamler, J.S. Nitrosative stress: Metabolic pathway involving the flavohemoglobin.
     Proc. Natl. Acad. Sci. USA **1998**, *95*, 14100–14105. [CrossRef] [PubMed]

27. Robinson, J.L.; Brynildsen, M.P. A kinetic platform to determine the fate of nitric oxide in *Escherichia coli*.
     PLoS Comput. Biol. **2013**, *9*, e1003049. [CrossRef] [PubMed]

28. Robinson, J.L.; Adolfsen, K.J.; Brynildsen, M.P. Deciphering nitric oxide stress in bacteria with quantitative
     modeling. *Curr. Opin. Microbiol.* **2014**, *19*, 16–24. [CrossRef] [PubMed]

29. Svensson, L.; Poljakovic, M.; Save, S.; Gilberthorpe, N.; Schon, T.; Strid, S.; Corker, H.; Poole, R.K.; Persson, K.
     Role of flavohemoglobin in combating nitrosative stress in uropathogenic *Escherichia coli*—Implications for
     urinary tract infection. *Microb. Pathog.* **2010**, *49*, 59–66. [CrossRef] [PubMed]

30. Bang, I.S.; Liu, L.; Vazquez-Torres, A.; Crouch, M.L.; Stamler, J.S.; Fang, F.C. Maintenance of nitric oxide and
     redox homeostasis by the *Salmonella* flavohemoglobin hmp. *J. Biol. Chem.* **2006**, *281*, 28039–28047. [CrossRef]
     [PubMed]

31. Stevanin, T.M.; Poole, R.K.; Demoncheaux, E.A.G.; Read, R.C. Flavohemoglobin Hmp protects *Salmonella enterica* serovar typhimurium from nitric oxide-related killing by human macrophages. *Infect. Immun.* **2002**, *70*, 4399–4405. [CrossRef] [PubMed]

32. Sebbane, F.; Lemaitre, N.; Sturdevant, D.E.; Rebeil, R.; Virtaneva, K.; Porcella, S.F.; Hinnebusch, B.J. Adaptive response of *Yersinia pestis* to extracellular effectors of innate immunity during bubonic plague. *Proc. Natl. Acad. Sci. USA* **2006**, *103*, 11766–11771. [CrossRef] [PubMed]

33. Richardson, A.R.; Dunman, P.M.; Fang, F.C. The nitrosative stress response of *Staphylococcus aureus* is required for resistance to innate immunity. *Mol. Microbiol.* **2006**, *61*, 927–939. [CrossRef] [PubMed]

34. Stern, A.M.; Hay, A.J.; Liu, Z.; Desland, F.A.; Zhang, J.; Zhong, Z.T.; Zhu, J. The NorR regulon is critical for *Vibrio cholerae* resistance to nitric oxide and sustained colonization of the intestines. *mBio* **2012**, *3*, e00013-12. [CrossRef] [PubMed]

35. Vareille, M.; de Sablet, T.; Hindre, T.; Martin, C.; Gobert, A.P. Nitric oxide inhibits Shiga-toxin synthesis by enterohemorrhagic *Escherichia coli*. *Proc. Natl. Acad. Sci. USA* **2007**, *104*, 10199–10204. [CrossRef] [PubMed]

36. Branchu, P.; Matrat, S.; Vareille, M.; Garrivier, A.; Durand, A.; Crepin, S.; Harel, J.; Jubelin, G.; Gobert, A.P. NsrR, GadE, and GadX interplay in repressing expression of the *Escherichia coli* O157: H7 LEE pathogenicity island in response to nitric oxide. *PLoS Pathog.* **2014**, *10*, e1003874. [CrossRef] [PubMed]

37. Schomburg, I.; Chang, A.; Placzek, S.; Sohngen, C.; Rother, M.; Lang, M.; Munaretto, C.; Ulas, S.; Stelzer, M.; Grote, A.; *et al.* BRENDA in 2013: Integrated reactions, kinetic data, enzyme function data, improved disease classification: New options and contents in BRENDA. *Nucleic Acids Res.* **2013**, *41*, D764–D772. [CrossRef] [PubMed]

38. Gardner, P.R.; Gardner, A.M.; Martin, L.A.; Dou, Y.; Li, T.; Olson, J.S.; Zhu, H.; Riggs, A.F. Nitric-oxide dioxygenase activity and function of flavohemoglobins. sensitivity to nitric oxide and carbon monoxide inhibition. *J. Biol. Chem.* **2000**, *275*, 31581–31587. [CrossRef] [PubMed]

39. Bowman, L.A.; McLean, S.; Poole, R.K.; Fukuto, J.M. The diversity of microbial responses to nitric oxide and agents of nitrosative stress close cousins but not identical twins. *Adv. Microb. Physiol.* **2011**, *59*, 135–219. [PubMed]

40. Lancaster, J.R., Jr. Nitroxidative, nitrosative, and nitrative stress: Kinetic predictions of reactive nitrogen species chemistry under biological conditions. *Chem. Res. Toxicol.* **2006**, *19*, 1160–1174. [CrossRef] [PubMed]

41. Lim, C.H.; Dedon, P.C.; Deen, W.M. Kinetic analysis of intracellular concentrations of reactive nitrogen species. *Chem. Res. Toxicol.* **2008**, *21*, 2134–2147. [CrossRef] [PubMed]

42. Robinson, J.L.; Brynildsen, M.P. Model-driven identification of dosing regimens that maximize the antimicrobial activity of nitric oxide. *Metab. Eng. Commun.* **2014**, *1*, 12–18. [CrossRef]

43. Robinson, J.L.; Brynildsen, M.P. An ensemble-guided approach Identifies ClpP as a major regulator of transcript levels in nitric oxide-stressed *Escherichia coli*. *Metab. Eng.* **2015**, *31*, 22–34. [CrossRef] [PubMed]

44. Campellone, K.G.; Giese, A.; Tipper, D.J.; Leong, J.M. A tyrosine-phosphorylated 12-amino-acid sequence of enteropathogenic *Escherichia coli* Tir binds the host adaptor protein Nck and is required for Nck localization to actin pedestals. *Mol. Microbiol.* **2002**, *43*, 1227–1241. [CrossRef] [PubMed]

45. Zaslaver, A.; Bren, A.; Ronen, M.; Itzkovitz, S.; Kikoin, I.; Shavit, S.; Liebermeister, W.; Surette, M.G.; Alon, U. A comprehensive library of fluorescent transcriptional reporters for *Escherichia coli*. *Nat. Methods* **2006**, *3*, 623–628. [CrossRef] [PubMed]

46. Datsenko, K.A.; Wanner, B.L. One-step inactivation of chromosomal genes in *Escherichia coli* K-12 using PCR products. *Proc. Natl. Acad. Sci. USA* **2000**, *97*, 6640–6645. [CrossRef] [PubMed]

47. Baba, T.; Ara, T.; Hasegawa, M.; Takai, Y.; Okumura, Y.; Baba, M.; Datsenko, K.A.; Tomita, M.; Wanner, B.L.; Mori, H. Construction of *Escherichia coli* K-12 in-frame, single-gene knockout mutants: The Keio collection. *Mol. Syst. Biol.* **2006**, *2*. [CrossRef] [PubMed]

48. Adolfsen, K.J.; Brynildsen, M.P. A kinetic platform to determine the fate of hydrogen peroxide in *Escherichia coli*. *PLoS Comput. Biol.* **2015**, *11*, e1004562. [CrossRef] [PubMed]

49. Denicola, A.; Souza, J.M.; Radi, R.; Lissi, E. Nitric oxide diffusion in membranes determined by fluorescence quenching. *Arch. Biochem. Biophys.* **1996**, *328*, 208–212. [CrossRef] [PubMed]

50. Kelm, M. Nitric oxide metabolism and breakdown. *Biochim. Biophys. Acta* **1999**, *1411*, 273–289. [CrossRef]

51. Zamora-Sillero, E.; Hafner, M.; Ibig, A.; Stelling, J.; Wagner, A. Efficient characterization of high-dimensional parameter spaces for systems biology. *BMC Syst. Biol.* **2011**, *5*, 142. [CrossRef] [PubMed]

52. Akaike, H. Information theory and extension of the maximum likelihood principle. In Proceedings of the 2nd International Symposium on Information Theory, Tsahkadsor, Armenia, 2–8 September 1971; Petrov, B.N., Csaki, F., Eds.; Akadémiai Kiadó: Budapest, Hungary, 1973; pp. 267–281.

53. Hurvich, C.M.; Tsai, C.L. Regression and time-series model selection in small samples. *Biometrika* **1989**, *76*, 297–307. [CrossRef]

54. Turkheimer, F.E.; Hinz, R.; Cunningham, V.J. On the undecidability among kinetic models: From model selection to model averaging. *J. Cereb. Blood Flow Metab.* **2003**, *23*, 490–498. [CrossRef] [PubMed]

55. Burnham, K.P.; Anderson, D.R. *Model Selection and Multimodel Inference: A Practical Information-Theoretic Approach*, 2nd ed.; Springer: New York, NY, USA, 2002.

56. Caspi, R.; Billington, R.; Ferrer, L.; Foerster, H.; Fulcher, C.A.; Keseler, I.M.; Kothari, A.; Krummenacker, M.; Latendresse, M.; Mueller, L.A.; *et al.* The MetaCyc database of metabolic pathways and enzymes and the BioCyc collection of pathway/genome databases. *Nucleic Acids Res.* **2015**, *36*, D623–D631. [CrossRef] [PubMed]

57. Hausladen, A.; Gow, A.; Stamler, J.S. Flavohemoglobin denitrosylase catalyzes the reaction of a nitroxyl equivalent with molecular oxygen. *Proc. Natl. Acad. Sci. USA* **2001**, *98*, 10108–10112. [CrossRef] [PubMed]

58. Mills, C.E.; Sedelnikova, S.; Soballe, B.; Hughes, M.N.; Poole, R.K. *Escherichia coli* flavohaemoglobin (Hmp) with equistoichiometric FAD and haem contents has a low affinity for dioxygen in the absence or presence of nitric oxide. *Biochem. J.* **2001**, *353*, 207–213. [CrossRef] [PubMed]

59. Stevanin, T.M.; Ioannidis, N.; Mills, C.E.; Kim, S.O.; Hughes, M.N.; Poole, R.K. Flavohemoglobin Hmp affords inducible protection for *Escherichia coli* respiration, catalyzed by cytochromes *bo'* or *bd*, from nitric oxide. *J. Biol. Chem.* **2000**, *275*, 35868–35875. [CrossRef] [PubMed]

60. Tilden, J.; Young, W.; McNamara, A.M.; Custer, C.; Boesel, B.; LambertFair, M.; Majkowski, J.; Vugia, D.; Werner, S.B.; Hollingsworth, J.; *et al.* A new route of transmission for *Escherichia coli*: Infection from dry fermented salami. *Am. J. Public Health* **1996**, *86*, 1142–1145. [CrossRef] [PubMed]

61. Tuttle, J.; Gomez, T.; Doyle, M.P.; Wells, J.G.; Zhao, T.; Tauxe, R.V.; Griffin, P.M. Lessons from a large outbreak of *Escherichia coli* O157:H7 infections: Insights into the infectious dose and method of widespread contamination of hamburger patties. *Epidemiol. Infect.* **1999**, *122*, 185–192. [CrossRef] [PubMed]

62. Wong, C.S.; Jelacic, S.; Habeeb, R.L.; Watkins, S.L.; Tarr, P.I. The risk of the hemolytic-uremic syndrome after antibiotic treatment of *Escherichia coli* O157:H7 infections. *N. Engl. J. Med.* **2000**, *342*, 1930–1936. [CrossRef] [PubMed]

63. Orth, D.; Grif, K.; Zimmerhackl, L.B.; Wurzner, R. Prevention and treatment of enterohemorrhagic *Escherichia coli* infections in humans. *Expert Rev. Anti Infect Ther.* **2008**, *6*, 101–108. [CrossRef] [PubMed]

64. Gardner, A.M.; Martin, L.A.; Gardner, P.R.; Dou, Y.; Olson, J.S. Steady-state and transient kinetics of *Escherichia coli* nitric-oxide dioxygenase (flavohemoglobin). The B10 tyrosine hydroxyl is essential for dioxygen binding and catalysis. *J. Biol. Chem.* **2000**, *275*, 12581–12589. [CrossRef] [PubMed]

65. Hyduke, D.R.; Jarboe, L.R.; Tran, L.M.; Chou, K.J.; Liao, J.C. Integrated network analysis identifies nitric oxide response networks and dihydroxyacid dehydratase as a crucial target in *Escherichia coli*. *Proc. Natl. Acad. Sci. USA* **2007**, *104*, 8484–8489. [CrossRef] [PubMed]

66. Pullan, S.T.; Gidley, M.D.; Jones, R.A.; Barrett, J.; Stevanin, T.A.; Read, R.C.; Green, J.; Poole, R.K. Nitric oxide in chemostat-cultured *Escherichia coli* is sensed by Fnr and other global regulators: Unaltered methionine biosynthesis indicates lack of S nitrosation. *J. Bacteriol.* **2007**, *189*, 1845–1855. [CrossRef] [PubMed]

9

Effect of Deep Drying and Torrefaction Temperature on Proximate, Ultimate Composition, and Heating Value of 2-mm Lodgepole Pine (*Pinus contorta*) Grind

Jaya Shankar Tumuluru

Idaho National Laboratory, 750 University Blvd., Energy Systems Laboratory, PO: Box: 1625, Idaho Falls, ID 83415, USA; jayashankar.tumuluru@inl.gov

Academic Editor: Gou-Jen Wang

Abstract: Deep drying and torrefaction compose a thermal pretreatment method where biomass is heated in the temperature range of 150–300 °C in an inert or reduced environment. The process parameters, like torrefaction temperature and residence time, have a significant impact on the proximate, ultimate, and energy properties. In this study, torrefaction experiments were conducted on 2-mm ground lodgepole pine (*Pinus contorta*) using a thermogravimetric analyzer. Both deep drying and torrefaction temperature (160–270 °C) and time (15–120 min) were selected. Torrefied samples were analyzed for the proximate, ultimate, and higher heating value. The results indicate that moisture content decreases with increases in torrefaction temperature and time, where at 270 °C and 120 min, the moisture content is found to be 1.15% (w.b.). Volatile content in the lodgepole pine decreased from about 80% to about 45%, and ash content increased from 0.77% to about 1.91% at 270 °C and 120 min. The hydrogen, oxygen, and sulfur content decreased to 3%, 28.24%, and 0.01%, whereas the carbon content and higher heating value increased to 68.86% and 23.67 MJ/kg at 270 °C and 120 min. Elemental ratio of hydrogen to carbon and oxygen to carbon (H/C and O/C) calculated at 270 °C and a 120-min residence time were about 0.56 and 0.47. Based on this study, it can be concluded that higher torrefaction temperatures ⩾230 °C and residence time ⩾15 min influence the proximate, ultimate, and energy properties of ground lodgepole pine.

Keywords: lodgepole pine; torrefaction; thermogravimetric analysis; proximate and ultimate composition

1. Introduction

In the recent United Nations Paris Framework Convention on Climate Change (PFCCC), there is a call to mitigate the global annual emissions of greenhouse gases by 2020 in order to reduce the global average temperature increase to less than 2 °C [1]. Renewable energies represent a diversity of sources that can help to maintain the equilibrium of different ecosystems. Among these, biomass is considered carbon neutral due to the carbon dioxide released during its conversion that is already part of the carbon cycle [2]. According to the Kyoto Protocol [3], increasing the use of biomass helps to reduce carbon dioxide emissions and to reduce its negative impact on the environment. According to the U.S. Department of Energy (DOE) [4], about a billion tons of biomass is available in the United States for energy applications. This enhances the ability of the United States to include biomass as a sustainable and significant part of domestic energy production.

1.1. Biomass Limitations for Solid and Liquid Fuel Applications

Energy from biomass can be produced using thermochemical (direct combustion, gasification, and pyrolysis), biological (anaerobic digestion and fermentation), and chemical (esterification) technologies.

Out of all of these technologies, combustion of biomass can provide a direct near-term energy solution. The inherent physical (particle size and density) and chemical characteristics (proximate, ultimate, and energy properties) of raw biomass restrict its use in higher percentages for direct-combustion applications. Furthermore, grinding raw biomass with high moisture content is very challenging due to its fibrous nature. The study of Tumuluru et al. [5] indicated that high moisture in biomass increases the grinding energy and negatively impacts the particle size distribution. Additionally, higher moisture in the biomass will result in plugging the grinder screens and the reactor and bridging of the particles in the conveyors. In terms of chemical composition, the raw biomass has a higher oxygen and hydrogen content and a lower carbon and calorific value as compared to fossil fuels [6]. All of these limitations do not make biomass a great candidate for production of solid and liquid biofuels. In their studies, Tumuluru et al. [7] researched the impact of feedstock supply system unit operations and the effects on feedstock quality and cost. The same authors suggested that new harvesting methods and mechanical, chemical, thermal pretreatment technologies will help to improve the biomass physical and chemical properties. These improvements will make biomass meet the specifications in terms of density, particle size, ash composition, and carbohydrate content for both biochemical and thermochemical conversion applications. Various thermal methods, such as dry and wet torrefaction (hydrothermal carbonization) and steam explosion, as well as chemical pretreatment techniques, such as ionic, acid, alkali, and ammonia fiber expansion, were investigated to understand how they improve the biomass specifications for biofuels production [8–13]. Recent studies by Sarkar et al. [14] and Yang et al. [15] on using torrefied and torrefied-densified switchgrass for both gasification and pyrolysis applications have indicated that higher quality syngas and bio-oil are produced when compared to the ones produced using raw switchgrass. In their studies, Hoover et al. [16], Ray et al. [17], and Aston et al. [18] indicate that chemical pretreatments (i.e., ammonia fiber expansion (AFEX), acid, and alkali) and further densification help to increase the liquid fuels production through biochemical conversion.

1.2. Dry Torrefaction

Dry torrefaction is a thermal pretreatment technique that is used to improve the physical and chemical properties of biomass. Torrefaction is defined as slowly heating the biomass in an inert atmosphere to a maximum temperature of 300 °C [19,20], which will result in a solid uniform product with lower moisture and higher energy content when compared to raw biomass. Various biomass reactions occur during torrefaction, such as devolatilization and carbonization of hemicellulose, depolymerization and devolatilization/softening of lignin and depolymerization, and devolatilization of cellulose. These reactions result in changes in biomass moisture, chemical composition, and energy content. The feedstock and process variables that influence the torrefaction process are (1) feedstock type, (2) particle size, (3) temperature, (4) residence time, and (5) heating rate [13]. Figure 1 indicates the change in the biomass components at different temperature regimes [13].

Tumuluru et al. [13] divide thermal pretreatment regimes into three zones: non-reactive 50–150 °C, reactive 150–200 °C, and destructive drying 200–300 °C. According to Tumuluru and Hess [21] and Tumuluru et al. [13], the reactive drying temperature regime is also called deep drying, whereas the destructive drying temperature is called the torrefaction regime. The initial heating of biomass up to 150 °C (non-reactive drying zone) removes the unbound water. Increasing the temperature from 150–200 °C removes most of the bound water by a thermo-condensation process [22]. Furthermore, increasing the temperature to >200 °C will result in decomposition, devolatilization, and carbonization reactions. Most of the biomass hemicellulose undergoes decomposition reactions, which result in significant changes in color, chemical composition, and physical characteristics. At temperatures >280 °C, the process is completely exothermic, which results in significant increases in the production of CO_2, phenols, acetic acid, and other higher hydrocarbons. During torrefaction, hemicellulose degradation causes the destruction of OH functional groups and will leave the material hydrophobic. The degree of hydrophobicity depends on the torrefaction temperature. Hydrophobicity will help biomass to absorb less moisture during its storage in different environments, therefore making it more stable against fungal and microbial attacks.

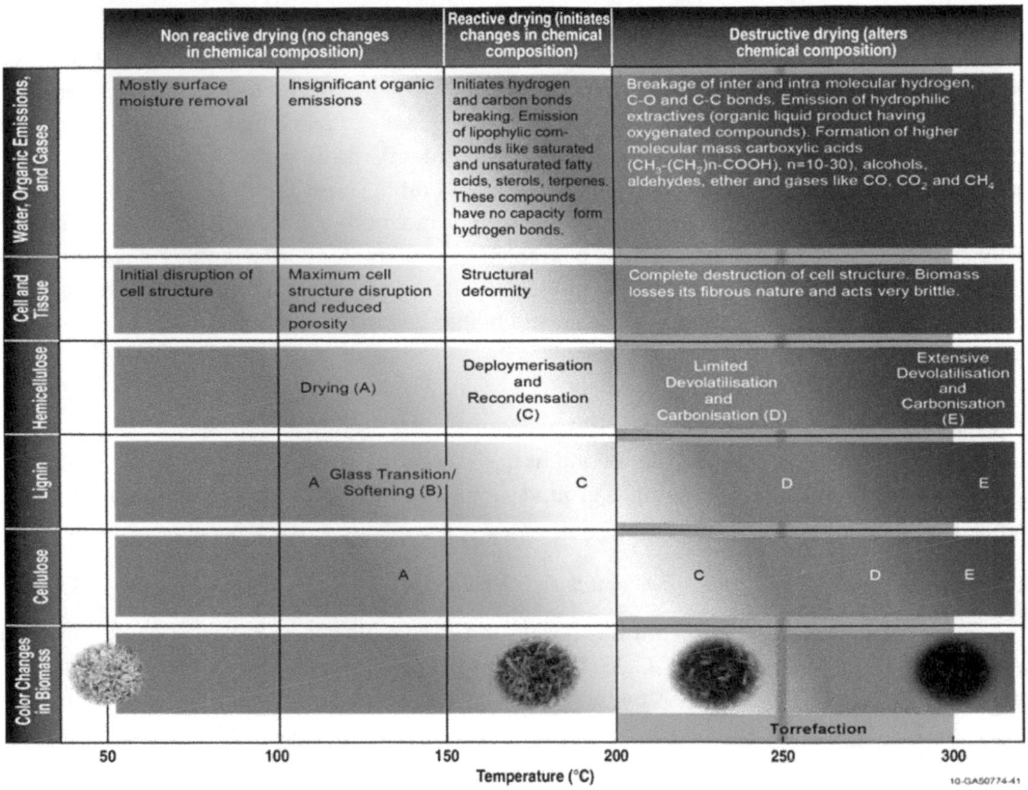

		Non reactive drying (no changes in chemical composition)		Reactive drying (initiates changes in chemical composition)	Destructive drying (alters chemical composition)	
Water, Organic Emissions, and Gases		Mostly surface moisture removal	Insignificant organic emissions	Initiates hydrogen and carbon bonds breaking. Emission of lipophilic compounds like saturated and unsaturated fatty acids, sterols, terpenes. These compounds have no capacity form hydrogen bonds.	Breakage of inter and intra molecular hydrogen, C-O and C-C bonds. Emission of hydrophilic extractives (organic liquid product having oxygenated compounds). Formation of higher molecular mass carboxylic acids $(CH_3-(CH_2)n\text{-}COOH)$, n=10-30), alcohols, aldehydes, ether and gases like CO, CO_2 and CH_4	
Cell and Tissue		Initial disruption of cell structure	Maximum cell structure disruption and reduced porosity	Structural deformity	Complete destruction of cell structure. Biomass losses its fibrous nature and acts very brittle.	
Hemicellulose			Drying (A)	Deploymerisation and Recondensation (C)	Limited Devolatilisation and Carbonisation (D)	Extensive Devolatilisation and Carbonisation (E)
Lignin		A	Glass Transition/ Softening (B)	C	D	E
Cellulose			A		C	D E
Color Changes in Biomass						Torrefaction
	50	100	150	200	250	300

Temperature (°C)

10-GA50774-41

Figure 1.   Non-reactive, reactive, and destructive drying temperature impact on the biomass components (adapted from [13]).

1.3. Deep Drying and Torrefaction Parameters

The biomass parameters commonly used in deep drying and torrefaction are reaction temperatures of 160–300 °C, heating rates of <50 °C/min, and residence times <30 min [13]. The absence or limiting of oxygen presence to <7% in the reactor by avoiding oxidation and auto-ignition during its heating is of great importance. Many researchers have worked on the torrefaction of agricultural and woody biomass using a thermogravimetric analyzer (TGA) [6,23,24]. These researchers have successfully used TGA, on both woody and herbaceous biomass, to understand the effect of temperature and residence time on torrefaction kinetics and the chemical, proximate, and energy properties. Tumuluru [6] studied torrefaction of corn stover and switchgrass using TGA, concluding that the torrefaction temperature has the most significant effect compared to residence time on the chemical and energy properties.

1.4. Objectives

In the past several decades, lodgepole pine (*Pinus contorta*) has gained much importance due to its high quality timber and yields [25]. This is the third largest timber type in the Western United States after ponderosa pine and Douglas fir. Lodgepole pine is available in larger quantities, which leads to a growing interest in its suitability as a feedstock for bioenergy applications [25–27]. In the present study, lodgepole pine was selected to understand the changes torrefaction can bring in terms of the proximate, ultimate, and energy properties. In the present research, a smaller particle size of 2 mm was selected to conduct the torrefaction studies. Studies conducted by Bridgewater *et al.* [28] indicate that for efficient thermo-chemical conversion, such as pyrolysis, particle sizes of approximately 2-mm are necessary. In general, bigger particle sizes result in secondary reactions, leading to the formation of more tars, whereas smaller particle sizes due to less time in the reaction zone result in minimizing the secondary reactions and, in turn, result in less tar formation.

The overall objective of this work is to conduct thermal pretreatment studies (deep drying and torrefaction) over a wide range of temperatures and residence times using the TGA. The specific objectives of this study are (1) to understand the effect of deep drying and torrefaction temperature in the range of 160–270 °C and a residence time of 15–120 min on the proximate composition (*i.e.*, moisture content, volatile content, fixed carbon content, and ash content), the ultimate composition (*i.e.*, carbon, hydrogen, sulfur, nitrogen, oxygen, and H/C and O/C ratios), and the higher heating value of a 2-mm lodgepole pine grind; (2) to understand the significance of the torrefaction temperature and the residence time with respect to the ultimate composition, proximate composition, and higher heating values; and (3) to develop linear regression models for the torrefaction parameters with respect to the proximate composition, ultimate composition, and higher heating values.

2. Materials and Methods

Clean lodgepole pine (*Pinus contorta*) wood chips were used in the present study. The wood chips were dried to <5% (w.b.) moisture content in a laboratory oven and were further ground in a Willey mill fitted with a 2-mm screen. The ground samples were double bagged, stored in an air-tight container, and used to conduct the torrefaction tests.

2.1. Torrefaction Studies Using the Thermogravimetric Analyzer

Deep drying and torrefaction studies on ground lodgepole pine was carried out using TGA (Figure 2). The LECO (Model No: TG 701) TGA was used to conduct the torrefaction tests. A N_2 atmosphere was used in this study. A method file was developed to carry out the torrefaction studies using TGA [24]. TGA also measures mass loss as total moisture, ash, volatile content, or loss-on-ignition. The LECO C/H/N analyzer was used for estimating the carbon, hydrogen, and nitrogen of the raw and torrefied samples. The American Society for Testing and Materials (ASTM standard) methods were used for estimating the chemical composition (Table 1). The higher heating values of raw and torrefied samples were measured using a bomb calorimeter. The reported chemical composition and energy property are an average of three measurements. Experiments were conducted in a temperature range of 160–270 °C and residence times of 15–120 min. Table 2 indicates the experimental design followed for conducting deep drying and torrefaction studies.

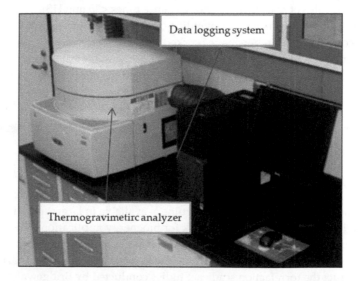

Figure 2. LECO Model 701 thermogravimetric analyzer [6].

Table 1. Methods to measure the chemical properties and higher heating values.

S. No.	Chemical Composition Proximate	Procedure
1	Moisture	ASTM D3173 [29]
2	Ash	ASTM D 3174 [30]
3	Volatiles	ASTM D3175 [31]
4	Fixed carbon	Fixed carbon calculated by the difference method
	Ultimate Composition	
1	Moisture	ASTM D3173 [29]
2	Carbon	ASTM D3178 [32]
3	Hydrogen	ASTM D3178 [32]
4	Nitrogen	ASTM D3179 [33]
5	Sulphur	ASTM D3177 [34]
7	Oxygen	Oxygen calculated by the difference method
8	H/C ratio	H/C: number of hydrogen atoms/number of carbon atoms = (%H/1)/(%C/12)
9	O/C ratio	O/C: number of oxygen atoms/number of carbon atoms = (%O/8)/(%C/12)
6	Higher heating value (HHV)	ASTM D5865 [35]

Table 2. Experimental design for the deep drying and torrefaction experiments.

Process	Temperatures (°C)	Residence Time (min)	Particle Size (mm)	Heating Rate (°C/min)
Deep drying	160, 180	15, 30, 60, and 120	2	10
Torrefaction	230, 270	15, 30, 60, and 120	2	10

2.2. Data Analysis: Analysis of Variance and Multiple Regression Analysis

The experimental data on the proximate composition, ultimate composition, and higher heating values obtained at different torrefaction temperatures (160, 180, 230 and 270 °C) and residence times (15, 30, 60 and 120 min) for a 2-mm lodgepole pine grind were used to draw the linear plots, develop multiple regression models, (Equation 1), and understand the significance of the torrefaction process variables with respect to chemical properties that were studied. Dell Statistica (Version 10) statistical software was used to develop the multiple regression models and analysis of variance (ANOVA) of the experimental data. The coefficient of determination was used to determine the model fit.

$$f(y) = b_0 + b_1x_1 + b_2x_2 \qquad (1)$$

where:

$$b_0, b_1 \text{ and } b_2 = \text{equation constants}$$

$$x_1 \text{ and } x_2 = \text{torrefaction temperature and torrefaction residence time}$$

3. Results

Figure 3 indicates the ground lodgepole pine samples that were torrefied at different temperatures and residence times. It is clear from the figure that when increasing the torrefaction temperature, the color of the biomass changes from brown to black. Tumuluru et al. [13] report that the biomass turns brown to black at 150–300 °C. This can be mainly attributed to the chemical compositional changes that occur in the biomass components, such as hemicellulose, lignin, and cellulose. The major reactions that change the biomass color are devolatilization and carbonization of hemicellulose, depolymerization and devolatilization/softening of lignin, and depolymerization and devolatilization of cellulose. Food industries use color as an indicator to understand the quality changes in products. For example, in a coffee-bean roasting process, the change in color is used as an indicator to define the degree of roasting. The chemical compositional changes are due to the breakage of hydrogen and carbon bonds. The disruption of most inter- and intra-molecular hydrogen bonds, C–C and C–O bonds, results in the formation of hydrophilic extractives, carboxylic acids, alcohols, aldehydes, ether, and gases (e.g., CO, CO_2, and CH_4). All of these chemical compositional changes within the biomass will significantly change its color. Branca et al. [36] indicate that the color of the wood chips changes when the biomass is torrefied at different temperatures and residence times. Nhuchhen *et al.* [37] observed a similar color change, where higher torrefaction temperatures resulted in a dark black product.

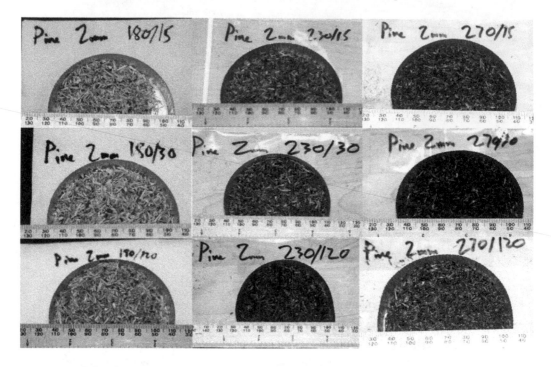

Figure 3. Color changes in torrefied lodgepole pine grind at different torrefaction temperatures and residence times.

3.1. Proximate, Ultimate Composition and Higher Heating Values

Proximate-composition measurement includes moisture content, volatiles' content, ash content, and fixed carbon, whereas fixed carbon is calculated based on the difference. The ultimate composition includes carbon, hydrogen, nitrogen, sulfur, and oxygen and the oxygen content is calculated by the difference method. The proximate, ultimate, and higher heating values of the 2-mm raw lodgepole pine grind are shown in Table 3.

Table 3. Proximate, ultimate, and higher heating values of the 2-mm lodgepole pine grind.

S. No.	Chemical Composition Proximate	(%)
1	Moisture	4.2
2	Ash	0.69
3	Volatiles	80.23
4	Fixed carbon	15.1
	Ultimate composition	
1	Carbon	52.23
2	Hydrogen	6.2
3	Nitrogen	0.47
4	Sulphur	0.022
5	Oxygen	41.23
6	H/C	1.42
7	O/C	0.59
8	Higher heating value (HHV)	19.37

3.2. Moisture Content

The initial moisture content of the 2-mm lodgepole pine grind was about 4.2% (w.b.). After torrefaction, the torrefied samples' moisture content had significantly decreased with the increase in torrefaction temperature and time. The lowest moisture content observed was about 1.15% (w.b.) at 270 °C and a 120-min residence time (Figure 4). The percent decrease observed in moisture content was about 23.09% from the initial value of about 4.2% (w.b.) to 3.23% (w.b.) at 160 °C for 15 min. Increasing the torrefaction temperature and time to 270 °C and 120 min, the moisture content of the samples reduced by about 72.61% from the initial value of 4.2%.. The decrease in moisture content from 160, 180, 230 and 270 °C and a 30-min residence time was about 29.04%, 43.80%, 56.90% and 68.57%. A similar decrease in moisture content was observed at other residence times, as well.

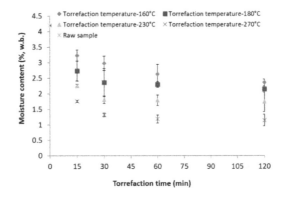

Figure 4. Moisture content in the lodgepole pine grind with respect to the torrefaction temperature and time.

3.3. Volatile Content

At lower torrefaction temperatures (160 and 180 °C) and different residence times, the decrease in volatile content in the torrefied biomass was relatively minimal (the maximum decrease was about 1.3% and 3.72% at 120 min with respect to the original value) (Figure 5). Increasing the torrefaction temperature to 230 and 270 °C and its residence time to 120 min reduced the volatile content to about 75.41% and 55.28% compared to its original value. The effect of torrefaction temperature showed a more significant effect when compared to the residence time. Carter et al. [38] reported similar volatile content (76.37% and 56.53%) in the lodgepole pine samples when torrefied at 225 and 275 °C for a 30-min residence time.

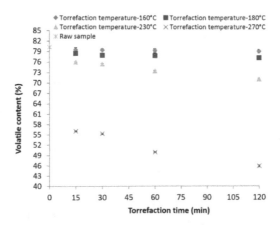

Figure 5. Volatile content in the lodgepole pine grind with respect to the torrefaction temperature and time.

3.4. Ash Content

The ash content in the torrefied material increased with the increase of torrefaction temperature and time (Figure 6). The changes in the ash content are mainly due to breakdown of carbon-hydrogen bonds, resulting volatile loss and further concentrating the ash content in the biomass. The highest ash content of about 1.91% was observed for samples torrefied at 270 °C and 120 min. The increase was about 176% with respect to the original value. The increase in relative ash content at different torrefaction temperatures (160, 180, 230 and 270 °C) and a 30-min residence time was 36.9%, 46.37%, 102.01% and 134.78% compared to its initial value of 0.69%.

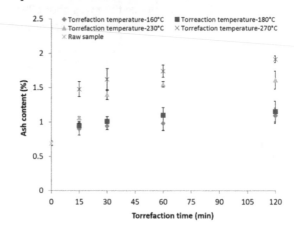

Figure 6. Ash content in the lodgepole pine grind with respect to the torrefaction temperature and time.

3.5. Fixed Carbon

Figure 7 indicates the changes in the fixed carbon content in the ground lodgepole pine torrefied at different temperatures and residence times. The initial fixed carbon of the raw biomass was 15.91%. Increasing the temperature to 160, 180, 230 and 270 °C and a 30-min residence time increased the fixed carbon content to about 18.43%, 18.79%, 22.34% and 41.01% (an increase of 15.83%, 18.01%, 40.41%, and 157.76% from its original value). The maximum fixed carbon observed at 270 °C and 120 min was about 50.48% (Figure 7). Fixed carbon content data matched closely with the data presented by Carter et al. [38], where at 225 and 275 °C for 30 min, the observed values were 22.75% and 42.49% for lodgepole pine. The results indicate that the increase is marginal at 160 and 180 °C at all residence times, but further increasing the temperature to 230 and 270 °C significantly increased the fixed carbon content.

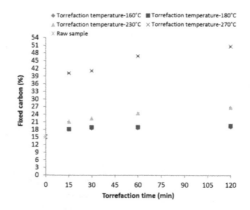

Figure 7. Fixed carbon content in the lodgepole pine grind with respect to the torrefaction temperature and time.

3.6. Ultimate Composition

Ultimate composition measurements include elemental carbon, hydrogen, sulfur, nitrogen, and oxygen by different methods. The ultimate composition of the 2-mm ground lodgepole pine raw material is 6.2% hydrogen, 52.3% carbon, 0.28% nitrogen, 0.02% sulfur, and 41.23% oxygen.

3.7. Carbon (%)

The initial elemental carbon in the 2-mm ground lodgepole pine sample was about 52.23%. Heating the biomass at 160 and 180 °C for a 30-min residence time increased the elemental carbon content to about 56%–57% (Figure 8). Further, increasing the temperature and time to 230 and 270 °C for 120 min increased the carbon content to about 62% and 68%. It is clear from the data that the increase in elemental carbon is significantly higher at temperatures ≥230 °C. It is also clear from the data that the increase in elemental carbon with the increase in residence time (from 15 to 120 min) is about 3%, whereas increasing the torrefaction temperature from 230 to 270 °C resulted in about an 8% rise for the residence times studied. Carter *et al.* [38] reports similar values of carbon content (64.17%) for pine when torrefied at 275 °C and a 30-min residence time.

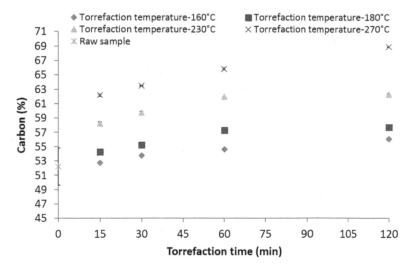

Figure 8. Carbon content in the lodgepole pine grind with respect to the torrefaction time and temperature.

3.8. Hydrogen (%)

The hydrogen content of the raw biomass samples observed was about 6.2%. The decrease in hydrogen content at 160 and 180 °C for different residence times was marginal, where a maximum decrease of about 5.85% was observed at 180 °C and 120 min (Figure 9). Increasing the torrefaction temperature and residence time to 230 and 120 min reduced the hydrogen content to about 5.54 and 3%, respectively. The results indicate that the decrease is more significant at higher temperatures of >230 °C. Furthermore, the data from the present study indicate that higher torrefaction temperatures and residence times played a major role in reducing the hydrogen content. At 270 °C and a 15-min residence time, the hydrogen content observed was 4.5%, whereas at a 120-min residence time, the final hydrogen content observed is about 3% at 270 °C. Carter *et al.* [38] observed similar trends where hydrogen was about 4.81% at 275 °C and 30 min; further increasing the time to 45 min reduced the hydrogen content to 3.47%.

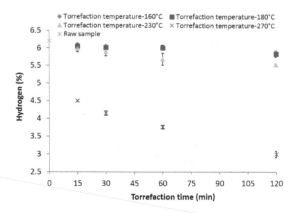

Figure 9. Hydrogen content in the lodgepole pine grind with respect to the torrefaction time and temperature.

3.9. Oxygen (%)

The oxygen content of the raw and torrefied samples was calculated based on the difference method. The initial oxygen content of the raw samples observed was 41.23%. At 160 and 180 °C, the oxygen content observed in the samples was in the range of 40.93%–37.04%, whereas at 230 °C, the oxygen content reduced to about 31.14% (a decrease of about 24.47% from the initial value). Further increasing the torrefaction temperature to 270 °C and the residence time to 120 min decreased the oxygen content in the sample to about 28.24% (a decrease of about 32% from the initial value). The data presented by Carter *et al.* [38] for oxygen content in the pine samples when torrefied at 275 °C for a 30-min residence time was 29.27%, which matched closely with the values reported in this research (32.24%). At 160, 180, 230, and 270 °C and a 30-min residence time, the observed oxygen values were 40.24%, 38.49%, 34.21% and 32.24% (Figure 10). The results indicate that torrefaction temperature had a higher impact on the oxygen content compared to the residence time.

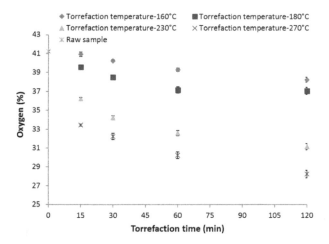

Figure 10. Oxygen content in the lodgepole pine grind with respect to the torrefaction time and temperature.

3.10. Nitrogen and Sulfur (%)

The initial nitrogen and sulfur content observed for the raw biomass samples was 0.47% and 0.02%. At lower temperatures of 160 and 180 °C and different residence times, the changes in nitrogen and sulfur content were marginal; however, increasing the torrefaction temperature to 230 and 270 °C at 30 min decreased the nitrogen and sulfur content to 0.25% and 0.015%. Furthermore, increasing the residence time to 120 min decreased the nitrogen and sulfur content to final values of 0.17% and 0.01%.

3.11. H/C and O/C Ratios

The initial hydrogen to carbon (H/C) and oxygen to carbon (O/C) ratios of the lodgepole pine sample were 1.42 and 0.59. The H/C ratio decreased when the torrefaction temperature and residence time was increased. The lowest H/C and O/C ratios observed were 0.56 and 0.47 at a torrefaction temperature of 270 °C and a 120-min residence time (Figure 11). At 160 and 180 °C and a 30-min residence time, the H/C ratio was about 1.31 and 1.28 and the O/C ratio was about 0.99 and 0.69. Increasing the torrefaction temperature to 230 and 270 °C and the residence time to 30 min decreased the H/C ratio to 1.19 and 0.88 and the O/C ratio to near 0.70 and 0.41. Further increasing the residence time to 120 min at the same temperature decreased the H/C and O/C ratios. The change in the O/C ratio with respect to the torrefaction temperature is more significant than the torrefaction residence time.

3.12. van Krevelen Diagram

The van Krevelen diagram was drawn for the O/C and H/C ratios, for the raw and torrefied ground lodgepole pine, and is compared to different grades of commercially available coals (Central Appalachian, Illinois Basin, Powder River Basin) (Figure 11). According to this plot, torrefaction of ground lodgepole pine shifts the elemental ratios of H/C and O/C (0.78 and 0.36) closer to the commercially available coals, such as Illinois Basin and Powder River Basin coals. Figure 11 indicates that the high-quality coals have a lower ratio of H/C to O/C (mainly due to the lower oxygen content and higher carbon content) compared to the ground lodgepole pine. Torrefaction of lodgepole pine at 160, 180 and 230 °C at different residence times advanced the H/C and O/C ratios closer to the commercial coals. At a 270 °C torrefaction temperature and a 15–120-min residence time, the ground lodgepole pine moved close to Power River Basin and Illinois Basin coal. There was a major shift in the H/C and O/C values from 230 to 270 °C (Figure 11). The shift is mostly due to the significant increase in carbon content and the steep decrease in oxygen and hydrogen content. The breakage of inter- and intra-molecular hydrogen and carbon-oxygen and carbon-carbon bonds result in the emission of extractives and oxygenated compounds, which might have resulted in the major shift in the H/C and O/C values. Additionally, Dutta [39] reports similar elemental ratios of the H/C and O/C ratio values when pine was torrefied at a higher temperature of 270 °C and a 30-min residence time.

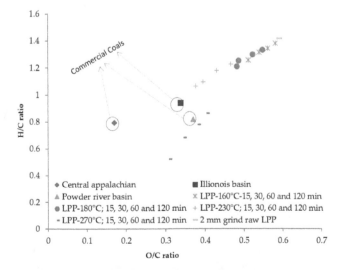

Figure 11. van Krevelen diagram for the raw, torrefied lodgepole pine grind, and the commercial coals.

3.13. Higher Heating Value (MJ/kg)

The initial higher heating value observed for the 2-mm ground lodgepole pine was 19.45 MJ/kg. Increasing the torrefaction temperature to 270 °C and the time to 30 min significantly increased the heating value to about 23.67 MJ/kg (Figure 12). At lower temperatures of 160 and 180 °C and a 120-min residence time, the heating value increase was found to be marginal (there is about a 1 MJ/kg increment). Increasing the torrefaction temperature from 180 to 230 °C for residence times of about 30 min, the increase was about 2 MJ/kg (21.34 MJ/kg), whereas at 270 °C, the increase was about 4 MJ/kg (23.21 MJ/kg) compared to the higher heating value of raw lodgepole pine grind. The maximum higher heating value observed at 270 °C and a 120-min residence time was about 21.57 MJ/kg. Peng *et al.* [40] report that torrefying lodgepole pine at 250 and 300 °C at 30 min results in higher heating values in the range of 20.58–23.02 MJ/kg.

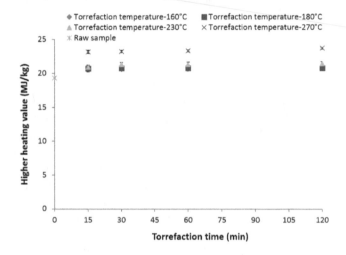

Figure 12. Higher heating value in the lodgepole pine grind with respect to the torrefaction time and temperature.

4. Analysis of Variance and Multiple Regression Models

The experimental data were further analyzed to understand the significance of the process variables, with respect to the proximate and ultimate composition and higher heating values. Multiple regression models were also developed for proximate, ultimate composition, and energy data. Table 4 shows the multiple regression equations fitted for the experimental data, and Table 5 indicates the significance of the process variables based on the ANOVA. The regression equations developed were found to be statistical significant at $p < 0.001$. The regression equations indicated that in the case of proximate composition, the torrefaction temperature and residence time were positively correlated for fixed carbon and ash content, whereas for moisture and volatiles, there was a negative correlation. Furthermore, the regression equations indicated that the coefficient of torrefaction temperature was higher when compared to the torrefaction residence time. In the case of the ultimate composition, hydrogen, nitrogen, oxygen, sulfur, the H/C ratio, and the O/C ratio were negatively correlated, whereas in the case of carbon content and higher heating values, they are positively correlated with the torrefaction temperature and residence time.

ANOVA analysis indicates that the moisture content, ash content, and fixed carbon were influenced by both the torrefaction temperature and residence time at $p < 0.001$. In the case of volatile content, the torrefaction temperature, at $p < 0.001$, and not by residence time, had a significant effect. In the case of the ultimate composition, hydrogen content was influenced by torrefaction temperature at $p < 0.001$, whereas residence time was found to be insignificant. Carbon, oxygen, and sulfur content were influenced by both torrefaction temperature and residence time at $p < 0.001$, whereas nitrogen content was influenced by torrefaction time at $p < 0.01$. The H/C and O/C ratios are influenced by the torrefaction temperature at $p < 0.001$, whereas the residence time influenced the H/C ratio at $p < 0.05$ and not the O/C ratio. In the case of a higher heating value, the torrefaction temperature was influenced at $p < 0.001$, and the residence time was not significant.

Bates and Ghoniem [41] indicate that torrefaction temperatures of 250 and 300 °C and residence times of 15–60 min result in 16%–30% solid mass loss; increasing the residence time further results in an additional 42%–48% mass loss. They have concluded that mass loss is faster in the first stage of torrefaction, which is primarily attributable to the decomposition of hemicellulose (with an increasing contribution from cellulose decomposition at higher temperatures). The non-condensable products, such as CO_2 and CO, reach the peak value at residence times of 10 min and then start to decline. The same authors also indicate that the amount of methanol and lactic acid, which is produced during the decomposition of acetoxy- and methoxy-groups, increased up to 10 min, and then it remained unchanged. The observations in this study indicate that the coefficient of torrefaction temperature is higher when compared to the torrefaction residence time. Additionally, the analysis of variance supports this observation, where torrefaction temperature has more significant impacts on the proximate and ultimate composition and its higher heating value when compared to torrefaction residence time. Nhuchhen [37] work on torrefaction of biomass indicate that the net effect of the residence time is not as prominent as the temperature; the same authors also concluded that in a rotary torrefaction system, the torrefaction temperature is more prominently impactful when compared to angular speed and the inclination of the dryer.

Table 4. Multiple regression equations for the chemical and energy properties.

S. No.	Chemical Composition and higher heating value	Multiple Regression Equation	Coefficient of Determination (R^2)	Statistical Significance
	Proximate composition			
1	Moisture content (%, w.b.)	$y = 5.013 - 0.0124x_1 - 0.00515x_2$	0.93	$p < 0.001$
2	Ash (%)	$y = -0.2718 + 0.0066x_1 + 0.0029x_2$	0.93	$p < 0.001$
3	Volatiles (%)	$y = 121.8021 - 0.2322x_1 - 0.0409x_2$	0.79	$p < 0.001$
4	Fixed carbon (%)	$y = -22.9741 + 0.2236x_1 + 0.0431x_2$	0.80	$p < 0.001$
	Ultimate composition			
1	Hydrogen (%)	$y = 9.4969 - 0.018x_1 - 0.0054x_2$	0.72	$p < 0.001$
2	Carbon (%)	$y = 36.3531 + 0.09716x_1 + 0.0397x_2$	0.97	$p < 0.001$
3	Oxygen (%)	$y = 54.344 - 0.0800x_1 - 0.03478x_2$	0.96	$p < 0.001$
4	Nitrogen (%)	$y = 0.7163 - 0.00183x_1 - 0.00069x_2$	0.89	$p < 0.001$
5	Sulfur (%)	$y = 0.03492 - 0.000071x_1 - 0.000047x_2$	0.89	$p < 0.001$
6	H/C ratio	$y = 2.158 - 0.004436x_1 - 0.00179x_2$	0.83	$p < 0.001$
7	O/C ratio	$y = 1.4845 - 0.0039x_1 - 0.00096x_2$	0.72	$p < 0.05$
8	Higher heating value (MJ/kg)	$y = 16.462 + 0.02321x_1 + 0.0026x_2$	0.82	$p < 0.001$

Table 5. Analysis of variance for the chemical and energy properties.

S. No.	Chemical Composition and higher heating value	Process Variables	
		Torrefaction Temperature (x_1)	Torrefaction Residence Time (x_2)
	Proximate composition		
1	Moisture content (%, w.b.)	(−) ***	(−) ***
2	Ash (%)	(+) ***	(+) ***
3	Volatile content (%)	(−) ***	ns
4	Fixed carbon (%)	(+) ***	(+) ***
	Ultimate composition		
5	Hydrogen (%)	(−) ***	ns
6	Carbon (%)	(+) ***	(+) ***
7	Nitrogen (%)	(−) ***	(−) **
8	Oxygen (%)	(−) ***	(−) ***
9	Sulfur (%)	(−) ***	(−) ***
10	H/C ratio	(−) ***	(−) *
11	O/C ratio	(−) ***	ns
12	Higher heating value (MJ/kg)	(+) ***	ns

Note: * $p < 0.05$; ** $p < 0.01$; *** $p < 0.001$; ns: non-significant.

5. Discussion

The moisture content of lodgepole pine biomass reduces both at deep drying and torrefaction temperatures. If the target is to reduce the moisture content of the biomass, deep drying at 160 and 180 °C for <30 min will help to reduce most of the moisture in the lodgepole pine. At deep drying temperature, the loss of moisture content can be due to dehydration reactions with less change to the chemical composition of the biomass. At higher torrefaction temperatures between ⩾180 and ⩽350 °C, the loss of moisture and volatiles can be due to the devolatilization and carbonization of hemicellulose and cellulose [24]. Bergman and Kiel [19] and Prins [42] report that drying and depolymerization occur between 225 and 325 °C for hemicelluloses. According to Bridgeman *et al.* [24], the loss of moisture is due to the evaporation and dehydration reactions between the organic molecules, which result in the release of organic and inorganic products from the biomass. The organics that are typically released during torrefaction include sugars, poly-sugars, acids, alcohols, furans, and ketones [13]. The present study indicated that the change in the chemical composition at deep drying temperature (160 and 180 °C) is minimal compared to torrefaction temperatures of 230 and 270 °C.

According to Tumuluru *et al.* [13], at deep drying temperatures of 160 and 180 °C, the changes in the biomass are due to dehydration reactions, whereas at torrefaction temperature of >230 °C, the major reactions are devolatilization and carbonization, causing significant changes in the chemical composition when compared to the raw material. Bates and Ghoniem [41], in their study, indicate that at a lower temperature of 200 °C, the composition is similar to non-torrefied willow, whereas at an increased temperature, it will increase the mass fraction of carbon, while those of hydrogen and oxygen decrease. The authors' two-step model indicates that in the first stage, oxygenated species, such as water, acetic acid, and carbon dioxide, are released, and at the second stage, lactic acid, methanol, and acetic acid were released. Medic *et al.* [43], in their studies on the effects of torrefaction process parameters on biomass feedstock upgrading, conclude that corn stover undergoes changes in chemical composition and energy content during the torrefaction process. These authors also indicate that at higher torrefaction temperatures, the biomass can be characterized in several ways, such as a mass loss of up to 45%, a decrease in the O/C ratio from 1.11 to 0.6, and an increase in the energy density of about 19%.

Tumuluru *et al.* [13], in their review on the torrefaction of biomass, indicate that at >250 °C, the biomass undergoes extensive devolatilization and carbonization. These reactions result in the loss of volatiles and an increase in the carbon content. Furthermore, it is observed that the change in the ash content is related to the loss of other biomass components, such as moisture and volatiles. According to Park *et al.* [44], during torrefaction, volatile loss is due to thermal breakdown of carbohydrate fractions, which results in the accumulation of the residual ash after torrefaction. The same authors further indicate that these changes increase the fixed carbon content of the biomass. Ultimate composition data indicate that oxygen is reduced during torrefaction and the reduction is increased at higher torrefaction temperatures. In the present study, changes in hydrogen and oxygen were marginal at a lower temperature of 180 °C, when compared to torrefaction temperatures of 230 and 270 °C. This observation is corroborated with the findings of other researchers [13,45–47]. The decrease of oxygen content is mainly due to the dehydration reaction, which produces water vapor and releases CO and CO_2. During torrefaction, the loss of volatiles, gases, and water also results in a decrease in hydrogen content. The decrease in oxygen and hydrogen content and the increase of carbon content (from approximately 51%–66%), at a 270 °C torrefaction temperature and a 30-min residence time, results in lowering the atomic O/C ratio (from 0.63 to 0.31). According to Tumuluru *et al.* [46], torrefaction temperatures of >300 °C may not be needed, because there is a significant loss of higher energy content volatiles. Furthermore, it may increase the relative ash content of the biomass.

According to the van Krevelen diagram, the substantial change in the atomic ratio of carbon, hydrogen, and oxygen makes torrefied material act more like coal, making the torrefied material more suitable for co-firing applications. Lodgepole pine torrefied at 270 °C and at different residence times reduced the H/C and O/C ratios and moved them closer to high-quality coal, such as Powder River

Basin and Illinois Basin coals. The lower O/C ratio observed at higher torrefaction temperatures can be due to the release of oxygen-rich compounds (e.g., CO, CO_2, and H_2O), whereas lower H/C ratios can be due to the formation of CH_4 and C_2H_6 during torrefaction. The decrease of the H/C and O/C ratios with an increase in the torrefaction temperature and residence time results in fuels that produce less smoke, less water-vapor formation, and lower energy loss during the combustion and gasification processes [6].

The heating value of biomass is an important quality attribute for energy generation applications [2]. The increase in torrefaction temperature and residence time increases the calorific value (higher heating value), and higher temperatures and residence times result in the loss of more volatiles and increased energy density. The increase in the heating value is due to the decrease in moisture content and an increase of carbon content in the samples. Zanzi et al. [22] and Nimlos et al. [48], in their report, confirm that increasing the torrefaction temperature increases the higher heating value. Tumuluru et al. [12] indicate that at torrefaction temperatures of >300 °C, there is a significant loss of higher energy content volatiles within the biomass.

The feedstock variable that impacts the proximate, ultimate, and energy properties of the torrefied material is the particle size used for torrefaction. Smaller particle sizes used in the present study might have been more reactive to torrefaction temperature due to the higher surface area. According to Nhuchhen [49] and Wang et al. [50], particle size impacts the devolatilization reactions. These authors further indicate that a smaller particle size results in a greater intra-particle effect and heat transfer. Another important torrefaction parameter that can impact the resolution of the experimental data is the ramp time. This can impact the torrefaction reaction kinetics of the various components, such as lignin, hemicellulose, and cellulose. Current studies indicate the torrefaction of ground lodgepole pine at 270 °C and a 15–30-min residence time improves the chemical composition and the higher heating value, making lodgepole pine more suitable for bio-power generation.

6. Conclusions

The research presented was carried out to understand the effect of torrefaction temperature and residence time on the proximate composition, ultimate composition, and higher heating values of a 2-mm lodgepole pine grind. Based on this research the following conclusions are drawn:

- The changes in proximate and ultimate composition were marginal at deep drying temperatures of 160 and 180 °C, whereas at torrefaction temperatures of 230 and 270 °C, the changes were significant.
- Increasing the torrefaction temperature to 270 °C and the residence time to 30 min significantly decreases the moisture content, hydrogen content, and oxygen content and increases the carbon and heating value.
- At 270 °C and a 120-min residence time, carbon content increases to 69.86%, while oxygen and hydrogen content decrease to 28.24% and 3%, whereas the volatile content decreases to 45.81%.
- The H/C and O/C ratios of the raw samples are about 1.42 and 0.59, whereas at 270 °C and 120 min, the H/C and O/C ratio decreases to 0.56 and 0.47.
- The changes in these chemical compositions are attributed to the devolatilization of hemicellulose, which will typically happen at torrefaction temperatures of >200 °C. These reactions result in the formation of water, carbon monoxide, and carbon dioxide and influence the hydrogen and carbon content of the biomass.
- The heating value increased from its initial value of about 19.41 MJ/kg to about 23.67 MJ/kg at 270 °C and a 120-min residence time. At lower torrefaction temperatures of 160–180 °C, the increase in the heating value is marginal.

Acknowledgments: This work was supported by the DOE Office of Energy Efficiency and Renewable Energy under DOE Idaho Operations Office Contract DE-AC07-05ID14517. Accordingly, the U.S. Government retains and the publisher, by accepting the article for publication, acknowledges that the U.S. Government retains a nonexclusive, paid-up, irrevocable, worldwide license to publish or reproduce the published form of this manuscript or to allow others to do so, for U.S. Government purposes.

Conflicts of Interest: The author declare no conflict of interest.

References

1. Framework Convention on Climate Change (FCCC), United Nations. Adoption of the Paris Agreement. In Proceedings of the Conference of Parties, Twenty-First Session, Paris, France, 11 December 2015.

2. Arias, B.R.; Pevida, C.G.; Fermoso, J.D.; Plaza, M.G.; Rubiera, F.G.; Pis Martinez, J.J. Influence of torrefaction on the grindability and reactivity of woody biomass. *Fuel Process. Technol.* **2008**, *89*, 169–175. [CrossRef]

3. United Nations. *Kyoto Protocol to the United Nations Framework Convention on Climate Change*; United Nations: Kyoto, Japan, 1998; pp. 1–20.

4. U.S. Department of Energy. *U.S. Billion-Ton Update: Biomass Supply for a Bioenergy and Bioproducts Industry*; Perlack, R.D., Stokes, B.J., Eds.; U.S. Department of Energy, Oak Ridge National Laboratory: Oak Ridge, TN, USA, 2011; p. 227.

5. Tumuluru, J.S.; Tabil, L.G.; Song, Y.; Iroba, K.L.; Meda, V. Grinding energy and physical properties of chopped and hammer-milled barley, wheat, oat and canola straws. *Biomass Bioenergy* **2014**, *60*, 58–67. [CrossRef]

6. Tumuluru, J.S. Comparison of Chemical Composition and Energy Property of Torrefied Switchgrass and Corn Stover. *Front. Energy Res.* **2015**, *3*, 46. [CrossRef]

7. Tumuluru, J.S.; Searcy, E.; Kenney, K.L.; Smith, W.A.; Gresham, G.; Yancey, N. Impact of feedstock supply systems unit operations on feedstock cost and quality for bioenergy applications. In *Valorization of lignocellulosic Biomass in a Biorefinery: From Logistic to Environmental and Performance Impact*; Kumar, R., Singh, S., Balan, V., Eds.; Nova Science Publishers: New York, NY, USA, 2016.

8. Karki, B.; Muthukumarappan, K.; Wang, Y.J.; Dale, B.; Balan, V.; Gibbons, W.R.; Karunanithy, C. Physical characteristics of AFEX-pretreated and densified switchgrass, prairie cord grass, and corn stover. *Biomass Bioenergy* **2015**, *78*, 164–174. [CrossRef]

9. Lynam, J.G.; Reza, M.T.; Vasquez, V.R.; Coronella, C.J. Pretreatment of rice hulls by ionic liquid dissolution. *Bioresour. Technol.* **2012**, *114*, 629–636. [CrossRef] [PubMed]

10. Reza, M.T.; Lynam, J.G.; Uddin, M.H.; Coronella, C.J. Hydrothermal carbonization: Fate of inorganics. *Biomass Bioenergy* **2013**, *49*, 86–94. [CrossRef]

11. Singh, S.; Cheng, G.; Sathitsuksanoh, N.; Wu, D.; Varanasi, P.; George, A.; Balan, V.; Gao, X.; Kumar, R.; Dale, B.E.; *et al.* Comparison of different biomass pretreatment techniques and their impact on chemistry and structure. *Front. Energy Res.* **2015**, *2*, 1–12. [CrossRef]

12. Tumuluru, J.S.; Hess, J.R.; Boardman, R.D.; Wright, C.T.; Westover, T.L. Formulation, pretreatment, and densification options to improve biomass specifications for co-firing high percentages with coal. *Ind. Biotechnol.* **2012**, *8*, 113–132. [CrossRef]

13. Tumuluru, J.S.; Sokhansanj, S.; Hess, J.R.; Wright, C.T.; Boardman, R.D. A review on biomass torrefaction process and product properties for energy applications. *Ind. Biotechnol.* **2011**, *7*, 384–401. [CrossRef]

14. Sarkar, M.; Kumar, A.; Tumuluru, J.S.; Patil, K.N.; Bellmer, D.D. Gasification performance of switchgrass pretreated with torrefaction and densification. *Appl. Energy* **2014**, *127*, 194–201. [CrossRef]

15. Yang, Z.; Sarkar, M.; Kumar, A.; Tumuluru, J.S.; Huhnke, R.L. Effects of torrefaction and densification on switchgrass pyrolysis products. *Bioresour. Technol.* **2014**, *174*, 266–273. [CrossRef] [PubMed]

16. Hoover, A.N.; Tumuluru, J.S.; Teymouri, F.; Moore, J.; Gresham, G. Effect of pelleting process variables on physical properties and sugar yields of ammonia fiber expansion pretreated corn stover. *Bioresour. Technol.* **2014**, *164*, 128–135. [CrossRef] [PubMed]

17. Ray, A.E.; Hoover, A.N.; Nagle, N.; Chen, X.; Gresham, G.L. Effect of pelleting on the recalcitrance and bioconversion of dilute-acid pretreated corn stover under low- and high-solids conditions. *Biofuels* **2013**, *4*, 271–284. [CrossRef]

18. Aston, J.E.; Tumuluru, J.S.; Lacey, J.A.; Thompson, D.N.; Thompson, V.S.; Fox, S. Alkaline deacetylation of corn stover: Effects on feedstock quality. In Proceedings of the 2015 AICHE Annual Meeting, Salt Lake City, UT, USA, 8–13 November 2015.

19. Bergman, P.C.A.; Kiel, J.H.A. Torrefaction for Biomass Upgrading. ECN-RX-05-180. In Proceedings of the 14th European Biomass Conference & Exhibition, Paris, France, 17–21 October 2005.

20. Fonseca, F.F.; Luengo, C.A.; Bezzon, G.; Soler, P.B. Bench unit for biomass residues torrefaction. In Proceedings of the Conference on Biomass for Energy and Industry, Würzburg, Germany, 8–11 June 1998.

21. Tumuluru, J.S.; Hess, J.R. New market potential: Torrefaction of woody biomass. Available online: http://www.iom3.org/materials-world-magazine/feature/2015/jun/02/new-market-potential-torrefaction-woody-biomass (accessed on 27 February 2016).

22. Zanzi, R.; Ferro, D.T.; Torres, A.; Soler, P.B.; Bjornbom, E. Biomass torrefaction. In Proceedings of the 6th Asia-Pacific International Symposium on Combustion and Energy Utilization, Kuala Lumpur, Malaysia, 20–22 May 2002.

23. Chen, W.H.; Kuo, P.C.E. A study on torrefaction of various biomass materials and its impact on lignocellulosic structure simulated by a thermogravimetry. *Energy* **2010**, *35*, 2580–2586. [CrossRef]

24. Bridgeman, T.G.; Jones, J.M.; Shield, I.; Williams, P.T. Torrefaction of reed canary grass, wheat straw and willow to enhance solid fuel qualities and combustion properties. *Fuel* **2008**, *87*, 844–856. [CrossRef]

25. Johnston, D.C. Estimating Lodgepole Pine Biomass. Theses, Dissertations, Professional Papers, Paper 2237, Missoula, MT, USA, 1977. Available online: http://scholarworks.umt.edu/etd (accessed on 1 June 2016).

26. Parisa, Z.; Sokhansanj, S.; Bi, X.; Lim, S.J.; Mani, S.; Melin, S.; John, K. Density, heating value, and composition of pellets made from Lodgepole Pine (Pinus concorta Douglas) infested with mountain pine beetle (Dendroctonus ponderosae Hopkins). *Can. Biosyst. Eng.* **2008**, *50*, 3.47–3.55.

27. Backlund, I. Cost-effective Cultivation of Lodgepole Pine for Biorefinery Applications. In *Faculty of Forest Sciences Department of Forest Biomaterials and Technology*; Swedish University of Agricultural Sciences Umeå: Uppsala, Switzerland, 2013.

28. Bridgewater, A.V.; Czernik, S.; Piskorz, J. An overview of fast pyrolysis. In *Progress in Thermochemical Biomass Conversion*; Blackwell Science Ltd.: Oxford, UK, 2001; pp. 977–997.

29. *ASTM Standard D3173-86, A. Standard Test Method for Moisture in the Analysis Sample of Coal and Coke*; ASTM International: West Conshohocken, PA, USA, 1996; Available online: http://www.astm.org/DATABASE.CART/HISTORICAL/D3173-87R96.htm (accessed on 6 January 2016).

30. *ASTM Standard D3174, A. Standard Test Method for Ash in the Analysis Sample of Coal and Coke from Coal*; ASTM International: West Conshohocken, PA, USA, 2002; Available online: http://www.astm.org/Standards/D3174 (accessed on 6 January 2016).

31. *ASTM Standard D3175, A. Standard Test Method for Volatile Matter in the Analysis Sample of Coal and Coke*; ASTM International: West Conshohocken, PA, USA, 2007; Available online: http://www.astm.org/Standards/D3175 (accessed on 6 January 2016).

32. *ASTM Standard D3178, A. Standard Test Methods for Carbon and Hydrogen in the Analysis Sample of Coal and Coke*; ASTM International: West Conshohocken, PA, USA, 2002; Available online: http://www.astm.org/Standards/D3178.htm (accessed on 6 January 2016).

33. *ASTM Standard D3179, A. Standard Test Methods for Nitrogen in the Analysis Sample of Coal and Coke*; ASTM International: West Conshohocken, PA, USA, 2002; Available online: https://www.astm.org/Standards/D3179.htm (accessed on 6 January 2016).

34. *ASTM Standard D3177, A. Standard Test Methods for Total Sulfur in the Analysis Sample of Coal and Coke*; ASTM International: West Conshohocken, PA, USA, 2007; Available online: http://www.astm.org/Standards/D3177.htm (accessed on 6 January 2016).

35. *ASTM Standard D5865-10a., A.S. Standard Test Method for Gross Calorific Balue of Coal and Coke*; ASTM International: West Conshohocken, PA, USA, 2010; Available online: http://www.astm.org/Standards/D5864-10a (accessed on 6 January 2016).

36. Branca, C.; di Blasi, C.; Galgano, A.; Brostrom, M. Effects of the Torrefaction Conditions on the Fixed-Bed Pyrolysis of Norway Spruce. *Energy Fuel.* **2014**, *28*, 5882–5891. [CrossRef]

37. Nhuchhen, D.R.; Basu, P.; Acharya, B. Torrefaction of poplar in a continouus two-stage, indirectly heated totary torrefier. *Energy Fuel* **2016**, *30*, 1027–1038.

38. Carter, C.L.; Abdoulmoumine, N.; Kulkarni, A.; Adhikari, S.; Fasina, O. Physicochemical properties of thermally treated biomass and energy requirement for torrefaction. *ASABE* **2013**, *53*, 1093–1100.

39. Dutta, A. *Torrefaction and Other Processing Options*; Guelph, U.O., Ed.; University of Guelph: Guelph, ON, Canada, 2011; Available online: http://www.ofa.on.ca/uploads/userfiles/files/animesh%20dutta.pdf (accessed on 24 February 2016).

40. Peng, J.H.; Bi, H.T.; Sokhansanj, S.; Lim, J.C. A study of particle size effect on biomass torrefaction and densification. *Energy Fuel* **2012**, *26*, 3826–3839. [CrossRef]

41. Bates, R.B.; Ghoniem, A.F. Biomass torrefaction: Modeling of volatile and solid product evlution kinetics. *Bioresour. Technol.* **2012**, *124*, 460–469. [CrossRef] [PubMed]

42. Prins, M.J. *Thermodynamic Analysis of Biomass Gasification and Torrefaction*; Technische Universiteit Eindhoven: Eindhoven, The Netherlands, 2005.

43. Medic, D.; Darr, M.; Shah, A.; Potter, B.; Zimmerman, J. Effects of torrefaction process parameters on biomass feedstock upgrading. *Fuel* **2012**, *91*, 147–154. [CrossRef]

44. Park, J.; Meng, J.; Lim, K.H.; Rojas, O.J.; Park, S. Transformation of lignocellulocis biomass during torrefaction. *J. Anal. Appl. Pyrolysis* **2013**, *100*, 199–206. [CrossRef]

45. Poudel, J.; Oh, S.C. A kinetic analysis of wood degradation in supercritical alcohols. *Ind. Eng. Chem. Res.* **2012**, *51*, 4509–4514. [CrossRef]

46. Tumuluru, J.S.; Boardman, R.D.; Wright, C.T.; Hess, J.R. Some chemical compositional changes in Miscanthus and white oak sawdust samples during torrefaction. *Energies* **2012**, *5*, 3928–3947. [CrossRef]

47. Tumuluru, J.S.; Boardman, R.D.; Wright, C.T. Response surface analysis of elemental composition and energy properties of corn stover during torrefaction. *J. Biobased Mater. Bioenergy* **2012**, *6*, 25–35. [CrossRef]

48. Nimlos, M.; Brooking, E.; Looker, M.J.; Evans, R.J. Biomass torrefaction studies with a molecular beam mass spectrometer. *Am. Chem. Soc.* **2003**, *48*, 590.

49. Nhuchhen, D.R. *Studies on Advanced Means of Biomass Torrefaction*; Dalhousie University: Halifaxm, NS, Canada, 2016.

50. Wang, M.J.; Huang, Y.F.; Chiueh, P.T.; Kuan, W.H.; Lo, S.L. Microwave-induced torrefaction of rice husk and sugarcane residues. *Energy* **2012**, *37*, 177–184. [CrossRef]

10

A Novel Cellulase Produced by a Newly Isolated *Trichoderma virens*

Rong Zeng [1,†], Xiao-Yan Yin [2,†], Tao Ruan [2], Qiao Hu [2], Ya-Li Hou [2], Zhen-Yu Zuo [2], Hao Huang [2] and Zhong-Hua Yang [2,*]

[1] College of Chemistry and Chemical Engineering, Hubei University, Wuhan 430062, China; rongzengce@163.com

[2] College of Chemical Engineering and Technology, Wuhan University of Science and Technology, Wuhan 430081, China; yinxiaoyan0516@163.com (X.-Y.Y.); 15671627439@163.com (T.R.); huqiao294873270@163.com (Q.H.); houyali88@163.com (Y.-L.H.); zuozhenyu@wust.edu.cn (Z.-Y.Z.); hhzy310@163.com (H.H.)

* Correspondence: yangzh@wust.edu.cn

† The authors contributed equally to this work and should be considered co-first authors.

Academic Editor: Daniel G. Bracewell

Abstract: Screening and obtaining a novel high activity cellulase and its producing microbe strain is the most important and essential way to improve the utilization of crop straw. In this paper, we devoted our efforts to isolating a novel microbe strain which could produce high activity cellulase. A novel strain *Trichoderma virens* ZY-01 was isolated from a cropland where straw is rich and decomposed, by using the soil dilution plate method with cellulose and Congo red. The strain has been licensed with a patent numbered ZL 201210295819.6. The cellulase activity in the cultivation broth could reach up to 7.4 IU/mL at a non-optimized fermentation condition with the newly isolated *T. virens* ZY-01. The cellulase was separated and purified from the *T. virens* culture broth through $(NH_4)_2SO_4$ fractional precipitation, anion-exchange chromatography and gel filtration chromatography. With the separation process, the CMC specific activity increased from 0.88 IU/mg to 31.5 IU/mg with 35.8 purification fold and 47.04% yield. Furthermore, the enzymatic properties of the cellulase were investigated. The optimum temperature and pH is 50 °C and pH 5.0 and it has good thermal stability. Zn^{2+}, Ca^{2+} and Mn^{2+} could remarkably promote the enzyme activity. Conversely, Cu^{2+} and Co^{2+} could inhibit the enzymatic activity. This work provides a new highly efficient *T. virens* strain for cellulase production and shows good prospects in practical application.

Keywords: *Trichoderma virens*; cellulase; strain isolation; straw; biomass

1. Introduction

With the development of agriculture, cellulose-rich straws from cropland have become one of the largest amounts of biomass. They should be comprehensively utilized as biological resources. Unfortunately, most crop straws, such as wheat straw and rice straw, are directly burned in cropland, which has caused serious atmospheric pollution in China [1]. If biological resources that are prevented from being directly burned in cropland can be utilized as carbohydrates, then the atmospheric pollution can be avoided and a huge amount of carbohydrates, such as fermentable sugars, can be provided. It is well known that the main component of straw is cellulose, which is the most common source of renewable carbon and energy on earth. Cellulose could be efficiently degraded to fermentable sugars with conventional biological processes. The fermentable sugars are the most important platform material, which could be further converted to ethanol (biofuel), single cell protein (SCP) and other chemicals by biotechnology [2–4]. In the process route, a high activity of cellulase is the key factor to this biological process of straw utilization as biological resources [5–7]. Screening and obtaining a novel high activity cellulase and its producing microbe strain is always the most important and essential way to improve the straw utilization. Moreover, cellulase has a very broad application in other fields such as food processing, oil extraction, agricultural industries, brewery, animal feed, textile and other fields like laundry, pulp, paper, detergent industry [8–12].

Cellulase is mainly produced by microbes, especially fungus, such as *Trichoderma*, *Aspergillus* and *Penicillium*. Nowadays, more and more *Trichoderma* fungi have been screened; however, there exists a lower-yield problem, so it is important to screen a strain that produces cellulase efficiently.

In this paper, we devoted our efforts to isolate a novel microbe strain which can produce high activity cellulase. Furthermore, the enzymatic properties of this new cellulase will also be evaluated.

2. Materials and Methods

2.1. Material and Media

The soil samples were obtained from land areas where straw is rich and some straw was decomposed. We collected soil samples from a wheat field (33°36′41″N and 116°56′17″E, Suzhou county, Anhui province, China), Yangtze River riverbank (30°38′14″N and 114°22′4″E, Wuhan City, Hubei province, China), and a paper mill (30°27′3″N, 114°11′59″E, Wuhan City, Hubei province, China).

The following mediums were used for enrichment, screening, identification and cultivation process. the enrichment medium was used to enrich the microbe, and it was composed of (in g/L): oatmeal extract (oatmeal 3.5 g adding water 30 mL and boil 20–30 min then filter to obtain the filtrate), agar 1.5, crystal violet 0.5, deionized water 100 mL. The Congo red medium was used to screen the target strain as the screening medium, it was composed of (in g/L): KH_2PO_4 1.0, $NaNO_3$ 3.0, KCl 0.5, $MgSO_4 \cdot 7H_2O$ 0.5, $FeSO_4 \cdot 7H_2O$ 0.01, CMC·Na 15, Agar 15–20 and Congo red 0.2. The identification medium was used to identify the target strain, which was composed of (in g/L): $NaNO_3$ 3.0, $MgSO_4$ 0.5, K_2HPO_4 1.0, KCl 0.5, $FeSO_4$ 0.01, sucrose or glucose 30.0. The fermentation medium was used to culture the target strain and produce cellulase, it was composed of (in g/L): yeast extract 1, CMC·Na 5, NaCl 20, K_2HPO_4 0.5 and $MgSO_4 \cdot 7H_2O$ 0.5.

2.2. Screening the Microbe for Cellulase and Microbe Identification

The soil sample (2.0 g) was suspended in 20 mL of sterile distilled water. Then 0.5 mL supernatant was inoculated on enrichment medium plates with coating inoculation, and cultured for three days at 30 °C. The colonies growing in the enrichment medium plates were inoculated in the Congo red medium plate with inoculating needle and further cultured on the third day at 30 °C. The colonies growing on the Congo red plate were the microbes which can excrete cellulase. The diameters of the colony and hydrolyzed circle can indicate the microbe's ability to produce cellulase [10]. The selected colonies were further purified by culture in new Congo red medium plate.

Morphological and molecular identification of the isolated microbes were done according to the Fungal Identification Manual [13]. The identification medium was used to culture the screened strain. A light microscope was used to observe its morphological features. Genomic DNA for molecular identification of the strain was extracted using the modified benzyl chloride method [14]. In order to identify the isolated microbe, the 18S rDNA was cloned by PCR (Initialization at 94 °C for 5 min, Denaturation at 94 °C for 30 s, Annealing at 54 °C for 1 min, Extension at 72 °C for 90 s, 30 cycles, Final elongation at 72 °C for 10 min, 4 °C to the end, PCR in 25 μL reaction system with pfu DNA polymerase) with the primers (Forward: 5′-GTAGTCATATGCTTGTCTC-3′, Reverse: 5′-TCCGCAGGTTCACCTACGGA-3′). The PCR-amplified products were purified by Gel Extraction kit (Omega Bio-Tek, Norcross, GA, USA) and sequenced by Sangon Biotech (Shanghai, China). The 18S rDNA was compared with sequences in nucleotide database (NCBI) using the BLAST algorithm. Multiple sequence alignment was carried out with CLUSTALW (Conway Institute UCD, Dublin, Ireland) multiple sequence alignment. The neighbour-joining phylogenetic analysis was constructed with MEGA v.5.0 software (Center for Evolutionary Medicine and Informatics, The Biodesign Institute, Tempe, AZ, USA).

2.3. Purification of the Cellulase from the Newly Isolated Microbe Culture Broth

The cellulase was produced by the newly isolated strain with fermentation medium at 30 °C for 4 days. The crude enzyme extraction was collected by centrifuging the cultivation broth, since cellulase is an extracellular enzyme. It was further purified by fractionation salting out (50%–70% saturation of ammonium sulfate), anion exchange chromatography (DEAE Sepharose Fast Flow, elution with 0–0.8 M·NaCl) and gel filtration chromatography (SephadexG-75). After purification, the purified product was assayed by SDS-PAGE.

2.4. Evaluation of Enzymatic Properties

The following enzymatic properties of the cellulase produced by the newly isolated microbe were evaluated. The optimum pH and pH stability, the optimum temperature and thermal stability and effects of metal ions to the enzymatic activity were investigated. Er mixed 0.5 mL enzyme with 0.5 mL substrate containing 0.05% (w/v) CMC-Na in citric acid buffer and incubated the mixture (pH = 5.0) for 1 h. The temperature range was from 30 °C to 90 °C, and pH was range from 2 to 9, and the concentration of metal ions was 0.01 mol/L. After incubation, we measured the cellulase activity, as described below.

2.5. Cellulase Activity Assay

The activity of the cellulase was assayed with CMC-Na activity measurement [15]. An enzyme sample of 0.5 mL was added into 1.5 mL of the reaction mixture (containing 0.05% (w/v) CMC·Na with pH 5.0 citric acid buffer) and incubated at 50 °C for 1 h. Then the reducing sugar released by the reaction was determined by DNS method. One unit of the CMC enzyme activity was defined as the amount of enzyme that catalyzed to produce 1 μmol of reduced sugar per minute with the reduction of CMC-Na.

3. Results

3.1. Isolation of Microbe for Cellulase and Microbe Identification

The purpose of isolation was to obtain a microbial strain that possessed ability to produce high activity cellulase. With the isolation procedure, four microbial strains were obtained and tested for cellulase activity, which can be released from cellulase to hydrolysis the CMC·Na and formed an obvious hydrolyzed circle. Table 1 describes the physiological characteristics of the isolated strains.

Table 1. Isolated strains for cellulase production.

Strain [1]	Colony Diameter/cm	Hydrolysised Circle Diameter/cm	Colony Color	Colony Morphology
ZY-01	2.30 ± 0.22	4.32 ± 0.26	Early phase white, Anaphase green	Edge irregular dentatus, Surface filiform
ZY-02	1.87 ± 0.41	2.82 ± 0.38	white	Edge irregular dentatus, Surface filiform
ZY-03	2.12 ± 0.49	3.02 ± 0.37	Front milk white, Back yellow	Edge irregular dentatus, Surface filiform
ZY-04	3.27 ± 0.43	fuzzy	Front yellow-white, Back red	Edge irregular dentatus, Surface smooth

[1] ZY-01: soil sample from wheat field, ZY-02 and ZY-03: soil sample from Yangtze River riverside, ZY-04: soil sample from paper mill.

In terms of the diameter size of the colony and hydrolysised circle that appeared on the plates, strain ZY-01 is the best strain for cellulase production. Since the strain possesses more cellulase production ability, then more cellulase activity can be released to hydrolysis of the CMC-Na and a larger hydrolysised circle and colony were formed. It can be confirmed that ZY-01 is a good high-producing strain. The cellulase production ability of this newly isolated ZY-01 strain was tested. The cellulase activity 7.4 IU/mL was obtained at a non-optimized fermentation condition.

The colony morphology and the morphological features of mycelium by light microscope both indicated that the screened ZY-01was a fungi. To further identify strain ZY-01, its 18S rDNA region was cloned by PCR and sequenced. Its nucleotide blast analysis was conducted in NCBI to find homologous sequences. The neighbor-joining phylogenetic tree, Figure 1, was constructed with these homologous sequences. The results show that strain ZY-01 is a *Trichoderma virens*. Furthermore, the morphological identification experiments also confirm that it is a *T. virens* according to the Fungal Identification Manual [13]. Then this newly isolated strain was named T. virens ZY-01 and deposited in the China Center for Type Culture Collection (CCTCC) with an accession number CCTCC M 2012205. Its 18S rDNA sequence was deposited at the GenBank database with the Accession No. JX121089. The cellulase productivity of this newly isolated *T. virens* is appealing compared to the previous reports [11,16–18]. According to these reports, the cellulase activity was about 7.8, 0.18, 1.43 and 7.51 IU/mL to the corresponding *T. reesei*, Mutated *T. viride*, *Fusarium oxysporum* H57-1 and *Rhizopus stolonifer* var. reflexus TP-02 at the optimized enzyme production conditions.

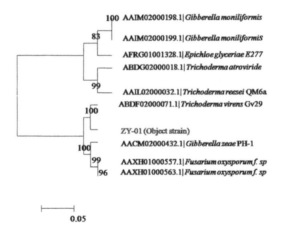

Figure 1. Neighbor-joining phylogenetic tree of 18S rDNA genes of strain ZY-01.

3.2. Purification of the Cellulase from the Newly Isolated T. virens ZY-01 Culture Broth

The cellulase was purified from the cultivation broth of *T. virens* ZY-01. With three complementary purification steps (fractional precipitation, anion exchange chromatography and gel filtration chromatography) Sephadex G-75 chromatography elution was distinguished by molecular weight of protein. The elution curves are shown in Figure 2. The CMCase activity mainly assembled at peak II.

As a result, cellulase was purified 35.8-fold with a 47.04% CMC enzymatic activity yield compared to the crude enzyme. The results of each separation step are given in Table 2.

Table 2. Results for purification of cellulose from *T. virens* ZY-01 culture broth.

Procedure	Protein Content/mg	Enzyme Activity/IU	Specific Activity IU/mg	Yield %	Purification Fold
Crude enzyme	197.3	174.1	0.88	–	–
Fractional precipitation	21.6	135.3	6.26	77.7	7.11
Anion exchange chromatography	15.8	106.4	6.73	61.1	7.65
Gel chromatography	2.6	81.9	31.5	47.04	35.8

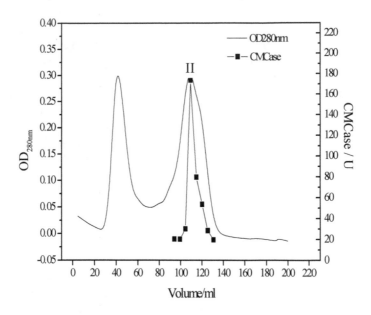

Figure 2. Elution profile gel filtration column chromatography.

The SDS-PAGE of the purification product was given in Figure 3. The SDS-PAGE result showed that there were three protein bands with molecular masses of 62 kDa, 58 kDa and 16 kDa. It indicates that this cellulase is a triple-subunit complex. In general, cellulase is multi-subunit complex. Tang and Balasubramanian respectively reported cellulases from *Rhizopus stolonifer* var. reflexus TP-02O and *Bacillus mycoides* S122C also had three subunits [17,18]. The molecular masses of the subunit are from 40 kDa to 64 kDa.

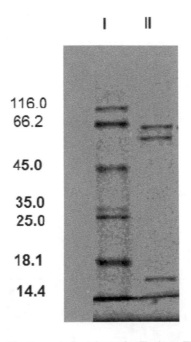

Figure 3. SDS-PAGE of the purification product from the *T. virens* ZY-01 cultivation broth. Lane I: protein ladder (β-galactosidase 116.0 kDa, bovine serum albumin 66.2 kDa, ovalbumin 45.0 kDa, lactate dehydrogenase 35.0 kDa, REase Bsp981 25.0 kDa, β-lactoglobulin 18.1 kDa, lysozyme 14.4 kDa), Lane II: purification product.

3.3. Enzymatic Properties of Cellulase Produced by T. virens ZY-01

The enzymatic properties are the fundamental bioinformation to an enzyme. They are important to it application. The following enzymatic properties of the new cellulase produced by T. virens ZY-01 were evaluated: the effect of reaction temperature, pH and metal ions to enzymatic activity, the thermal stability and pH stability. The corresponding results are present in Figures 4–8. Figure 4 showed that the CMCase activity was highest at 60 °C as protein degenerated gradually at a higher temperature. Figure 5 showed that the enzyme kept 90% of activity at 40 °C to 50 °C, the remained CMCase activity decreased obviously when higher than 50 °C. Figure 6 showed that the activity increased sharply from pH = 2.0 to 4.0, when pH is higher than 6.0, the activity decreased, so the optimum reaction pH is 6.0. Figure 7 showed that the CMCase activity of cellulase remained stable at the range of 5.0–6.0, the activity could keep 80%. If pH < 5.0 or pH > 6.0, the CMCase activity decreased, and the cellulase was sensitive to pH. Figure 8 showed that Zn^{2+}, Ca^{2+} and Mn^{2+} help cellulase while Co^{2+} and Cu^{2+} inhibit the enzyme activity.

The temperature could speed up the reaction, but the activity of cellulase would fade along with the increasing temperature. The space structure of the enzyme would be destroyed when over an acidic or basic environment, causing the change of conformation, and the loss of enzyme activity. The results showed that the optimum reaction temperature and pH is 60 °C and pH 5. As the component of cellulase is protein, its structure is unique, and metal ions carry more than one positive charge, its effect is stronger than proton. Besides, metal ions have a complexing action, which could maintain the concentration of solution at a stable stage. Therefore, metal ions can clearly affect cellulase activity. The results showed that Zn^{2+}, Ca^{2+} and Mn^{2+} are its activators; they can significantly promote its activity. The relative enzyme activity is respectively about 139.2%, 125.5% and 131.4% to the blank. In contrast, Co^{2+} and Cu^{2+} can obviously inhibit its activity. The relative enzyme activity is just 17.6% and 5.9% to the blank, respectively. It also can be seen that the stability of this cellulase is perfect. Even it was incubated at 50 °C for 1 h, the residual activity can maintain it at 90%. Since it was stored in pH 5 buffer for 24 h, its residual activity was about 85%. These enzymatic properties of the newly obtained cellulase are similar to the cellulase from Bacillus mycoides S122C [18].

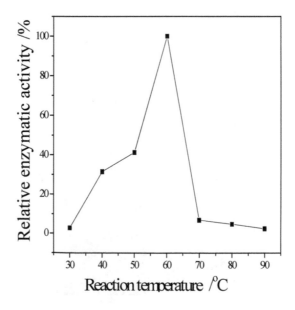

Figure 4. Effect of reaction temperature on enzymatic activity.

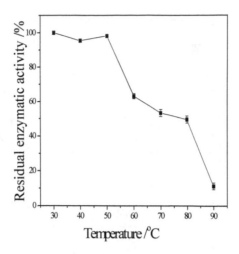

Figure 5. Thermal stability of the cellulose.

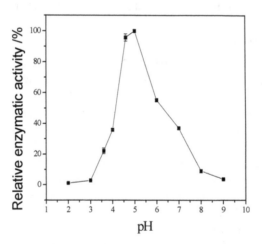

Figure 6. Effect of reaction pH on enzymatic activity.

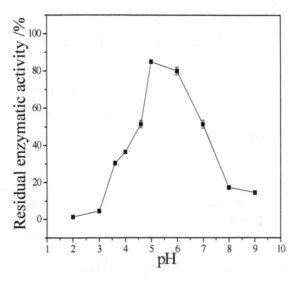

Figure 7. Stability of the cellulase to pH.

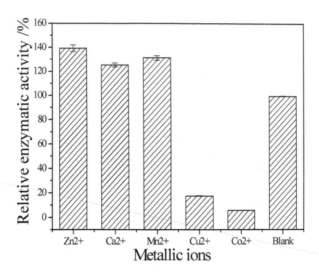

Figure 8. Effects of metal ions on enzymatic activity.

4. Discussion and Conclusions

Trichoderma virens produced enzyme activity is up to 31.5 U/mg. This is high compared with *Fusarium Oxysporum*, which produced cellulase, and its enzyme activity was only 1.43 IU/mL [19], and *Aspergillus sydowii*, the CMCase of which was 1.32 IU/mL [20].

A novel *Trichoderma virens* was isolated from straw cropland. It can secrete high activity cellulase. The cellulase is a triple-subunit complex with molecular masses of 62, 58 and 16 kDa of each subunit. The optimum temperature and pH of the cellulase are 50 °C and pH 5.0. Zn^{2+}, Ca^{2+} and Mn^{2+} could remarkably promote the enzyme activity. Conversely, Cu^{2+} and Co^{2+} are strong inhibitors to the cellulase. It possesses good stability in terms of thermal and pH factors. Based on these findings, it can be concluded that this newly obtained microbe is a good prospect for practical cellulase production and has good application for the new cellulase.

The *Trichoderma virens* ZY-01 was isolated for its high yield of cellulase, which is expected to be useful for hydrolysis of cellulosic and hemicellulosic substrates at proper temperatures, particularly for converting biomass into biofuels, to solve the energy crisis and the environmental pollution problem.

Acknowledgments: This study was supported by the National Natural Science Foundation of China (Grant No. 21376184), the Scientific Research Foundation for the Returned Overseas Chinese Scholars (State Education Ministry), the Foundation from Educational Commission of Hubei Province of China (Grant No. D20121108) and the Innovative Team of Bioaugmentation and Advanced Treatment on Metallurgical Industry Wastewater.

Author Contributions: Rong Zeng and Zhong-Hua Yang conceived and designed the experiments; Rong Zeng, Xiao-Yan Yin and Tao Ruan performed the experiments; Hao Huang and Ya-Li Hou analyzed the data; Zhen-Yu Zuo contributed reagents/materials/analysis tools; Qiao Hu and Zhong-Hua Yang wrote the paper.

References

1.   Jeya, M.; Zhang, Y.-W.; Kim, I.-W.; Lee, J.-K. Enhanced saccharification of alkali-treated rice straw by cellulase from *Trametes hirsuta* and statistical optimization of hydrolysis conditions by RSM. *Bioresour. Technol.* **2009**, *100*, 5155–5161. [CrossRef] [PubMed]

2.   Back, S.C.; Kwon, Y.J. Optimization of the pretreatment of rice straw hemicellulosic hydrolyzates for microbial production of xulytol. *Biotechnol. Bioprocess Eng.* **2007**, *12*, 404–409. [CrossRef]

3.   Delmer, D.P.; Haigler, C.H. The regulation of metabolic flux to cellulose, a major sink for carbon in plants. *Metab. Eng.* **2002**, *4*, 22–28. [CrossRef] [PubMed]

4.   Olsson, L.; Hahn-Hagerdal, B. Fermentation of lignocellulosic hydrolsates for ethanol production. *Enzym. Microb. Technol.* **1996**, *18*, 312–331. [CrossRef]

5.   Ramos, L.P.; Nahzad, M.; Saddler, J.N. Effect of enzymatic hydrolysis on the morphology and fine structure of pretreated cellusolic residues. *Enzym. Microb. Technol.* **1993**, *15*, 821–831. [CrossRef]

6.   Bothwell, M.; Daughhete, S.; Chaua, G. Binding capacilities for *Thermonmonspora fusca* E-3, and E-4. The E-3 binding domain and *Trichoderma reesei* CBHI on avicel and bacterial microcryrstallie cellulose. *Bioresour. Technol.* **1997**, *60*, 169–178. [CrossRef]

7.   Hii, K.-L.; Yeap, S.-P.; Mashitah, D.M. Cellulase production from palm oil mill effluent in Malaysia economical and technical perspectives. *Eng. Life Sci.* **2012**, *12*, 7–28. [CrossRef]

8.   Kuhad, R.C.; Gupta, R.; Singh, A. Microbial cellulases and their industrial applications. *Enzym. Res.* **2011**, *2011*, 280696. [CrossRef] [PubMed]

9.   Lin, Y.; Wei, W.; Yuan, P. Screening and identification of cellulase-producing strain of *Fusarium oxysporum*. *Procedia Environ. Sci.* **2012**, *12*, 1213–1219.

10. Assareh, R.; Shahbani, H.Z.; Akbari, K.N.; Aminzadeh, S.; Bakhshi, G.K. Characterization of the newly isolated *Geobacillus* sp. T1, the efficient cellulase-producer on untreated barley and wheat straws. *Bioresour. Technol.* **2012**, *120*, 99–105. [CrossRef] [PubMed]

11. Li, X.H.; Yang, H.J.; Roy, B.; Park, E.Y.; Jiang, L.J.; Wang, D.; Miao, Y.G. Enhanced cellulase production of the *Trichoderma viride* mutated by microwave and ultraviolet. *Microbiol. Res.* **2010**, *165*, 190–198. [CrossRef] [PubMed]

12. Zhao, Y.; Chen, G.; Zhou, W.; Hou, Y.; Yang, Z. Progress of cellulase and cellulase gene research. *Biotechnol. Bull.* **2013**, *2013*, 35–40. (In Chinese)

13. *Fungal Identification Manual*; Wei, J.C., Ed.; Shanghai Science and Technology Press: Shanghai, China, 1979. (In Chinese)

14. Xue, S.J.; Yue, T.L.; Yuan, Y.H.; Gao, Z.P. An improved method for extracting fungal DNA. *Food Res. Dev.* **2006**, *27*, 39–40. (In Chinese)

15. Dashtban, M.; Maki, M.; Leung, K.T.; Mao, C.; Qin, W. Cellulase activities in biomass conversion measurement methods and comparison. *Crit. Rev. Biotechnol.* **2010**, *30*, 302–309. [CrossRef] [PubMed]

16. Saravanan, P.; Muthuvelayudham, R.; Viruthagiri, T. Application of statistical design for the production of cellulase by *Trichoderma reesei* using mango peel. *Enzym. Res.* **2012**, *2012*, 157643.

17. Tang, B.; Pan, H.; Tang, W.; Zhang, Q.; Dinga, L.; Zhang, F. Fermentation and purification of cellulase from a novel strain *Rhizopus stolonifer* var. *reflexus* TP-02. *Biomass Bioenergy* **2012**, *36*, 366–372. [CrossRef]

18. Balasubramanian, N.; Toubarro, D.; Teixeira, M.; Simōs, N. Purification and biochemical characterization of a novel thermo-stable carboxymethyl cellulase from azorean isolate *Bacillus mycoides* S122C. *Appl. Biochem. Biotechnol.* **2012**, *168*, 2191–2204. [CrossRef] [PubMed]

19. Yuan, L.; Wang, W.; Pei, Y.; Lu, F. Screening and Identification of Cellulase-Producing Strain of Fusarium Oxysporum. *Procedia Environ. Sci.* **2011**, *12*, 1213–1219. [CrossRef]

20. Ketna, M.; Digantkumar, C.; Jyoti, D.; Anand, N.; Datta, M. Production of cellulase by a newly isolated strain of *Aspergillus sydowii* and its optimization under submerged fermentation. *Int. Biodeterior. Biodegrad.* **2013**, *78*, 24–33.

11

Metabolic Control in Mammalian Fed-Batch Cell Cultures for Reduced Lactic Acid Accumulation and Improved Process Robustness

Viktor Konakovsky [1], Christoph Clemens [2], Markus Michael Müller [2], Jan Bechmann [2], Martina Berger [2], Stefan Schlatter [2] and Christoph Herwig [1,*]

Academic Editors: Mark Blenner and Michael D. Lynch

[1] Institute of Chemical Engineering, Division of Biochemical Engineering, Vienna University of Technology, Gumpendorfer Strasse 1A 166-4, 1060 Vienna, Austria; vkonakovtuwien@gmail.com
[2] Boehringer Ingelheim Pharma GmbH & Co. KG Dep. Bioprocess Development, Biberach, Germany; christoph.clemens@boehringer-ingelheim.com (C.C.); markus_michael.mueller@boehringer-ingelheim.com (M.M.M.); jan.bechmann@boehringer-ingelheim.com (J.B.); martina.berger@boehringer-ingelheim.com (M.B.); stefan.schlatter@boehringer-ingelheim.com (S.S.)
* Correspondence: christoph.herwig@tuwien.ac.at

Abstract: Biomass and cell-specific metabolic rates usually change dynamically over time, making the "feed according to need" strategy difficult to realize in a commercial fed-batch process. We here demonstrate a novel feeding strategy which is designed to hold a particular metabolic state in a fed-batch process by adaptive feeding in real time. The feed rate is calculated with a transferable biomass model based on capacitance, which changes the nutrient flow stoichiometrically in real time. A limited glucose environment was used to confine the cell in a particular metabolic state. In order to cope with uncertainty, two strategies were tested to change the adaptive feed rate and prevent starvation while in limitation: (i) inline pH and online glucose concentration measurement or (ii) inline pH alone, which was shown to be sufficient for the problem statement. In this contribution, we achieved *metabolic control* within a defined target range. The direct benefit was two-fold: the lactic acid profile was improved and pH could be kept stable. Multivariate Data Analysis (MVDA) has shown that pH influenced lactic acid production or consumption in historical data sets. We demonstrate that a low pH (around 6.8) is not required for our strategy, as glucose availability is already limiting the flux. On the contrary, we boosted glycolytic flux in glucose limitation by setting the pH to 7.4. This new approach led to a yield of lactic acid/glucose (Y L/G) around zero for the whole process time and high titers in our labs. We hypothesize that a higher carbon flux, resulting from a higher pH, may lead to more cells which produce more product. The relevance of this work aims at feeding mammalian cell cultures safely in limitation with a desired metabolic flux range. This resulted in extremely stable, low glucose levels, very robust pH profiles without acid/base interventions and a metabolic state in which lactic acid was consumed instead of being produced from day 1. With this contribution, we wish to extend the basic repertoire of available process control strategies, which will open up new avenues in automation technology and radically improve process robustness in both process development and manufacturing.

Keywords: CHO cell culture; scale-down; fed batch; automation; Lactic acid control; pH; metabolic control; MVDA; uncertainty; online analyzer

1. Introduction

1.1. Problem Statement

A great amount of modern biopharmaceuticals such as monoclonal Antibodies (mAbs), fusion proteins, bi-specific antibodies, IgG and others are produced today by highly specialized cells in a fed-batch bioprocess. The Chinese hamster ovary (CHO), well known by cell banking institutes and authorities, is often engineered (*i.e.*, GS-CHO) to satisfy a tailored production purpose. Often a large-scale fed-batch or perfusion operation is required to meet the market demand. In such processes, the feeding regime was found to have a great impact on lactate production, which negatively affects the productivity [1]. As a consequence, variations in both quantity and quality during recombinant protein processes affect the profitability of the production plant markedly.

The lactic acid profile of a mammalian cell culture process affects the maximum achievable cell count and final product titer [2,3]. Manufacturing runs with a high and a low lactic acid profile have been linked with productivity of the process by using Multivariate Data Analysis (MVDA) methods [4,5]. It does not come as a big surprise that there is not only one but there are several approaches to decrease lactic acid. Some methods encompass genetic modifications of the clone [6–8], changes in medium composition [9–12], interventions on a genetic level [13,14], feeding strategy [15–18], modification of metabolic state [19–23] or the physico-chemical environment [24–30], to name a few.

As lactic acid is produced, pH falls—which can again impair cell proliferation [31] at critically low pH. Depending on the clone and cell type, the lower critical pH limit is typically between 6.6–6.8 in mammalian cell culture [24]. In several other studies, pH was also associated with interfering with product quality [32]; therefore, the way pH is controlled is still an interesting question today. Ideally, pH should be fixed during the whole process and should never need adjustment by either acidic or alkaline control agents. In practice this is not exactly the case. An increase of osmolality and stress to the cells is often the result of a suboptimal pH control strategy. Operation at low pH is observed to restrict the formation of lactic acid at the cost of inhibiting cellular proliferation while an alkaline pH has the opposite effect [24,25,33]. Therefore, many industrial processes start with high pH and shift to low pH during the culture as will be seen a bit later in this contribution. However, modifying the pH alone is not fast enough to prevent lactic acid build-up, as can be seen in the lactic acid profiles in our data—the metabolic state itself must be modified, and this can be done with an appropriate feeding strategy.

1.2. Feeding Strategies

The choice of an appropriate feeding strategy depends very much on the context; they may be based on well-defined industrial production runs, where little variation is anticipated and a historical run can be used as template for the next run. Alternatively, they are derived by in-line, at-line or on-line signals from the current process in real time, which is often the case in a bioprocess development environment [34–36]. Therefore, a methodology to deliver feed adaptively, accurately and safely in mammalian cell culture development is of very high relevance [37–40]. In order to control lactate production by adaptive feeding, the main understanding is that a limited metabolic state, mainly a carbon limitation, needs to be designed. The methodology is typically divided into data-driven and adaptive feeding approaches.

1.2.1. Data-Driven Feeding

Data driven approaches rely on knowledge gathered from historical data. Drapeau [16], Luan [15,41] and Gagnon *et al.* [42] published methods to control mammalian cell culture in glucose limitation (around 1 mM). As criteria for an increase or decrease of a previously determined feed rate from very similar historical runs, if too much lactic acid is produced, the feed is too high and needs to decrease, whereas when the pH is rising, the feed is too low and needs to increase. This intuitive and simple method may be limited if buffers are employed, which mask lactic acid production and may lead to periods of starvation [43]. Aehle *et al.* [44] described the control of a mammalian fed-batch cultivation by limiting glutamine instead of glucose availability, thus affecting carbon source utilization and reducing ammonia yields, to control the feed rate. Previous experiments under similar conditions or numerical optimization studies were used to establish a tcOUR (total cumulative oxygen uptake rate) profile. This profile was correlated *ab initio* to the historical viable biomass profile. Future repetitions of the experiment with different initial viable biomass concentrations led to a highly reproducible cell concentration estimation and also to the desired sub-maximal specific growth rate. In order to deliver the correct amount of feed in real time to the variable biomass, the glutamine per biomass yield (Y X/GLN) had to be known and proved to be fairly constant during the six days of cultivation.

Liu *et al.* [45] even suggested the feed-forward control of mammalian cell cultures. They derived a historical growth rate by linear calibration of previous growth curves and used it together with the yield of biomass per glucose (Y X/GLC) to set up a determined feed rate. Validation with three different cell lines led to an improvement of titer as well as IVCC (integrated viable cell concentration). This leads to the question of why data-driven strategies work so extremely well. One reason is that the validation experiments are short so that the verification does not face the difficulty of describing a considerable death phase, which leads to a reduced representation of a typical industrial process. Additionally, different clones in different scales often have different biomass and metabolite profiles. Differences in medium composition significantly diminish the usefulness of pre-determined feed profiles. Typically, even large datasets of one clone in scale 1 cannot predict the behavior of another clone in scale 1 or the same clone in scale 2. Therefore, a sufficient number of experiments are typically conducted *a priori*, which can build up enough confidence that a data-driven approach does not fail, which would be especially devastating in a manufacturing-scale run.

1.2.2. Adaptive Feeding Strategies

Usual orthogonal process development runs are too different to successfully apply a data-driven pre-calculated feed or biomass profile from historical data. For these reasons we need to focus on simple mechanistic relationships which allow automatic feed rate adaptation to the current and not a past process. A look into literature shows that there are several ways to accomplish this by deriving a feed in real time using in-line, at-line and on-line signals.

Zhou and Hu first mention signals which may be used to detect the metabolic state of mammalian cell cultures and derive a feeding profile on-line [46]. The turbidity probe allowed the conversion of optical density (OD) to cell concentration with a simple calibration curve as well as the oxygen uptake rate (OUR) which made it possible to detect a more direct measure for cellular activity. Noll and Biselli [47] used a capacitance probe to deliver feed to a fluidized bed culture. The yield X/GLN was constant during the whole time which allowed keeping glutamine at 0.45 mM with an online massflow dosing system. Dowd *et al.* [48] used capacitance as well to estimate viable cell concentrations in a perfusion process to control the nutrient feed rate. The feeding was adjusted automatically and was used to maximize the productivity. Zhou *et al.* [49] used the OUR measurement to adjust the nutrient feeding rate in a fed-batch culture. This was possible by applying a previously determined yield of glucose per oxygen (GLC/OX) consumption and led to the control of the process at very low glucose and amino acid levels. The strategy was applicable as long as a reliable calculation of the kLa (volumetric oxygen transfer coefficient) allowed determination of OUR. Europa *et al.* [50] used such a setup to show how controlling substrates at low levels not only led to reduced metabolite formation but also to a multiplicity of steady states, in which a more efficient metabolism was observed. Ozturk *et al.* [38] utilized an at-line analyzer to manipulate the perfusion rate of a culture in real time. Li *et al.* [51] proposed several other techniques to control mammalian cell culture, which are based on the cell's specific oxygen uptake rate (qOUR) during the process. In combination with substrate or metabolite measurements, qOUR was used as an indicator to change other process parameters on-line. Among those, feed rate, pH, temperature, stirring speed, and pO_2 were adapted when limitations, such as reduced qOUR or low substrate, could be detected. Lu [37] compared two methods to control the feed rate: in one instance, an auto sampler method was used to hold a particular glucose concentration, and in the other, a capacitance signal was used as a surrogate for cell growth to control the feed rate. In both cases, the target glucose concentration was non-limited between 4–6 g/L to prevent over- and under-feeding. Variable, specific consumption rates were observed over time, which are typical for operations at such high concentration ranges of substrate. As an online or offline measurement was available, it could be used to correct these deviations. Alternatively, previous experiments were shown to be useful to deduct the evolution of the specific consumption rates, with the drawback that this again makes the process dependent on historical data.

1.3. Challenges in Process Control

The aforementioned contributions describe the state of the art; however, there are still many gaps in knowledge which need to first be mentioned and then attempted to be closed. Many methods are based on some prior knowledge of the process. However, in reality there are many sources for deviations which complicate things considerably. Some sources of complication with which we were confronted in our labs and want to report are listed in Table 1.

Table 1. Selection of observed sources of complications in the development of feeding strategies in our labs.

Source	Influence
Clones	Metabolic needs may differ greatly, leading to the perpetual development of historical feeding profiles. In adaptive feeding regimes, clone-dependent differences of dielectric properties may complicate biomass estimation when capacitance probes are used, while turbidity probes may detect more or less cell debris in the decline phase, depending on which clone was used.
Scales	Especially on-line offgas/kLa–dependent control strategies may become very difficult to transfer because they depend on the aeration and stirrer cascade strategy (*i.e.*, constant or adaptively increasing gas flow to hold pO_2).
Assumptions	Constant yields (*i.e.*, GLC/OX, X/GLC, X/GLN, *etc.*) may change over time, leading to stoichiometric over- or under-feeding.
Media	Addition of growth-influencing components may change historical feeding profiles completely and make a direct comparison between experiments difficult as these changes have further implications on the process.
Process parameters	Changes in temperature, stirrer speed, pO_2, pH or pCO_2 levels may affect gas solubility, buffer capacity, offgas profiles, cellular stress level, and growth and may change the metabolic requirements for both adaptively or historically calculated feed rates.

We do not claim to offer solutions to all mentioned points but want to make the reader aware of their existence and propose solutions for some of them in our own work: the herewith presented adaptive feeding strategy was based on estimation of biomass in the current process, making it independent of historical data. The method we used in our labs to estimate biomass, and with it the adaptive feeding rate, has been shown in a previous contribution to be transferrable between clones and scales [52].

We coped with changing metabolic yields by limiting glucose concentrations and the glucose uptake rate to a desired range. The yield was still subject to variation, but could not exceed a particular range—the rate at which substrate was supplied corresponded to the desired consumption rate of the cells in real time, which made it controllable. The method is based on a first principle investigation using correlations between the specific rates such as glucose or lactic acid to determine the feed rate set-point. The metabolic response may differ from clone to clone in the slope of the yield Y L/G, but once established, this correlation will hold for one clone which makes the general methodology applicable to different clones. Uncertainty consisting of both metabolic state and biomass was considered by applying a range for the feed set-point; even though the yields and biomass error are subject to change, certain signals (*i.e.*, pH, among others) could be used to switch the feeding rate automatically to ensure sufficient supply of substrate for the whole process time.

1.4. Goal

The goal of this study was two-fold: to reduce lactic acid levels and to minimize the pH control actions by additional acid-base feeding using a dynamic process environment as encountered in a process of industrial relevance. This contribution therefore proposes a novel adaptive feeding strategy, which is based on estimating the viable cell concentration with a capacitance probe in glucose-limited growth conditions, targeting a low lactic acid/glucose yield (Y L/G) and taking pH variation into account in the control strategy. The workflow followed in this contribution was (i) to analyze historical process development data to get an understanding in which range specific glucose consumption shows a favorable Y L/G profile; (ii) MVDA to determine the contributions of important parameters influencing lactic acid production; and (iii) development of the feeding strategy based on a high and low uncertainty of the metabolic state and biomass estimation error. The results show that lactic acid build-up can be decreased by confining glucose flux to a particularly low range, regardless of the pH set-point. However, pH may have helped to keep glucose consumption higher than reference processes late in the process, which in turn may have had a positive effect on productivity. The challenge was finding a way to address this in a dynamic fed-batch process; therefore, our methodology aims at being directly transferrable from process development back to manufacturing conditions.

2. Experimental Section

2.1. Cell Lines

All data in this contribution was collected from one single engineered CHO clone (derived from CHO DG44), subsequently referred to as "Clone B". This clone was kindly provided by Boehringer Ingelheim (Ingelheim, Germany) for the necessary experiments at the VUT (Vienna University of Technology). Only Clone B data was used for data analysis purposes as the metabolic behavior is different from Clone A data. However, for completeness, the data to construct and validate the multivariate model for biomass estimation (see chapter MVDA) were recorded with "Clone A" clones. The biomass model was shown to be transferrable and also estimate Clone B data. Clone A data is not discussed further in this contribution, but more information could be found in our recent publication [52].

2.2. Available Dataset

A historical Clone B dataset consisting of 29 fed-batch fermentations from process development (2L and 80L) was kindly provided by Boehringer Ingelheim (Ingelheim, Germany) to generate process understanding using MVDA techniques. These data sets are also referred to as "Historical data". On this basis, fed-batch fermentations were mirrored in a scale-down model (2L), and two experiments, "R-30" and "R-31", were selected and described in detail in this contribution.

2.3. Media

Media for the seed train and fed-batches are proprietary in composition and subject to variations in starting levels of metabolites, growth factors, selection pressure, *etc.* All components were serum-free, animal-component-free, and chemically defined.

2.4. Process Setup

Clone B cells were cultivated in shake-flasks (Corning Inc., Corning, NY, USA) in incubators (Minitron, Infors, Bottmingen, Switzerland) with 5% partial CO_2 pressure at physiological temperature

(35–37 °C) on orbital shakers at 120 rpm (orbit 50 mm). Passaging was performed every third day in proprietary chemically defined media and the bioreactors (3.6L, Infors HT, Bottmingen, Switzerland) were inoculated at an initial seed density of 3×10^4 to more than 1×10^6 cells/mL. Process information was logged using the process management system Lucullus (PIMS, Lucullus, Biospectra, Switzerland). A capacitance probe (Biomass Sensor, Hamilton Bonaduz AG, Bonaduz, GR, Switzerland) in scanning mode was used with an excitation frequency ranging from 0.3 MHz to 10 MHz, the signal was collected in real time and recorded between one and more than 60 min steps. Offline concentration of total cells, viable cells and viability was measured using an image-based white/dark classification algorithm after automatic trypan blue staining integrated in the cell counter (Cedex HiRes, Roche, Basel, Switzerland). Main metabolite concentrations (Glucose, Lactic acid and IgG) were measured on-line and off-line using a photometric robot (CubianXC, Optocell, Bielefeld, Germany).

2.5. Online Enzymatic Analyzer

An enzymatic analyzer (CubianXC, Optocell, Germany) was coupled to the process by withdrawing sample via a ceramic membrane made of Al_2O_3 (pore size 0.2 μm, membrane area 17.8 cm^2, membrane thickness 1.6 mm; IBA, Heiligenstadt, Germany). The supernatant was removed from the reactor at certain intervals at a flow rate between 0.2–1.0 mL/min. The sampling intervals of the analyzer were tested between 3–12 h in experiment R-30 and a designated sampling interval of 6 h was set for the experiment R-31 to measure, among others, glucose and lactic acid concentrations. Purging the lines and sampling took a total of 30 min where a maximum of 12 mL cell-free supernatant was withdrawn per sample. Faster purging and more frequent sampling (up to 30 min intervals) was technically possible, but not required for our purpose. A more detailed description of the enzymatic robot's set-up and substrate and metabolite measurements is described elsewhere [53].

2.6. MVDA

2.6.1. Biomass Model

In brief, multivariate models perform better at capturing the declining phase of a culture than linear models. One of these was available to estimate biomass in historical Clone A runs. The resulting model was transferrable to another clone (here: Clone B) by adaptation of the slope of the multivariate model for this new clone. For more information please refer to our publication [52]. The model was constructed prior to the experiment with Datalab 3.5 software (kindly provided by Prof. Lohninger, Vienna University of Technology, Vienna, Austria) [54] and the transferred Clone A model was used to estimate Clone B biomass in real time.

2.6.2. Data Mining

Historical Clone B data from process development was harmonized by using the same calculation routine in Matlab (Mathworks, Natick, MA, USA) for all runs, including selected experiments R-30 and R-31 at the institute, to exclude possible variations after data treatment. In order to gather process understanding of how and by which importance the process parameters relate to each other, we used PLS-R (partial least squares regression) in Simca 13.0.3 (Umetrics, Sweden) to analyze the historical dataset. These parameters were chosen with considerations explained a little later in the results section and encompassed the specific rates and concentrations of glucose and lactic acid as well as pH, which was used in its non-logarithmic representation of H^+ concentration.

2.7. Process Control

2.7.1. Process Control Scheme

The recorded spectra were recorded every minute and directly translated into a real-time feed rate. The following script (Figure 1) controlled the process in the experiments. First, biomass was estimated using the coefficients calculated from a multivariate model. We used the assumption that glucose is limited and solely the feed rate is determining the volumetric glucose uptake rate. Both entities are combined to a specific glucose uptake rate, which is in turn used as a set-point to calculate a real-time feed rate. The calculation of specific substrate consumption qs was done by dividing the volumetric rate by the viable cell concentration. As derivative rates tend to be noisy, all specific rates were treated using Matlab's *rlowess* function [55]. The set-point of the feeding rate was corrected in real time by either of two factors: by an online enzymatic analyzer result (such as threshold glucose concentration), or by pH (for instance a pH upper or lower band), or a combination of both.

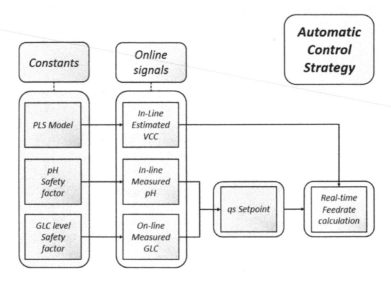

Figure 1. Automatic control strategy using exclusively real-time available signals. Constants (safety factors and the biomass model) were defined *a priori* which led to either a rather loose or very strict adherence to a desired qs set-point.

2.7.2. Feed Rate Calculation

We estimate biomass and feed stoichiometrically for the specific consumption rate q of a substrate s which the cells should ideally hold. A material balance for substrate glucose was used [56] to calculate qs and historical data analysis was performed to find a useful qs set-point which should be held over time. As qs is not a static value and subject to change, the question of which qs should be set was a very important one. We address it and the implications a bit later in the results section of our work. The feed set-point was controlled gravimetrically by a pump (Lambda Preciflow, Czech Republic) which received the adaptive set-point from a process management system Lucullus (PIMS, Lucullus, Biospectra, Switzerland) in real time. The adaptive feed rate (see Equation (1)) increased and decreased with biomass, which was estimated using a PLS model. The rate at which cells were fed stoichiometrically depended on the experiment and could be influenced by certain process events (Equation (2) and Table 2) to prevent starvation of the cells without having to rely on operator interventions.

The real-time automatic feed rate is described by Equation (1), where qs (which may be glucose or any other substrate) is kept constant.

$$Feed\left[\frac{ml}{h}\right] = \frac{VCC\left[\frac{c}{ml}\right] \cdot V_{Reactor}\left[mL\right] \cdot q_s\left[\frac{mg}{c \cdot h}\right]}{Feed\ concentration_{Substrate}\left[\frac{mg}{ml}\right]} \qquad (1)$$

We used a qGlc *range* instead of a single, fixed set-point, which depends on a previously defined safety factor (ranging here from −25% to +100%). As a switch between qs set-points we used inline pH and online available signals of substrates or metabolites measured by the online analyzer (here glucose); therefore, the qs we specified to be desirable to reduce Y L/G is multiplied by a factor which corresponds to the expected variation in the process. For both experiments, the factor (Equation (2)) by which qs changed was between 0.75 and 2.0 (1.0 minus safety factor of 25% and 1.0 plus safety factor of 100%) and if 1.0 is replaced by the actual feeding set-point, qs reads 8 (rounded up) to 30, depending on the experiment. The safety factor specification by either pH or online analyzer was distinct in both experiments and is explained in detail in Table 2.

$$qs_{adapted} = qs - safety\ factor\ f\ (pH,\ online\ analyzer) \cdot qs \tag{2}$$

The final equation (Equation (3)) as it was used in the control strategy now reads:

$$Feed \left[\frac{ml}{h}\right] = \frac{VCC\left[\frac{c}{ml}\right] \cdot V_{Reactor}\ [mL] \cdot q_{s\ adapted}\left[\frac{mg}{c \cdot h}\right]}{Feed\ concentration_{S,in}\left[\frac{mg}{ml}\right]} \tag{3}$$

Table 2. Control specifications: Initial qs set-point (SP) in [pg/ch] (picogram per cell per hour), correction by a defined safety factor in [%] under certain conditions (pH and Online Analyzer), and Target SP.

	Control Specifications			
	Switch Conditions		Target SP [pg/ch]	
Experiment	R-30	R-31	R-30	R-31
Initialization	Start feed after 2 h	Start feed after 2 h	−15	−10
	pH control [†]			
pH high	If pH ⩾ 7.1. increase qs by 100%	If pH ⩾ 7.4. increase qs by 25%	−30	−13
pH OK	If pH between 7.1 and 6.9 use the desired qs	If pH between 7.1 and 6.9 use the desired qs	−15	−10
pH low	If pH ⩽ 6.9 reduce qs by 25%	If pH ⩽ 6.9 reduce qs by 25%	−11	−8
	Online Analyzer control [‡]			
GLC low	If Gluc ⩽ 0.4 increase qs by 100%	Monitoring	−30	−10
GLC high [ʃ]	If Gluc > 0.4 use the desired qs	Monitoring	−15	−10

[†] The pH range was chosen to lie in the physiological range. However, it could be extended to conditions close to the maximum tolerance which may lie somewhere between pH 6.5 and up to 8.0 for mammalian cells [27]; [‡] The correction order of the set-point is as follows: first pH, then online analyzer. This is important because if the feed is already reduced by pH, it was not done so a second time by the online analyzer. The online analyzer in experiment R-31 had no purpose other than monitoring the metabolite concentrations; [ʃ] A high glucose concentration is the current status quo in most industrial mammalian cell culture processes.

2.7.3. Operation Window for qs

The set-point for feeding was not fixed but assumed several distinct values in two experiments, which are summarized in Table 2. The rationale for the range calculation of the set-point is based on a possible biomass estimation error and will be explained a little later in both Sections 3 and 4 including an exemplary calculation. In brief, the initial qs set-point was selected to lie in an area relevant for control purposes (reduced or negative Y L/G), in our case −10 and −15 p/cell*h.

3. Results and Discussion

3.1. MVDA for Assessing Lactate Metabolism

To capture the high dimensionality of a heterogeneous dataset, we employed PLS-R, a chemometrics tool [57,58] relatively straight-forward in its application [59]. The single most important feature we were interested in describing was the lactic acid metabolism. Therefore, only a handful of predictor variables were selected with the following considerations in mind:

1. We wanted to be able to set or influence the selected parameters easily, which is the case for the ones we selected, *i.e.*, pH.
2. We did not wish to become dependent on other parameters by including them in the analysis, which might then not be frequently analyzed, such as amino acids.
3. We did not want to include parameters or define new parameter ranges, which are fixed in a platform/manufacturing process, *i.e.*, pO_2 or the pH (acid/base) regulation strategy.

From the available variables and the considerations above, the following were always available in all processes, for the whole process time, and therefore the most straightforward to describe lactic acid production or consumption (qLac): (i) glucose concentration (GLC); (ii) specific glucose consumption (qGlc); (iii) pH in the form of hydrogen ion concentration (H^+); and (iv) lactic acid concentration (LAC).

A score scatter plot was used to identify all data inside a 95% Hotelling's confidence ellipsoid and used in the subsequent analysis (Figure 2A). A new PLS-R model was built without the outliers, and variable variation (R^2) and variable prediction (Q^2) were calculated. We were interested in the ranking of parameters according to their relevance to reduce qLac to get a better understanding on what to focus on in our experiments. The resulting fit did not play a role in accepting or rejecting the predictive quality of the results as we did not use the model in a quantitative way (Figure 2B). High LAC as well as H^+ prevented further increase of qLac [28], which could be mechanistically explained by chemical gradient action between the inside and outside of the cell [60,61]. The specific glucose consumption, meanwhile, was quite naturally indirectly proportional to lactic acid production and a much better predictor of metabolic behavior than, *i.e.*, GLC levels.

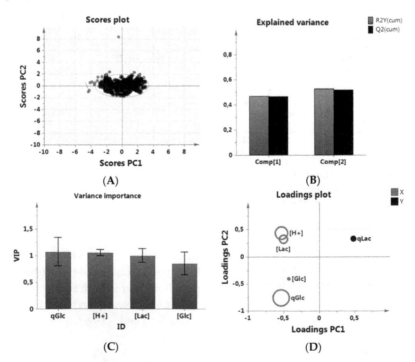

Figure 2. MVDA of the historical data set. (**A**) removal of outliers prior to analysis; (**B**) PLS-R model fit using two Principal Components, here Comp 1 and 2. R2Y(cum) indicates the explained variance, while Q2(cum) explains the predictive quality using one resp. two principal components; (**C**) VIP plot of the most important variables in the analysis sorted by relevance; (**D**) Loading scatter plot of data relationship between predictors (X) and predicted (Y) variable discriminated by VIP size.

A VIP (variance importance of the projection) analysis was used to simplify the multivariate data analysis by calculating a relative score for those predictor variables, which were used to describe the predicted variable qLac. A high VIP score (Figure 2C) resulted in a larger size representation of the given parameter in Figure 2D. The positions of predicted variable qLac together with the predictor variables explained the overall relationship of all parameters; H^+ and LAC were clustered closely to each other and indicate a closer relationship, while all parameters, including GLC and qGlc, were positioned opposite of qLac, indicating an inverse relationship (see the Appendix for more information on the importance of GLC in this analysis, in particular Figures A1 and A2). As GLC had the lowest VIP score, its importance may be understood to be low in this data set.

High GLC has an effect early in the culture while cells are "fit" to take up large amounts of glucose, so to speak. High GLC or overfeeding has almost no effect on cells later on—their maximum qs simply decreases over time, even if GLC was held perfectly constant over time. We can only speak for our experiments, but maybe this was already observed by other labs in which CHO cells were metabolically engineered to yield a reduced lactic acid profile. One would expect qLac to increase along with rising GLC levels, but from historical data this is not exactly the case, which implies that not GLC but the specific glucose consumption rate needs to be tightly controlled.

Summarizing our findings, we found that qGlc was by far the most important parameter to describe qLac behavior in process development runs, while H^+ as well as LAC apparently played a role as well, although much less pronounced. GLC was statistically correlated with qLac due to the nature of the clone to consume lactic acid also during high glucose concentrations, but this particular finding was of little practical use. Therefore, the following experiments featured a modification of the most important parameters, which are qGlc and H^+, in a glucose-limited state.

3.2. Lactic Acid Metabolism

The clone we investigated already featured a reduced LAC profile and was capable of lactic acid consumption as can be seen in Figure 3A,B. MVDA revealed that qLac can be expressed as a function of the parameters qGlc, LAC, H^+, but not necessarily GLC; hence, the feeding had to be based on qGLC and not on GLC itself to improve lactic acid profiles, the latter being the normal control entity in industrial processes.

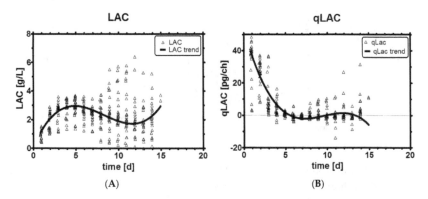

Figure 3. Historical LAC profiles (**A**); Historical qLac profiles (**B**).

A strong early consumption of glucose is the main reason for high LAC levels later on, as will be seen a bit later in chapter 3.6. Historical process development data, where glucose flux was limited and low lactic acid levels were observed during the process, supports this statement. Le [4] shows that early intervention in the process affects the process outcome, which may be, for instance, the lactic acid profile. This intervention was demonstrated in this contribution with an early limitation of glucose availability, especially while the cells are still growing well. The benefit was to prevent LAC buildup and avoid many challenges with LAC and pH in cell culture. As any limitation of substrate may lead to reduced cell counts, we have tried both a weak and a strong limitation of qGlc to improve the process.

3.3. Impact of pH on qGlc

The historical processes all featured a pH set-point with an upper and lower range, inside of which no control agents are yet used—as result, the pH can assume several values, depending on the current control set-point. This raises the question if pH could be kept somehow stable by design and de-correlated from the typically concomitantly occurring undesired rise in LAC levels at alkaline pH. We differentiate between two different pH states: "pH low" (6.6–6.9) and "pH high" (6.9–7.2).

Figure 4A,B show qGlc and on-line pH profiles of all 29 process development runs. The pH of all observations was split into two groups, higher or lower than pH 6.9, and plotted against qGlc. Under high pH conditions, a high qGLC range was observed, while at acidic pH, qGlc was more restricted (Figure 4C,D). As a result, high pH was often associated with high qGlc and *vice versa*. During the mid- and end-phase, pH did not impact or change qGlc much, while the effect that pH might exert on qGlc was strongest while the cells were growing well. Therefore, pH may be used indirectly to change qLac, as both qGlc and qLac are highly correlated (Figure 2D).

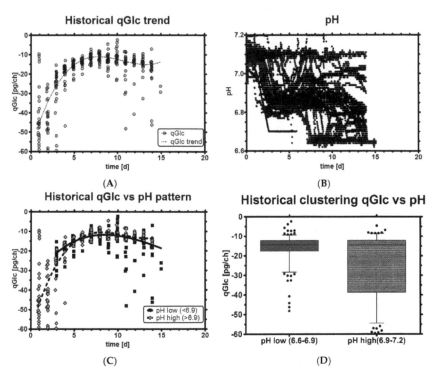

Figure 4. Historical qGlc profiles (**A**); On-line pH profile (**B**); Clustering of qGlc with a pH classifier (dark: low pH, bright: high pH) over time (**C**) and as box plot (**D**). A robust third-order polynomial function was applied in Graphpad Prism software to capture the general trend of a typical qGlc profile in (**A**) and qGlc in (**C**). No runs started with pH < 6.9 as pH initially falls and is then controlled.

3.4. Set-Point Selection for qGlc

The most important parameter which influences qLac and therefore LAC levels in the fermentation is qGlc. Other important parameters which also influence qLac, such as oxygen availability [62] or methods to improve pyruvate uptake into the tricarboxylic acid circle (TCA) [6], are acknowledged but out of scope of this publication. Historical data showed that the cells apparently never consumed less glucose than approximately 8 pg/ch (pg per cell per hour) and the gross of all measured qGlc lied in the range of 10–20 pg/ch. Plotting qGlc *versus* the yield qLac/qGlc (in short Y L/G) implies that, with the control of qGlc, a particular target yield might be achieved (Figure 5). This relationship may also be used to design a stoichiometric feed by trying to feed at a defined qGlc set-point in further experiments.

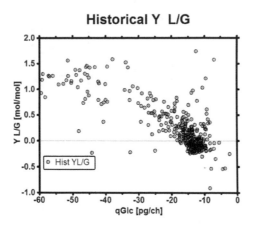

Figure 5. Historical yield qLac/qGlc (Y L/G) *versus* qGlc showing the consequence of particular qGlc on the yield.

3.5. *Impact of a Broad Range as qGlc Set-Point on the Lactic Acid Profile*

First, we attempted to hold the cell's qGlc set-point in a broad range between 10 and 30 pg/ch over the whole cultivation time by feeding at a target set-point rate which changes stoichiometrically with biomass. The feeding strategy is described in the experimental section (Table 2). For most of the time, the culture was controlled at a glucose level <0.36 g/L with a short excursion to up to 1 g/L when a high feeding set-point was automatically selected and the cells could not consume the substrate quickly enough before the next measurement.

In Figure 6A, it can be seen that the target specific rate of qGlc switched automatically between high and low set-point when certain conditions were fulfilled. Simply put, on-line pH was used as the main switch to affect the feed rate by changing the qGlc set-points between −10 and −30 pg/ch in real time. A second switch was the GLC level measured by an online analyzer, but pH had a higher hierarchy level as it was a faster available signal than the metabolite levels which were measured every couple of hours and so the associated events were rarely ever triggered. When the viable cell concentration reached a maximum, the high qGlc set-point was exchanged with the default set-point (10 resp. 15), because we know from experience that this clone does not require so much substrate so late in the cultivation.

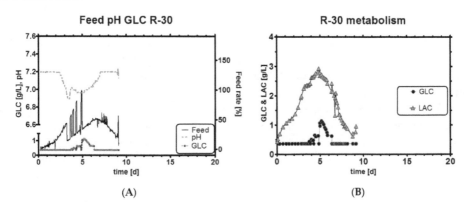

Figure 6. *Cont.*

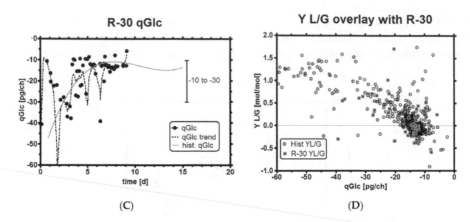

Figure 6. Experiment R-30, feeding profile with pH and GLC level (**A**); GLC and LAC concentrations (**B**); qGlc with trend and historical qGlc (**C**); overlay of resulting Y L/G profile with historical data (**D**).

Overall, this feeding strategy did not reduce LAC levels by much (Figure 6B) compared to historic reference runs (shown later). This was to be expected as we selected a purposely high qGlc set-point resulting in high glucose uptake rates. The qGlc trend of this run very much resembles the one from the historical runs which are overlaid in Figure 6C,D. As can be seen, qGlc falls well within the previously defined limits and shows the characteristics of a non-glucose-limited run (but *is* glucose- limited), which is relevant because it means two things: first, to achieve a lower qLac, a lower qGlc set-point must be selected (which could be relatively high so that the total carbon influx into TCA is not greatly reduced), and second, and perhaps more interestingly, a high GLC concentration is *not* necessarily required to keep the culture well. This is a very important statement because it implies that mammalian cell cultures may be operated in limitations if the biomass and qGlc can be estimated well. Both come with a particular error and, for this reason, a range rather than a fixed stoichiometric feeding point is necessary to keep the culture well supplied with enough nutrients over the whole process time. In the next chapter we briefly describe how a cell culture may be stoichiometrically fed if errors are rather low with experiment R-31.

3.6. Impact of a Tight Range as qGlc Set-Point on the Lactic Acid Profile

We select a second experiment, in which we used a lower qGlc set-point criterion and selected a very stringent band between −8 and −13 pg/ch as the operational range. The rationale for such a strong glucose inhibition was that lactic acid production should, in theory, be very low because Y L/G approaches zero for this qGlc, as seen in Figure 5. The experiment R-31 shows a few noteworthy differences when compared to any historical process development runs, and, in fact, any runs we have found so far in the literature: first, pH was extremely stable (Figure 7A) during the whole process time and no control action by either acid or base was required. If, as a consequence of this strategy, pH should begin to rise too much in a run, acids are easier to add than to remove from the system. During the whole fermentation time, the tight qGlc set-point was never breached by more than 25% (which corresponds to the error of our biomass estimation method). Even when overestimating biomass, the consequence resulted in a total increase of qGlc by roughly 3 pg/ch, equaling roughly 16 pg/ch at the time of the largest error. This implies that even though biomass estimation is not perfectly accurate and deviates within a certain error range (as observed for the capacitance signal [63], but also any other signal), the qGlc set-point range simply needs to be selected high enough. A possible biomass underestimation cannot threaten the culture and if the set-point is still low enough, the benefits from a reduced LAC buildup can be still fully reaped (see Section 4 for an example).

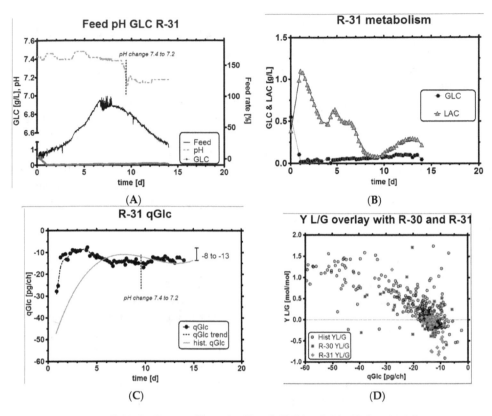

Figure 7. Experiment R-31, feeding profile with pH and GLC level (**A**); GLC and LAC concentrations (note that leftover GLC is consumed and results in a small LAC until it is consumed, but qs changes immediately to the target value once GLC is limited by the stoichiometric feeding strategy) (**B**); qGlc (including dotted trend line) and historical qGlc (full line) (**C**); overlay of resulting Y L/G profile with historical data (**D**). Both targets (qGlc and resulting lower Y L/G) can be better achieved due to the tighter control specification, when compared with run R-30. A comparison follows in Table 3.

Table 3. Statistical evaluation: in experiment R-30, a slight improvement of Y L/G and P/G compared to the historical runs can be seen. In experiment R-31, the improvements for Y L/G and P/G are most pronounced. The variation of qGlc could be massively reduced as can be seen in the low value for the standard deviation. N describes the number of available sampling events in the historical data and the experiments R-30 and R-31.

	qGlc [pg/ch]			Y L/G [-]			Y P/G [-]		
ID	Historical	R-30	R-31	Historical	R-30	R-31	Historical	R-30	R-31
MEAN	−22.39	−19.15	−13.92	0.41	0.18	0.03	0.29	0.31	0.36
SD	16.53	14.62	3.64	0.63	0.49	0.31	0.12	0.05	0.1
	$N_{All} = 667, N_{R-30} = 60, N_{R-31} = 56$								

Another observation was the potential effect that pH may have had on qGlc. In all historical experiments, qLac was strongly correlated with pH, but here, for the first time, we could successfully decouple both parameters and investigate what happens if a high pH meets low LAC. During biomass overestimation, we could see that qGlc is higher than the historical reference but within the previously selected qGlc range of −8 to −13 pg/ch (Figure 7C). After the viable cell concentration maximum was reached, a pH change of −0.2 units brought the pH down from 7.4 to 7.2 (Figure 7A). After a delay of one to two days, qGlc decreased slightly by *ca.* 15%, which cannot be seriously discussed in terms of statistical significance. However, the direction of qs change after pH change would fit well into the overall picture of pH affecting qs (pH acidic: decreased GLC influx, pH basic: more GLC influx into the cells). It may be too early to conclude that pH could be used to modulate the maximum possible qGlc so extremely easily (which depends, among others, also on amino acid availability). However, the data suggest that exactly this may be the case, so this claim might be worth being explored by other research groups in the future.

We claim to have reached the goal of this study to reduce lactic acid profiles successfully, as can be seen in Figure 7B, with a minimum amount of experiments. Even though we exposed cell cultures to an unusually high pH of 7.4, which is usually correlated with high qLac, our feeding strategy showed exceptionally low LAC levels by controlling qGlc tightly and accurately. The culture could be kept in a metabolically highly interesting state for a duration of 14 days in a dynamic fed-batch process. The resulting Y L/G was, therefore, around zero most of the time and also assumed negative values, which are found to correlate with high titers in cell culture (Figure 7D) [5,64].

3.7. Adaptive Feeding Using Real-Time Switches

3.7.1. Adaptive Feeding Using pH Correction

It is not in the nature of an error to announce an over- or under-estimation, so when exactly can the target set-point be switched on or off with an upper or lower range set-point, and how high or low can those possible set-points be defined? The answer to these questions lies in the enormous wealth of on-line signals which can be used exactly for this purpose [35,65]. In the first experiment, pH and the GLC level were used, and in another only pH was enough to switch the qGlc set-point to particular levels automatically. As the scale-down bioreactor model mirrors a manufacturing process with a floating pH dead band, we found pH suitable to automatically correct the feeding set-point. As a consequence, acid/base control for pH was not required at all. This may have had a positive impact on cellular physiology and, as a consequence, on the final titer. The strategy to include pH in set-point adaptation was adapted from Gagnon [42], but our control is more tightened due to the real-time and closed-loop approach instead of using predetermined rates. In the first experiment, we selected the switching criteria a little too broadly, and the set-point jumped at the smallest occasion, as can be seen in the feeding profile of run R-30 (Figure 6B). The reason for this was partially an unlucky first selection of the Boolean which often activated and deactivated the switch, but also the combined application of another switch from the online analyzer, as we had both signals at our disposal.

3.7.2. Adaptive Feeding Using an Online Metabolic Analyzer

Ozturk described how online analyzers were used to hold a particular substrate or metabolite level [38]; however, for our problem statement this was not necessary, as the low glucose level posed no threat to the culture due to robust stoichiometric feeding. The online analyzer was working well but was actually not required to keep lactic acid levels low in a dynamic fed-batch. Therefore, we restricted the use of the set-point switch in experiment R-31 to pH only (Figure 7B). We conclude that one well-chosen switching criterion is enough for the task to keep the control strategy simple and intuitive. Although signals are available every minute, we advise to reduce control action to every half hour or more—otherwise, tiny excursions of the control range lead to feed corrections in the same time which makes the feeding profile impossible to interpret without strong smoothing. However, process development may very much benefit from highly frequently measured concentrations (*i.e.*, in the hour range) and unlock precise event detection, *i.e.*, after pulsing or shifting [66,67], to develop process understanding which may then translate into better process development and control.

3.8. Comparison of Experiments with Historical Performance

Direct comparison of key differences between historical data and the experiments reveals the following findings: Selecting the qGlc band for a stoichiometric feeding strategy in a very high range from −10 to −30 pg/ch led to comparable concentrations (Figure 8A) and metabolic profiles (Figure 8C) to the reference runs, even though glucose was limited at levels around 0.36 g/L (with a temporary shoot-up up to 1.2 g/L GLC). The operation in limited GLC conditions did not lead to a reduction of growth at all, as can be seen in the exceptionally high cell concentration (Figure 8E), but as cells started dying very fast in this process, a high final product titer was finally not obtained in this fermentation (Figure 8F).

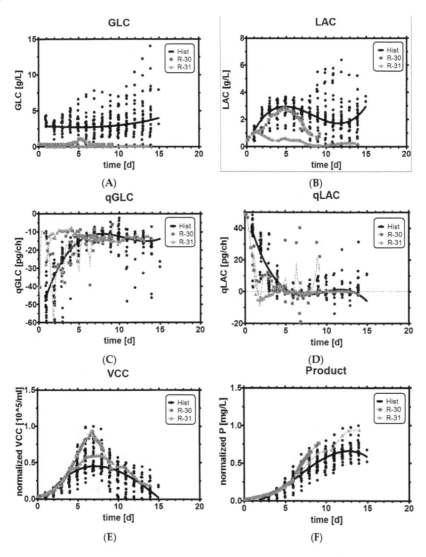

Figure 8. Benchmarking limited glucose control with historical data (black dots: historical data, squares: experiment R-30 with a broad qGlc set-point, triangles: experiment R-31 with a tight qGlc set-point. (**A**) and (**B**): metabolic profiles of glucose and lactic acid; (**C**) and (**D**): Specific glucose and lactic acid rates; (**E**) and (**F**): Viable cell concentration and product titer compared to historical data.

In contrast to a very broad selection of qGlc, a very tightly controlled qGlc band between −8 to −13 pg/ch resulted in almost immediate LAC uptake from the environment (Figure 8B,D), while at the same time limiting the availability of GLC to a stoichiometric feed (Figure 8A,C). As can be seen, such a strong substrate limitation showed an effect on the growth profile, and on the maximum cell concentration (Figure 8E), which was somewhat low but still higher than the average mammalian cell cultivation. However, the declining phase was much softer and flatter, which may have been the reason for the high final titer at the end of the process (Figure 8E,F) as cells were still productive instead of being in the process of quickly dying. To be more precise, we observed that the majority of the cells in the tight qGlc limitation did not turn into dead cells, but lysed, which could be seen in the high viability (80%) at the end of the process run time (Figure A3 in the Appendix). This could indicate that cells took a different way to die than under regular process conditions. In a speculative afterthought, development of drugs which improve resistance to either cell lysis or apoptosis [68–70] might help to prolong cultivation (and production) time. A lower maximum viable cell count (VCC), but at the same time higher titer, suggests that the culture may have been simply more productive or productive for longer than the references.

From the experiments shown, we cannot truly distinguish which of the two parameters (qGlc or pH) finally led to more product. We assume that qGlc had the higher impact, simply because there is no truly direct mechanistic link between productivity and pH. It may have been a combination of both effects, but this statement would require much more rigorous testing, *i.e.*, by running a DoE to determine the exact contributions and interactions (of H+, qGlc set-point, qGlc upper and lower range, Y L/G, other low GLC levels, *etc.*), which was far out of scope of this contribution. Table 3 summarizes the experimental findings with regard to qGlc, Y L/G and Y P/G. The strongest limitation of average qGlc over the whole process time (−62% compared to the reference) resulted in a decrease of the Y L/G yield to almost zero, while at the same time the product yield was increased by 25%. For this we only needed to estimate biomass using a capacitance probe, which is already an accepted standard in many development labs. Usually, an increase in product yield has to be critically regarded, especially if the final titer turns out to be lower overall than the reference, but this is very clearly not the case here.

4. Conclusions

4.1. Eliminating the Root Cause for High Lactic Acid Concentrations

Historical data was analyzed by MVDA techniques where the specific glucose uptake rate was identified as the most relevant parameter to reduce the specific lactic acid production rate and, as a consequence, the lactic acid profiles of the culture. A feeding strategy was developed on this basis to set up a target specific glucose consumption rate for the whole process runtime of a fed-batch. This target rate was set up by estimating cell count via a capacitance probe, which can be used either for process monitoring [71–74] or control purposes [47,75]. Feeding was realized by supplying the cells stoichiometrically in a range between 8 to 30 pg/ch in real time with feed solution instead of using a previously defined off-line feeding profile which cannot react to deviations in the present process. Glucose is shown to be efficiently held constant to levels around 2–3 times the Km of mammalian cell lines during the whole process time without having a negative effect on viable cell concentration, as long as the specific glucose consumption feeding rate is selected to be high. Although the main effects leading to a reduction of lactic acid happen in the first days, it is well worth mentioning the remainder of the experiment. The proposed strategy also keeps the culture in the decline phase in a stoichiometrically fed state. This is a very important finding because it eliminates otherwise-required operator interventions and makes it easy to integrate it in an automatized manufacturing environment. We demonstrate that, if a strong limitation of the specific glucose consumption rate is selected, a very positive effect on the stability of the pH signal was observed together with a complete prevention of lactic acid build-up.

4.2. Directions from MVDA

Multivariate data analysis (MVDA) techniques can be tricky to use and sometimes misleading, as we have seen with the parameter GLC. The historical data set was comprised of glucose levels which are far above the cell's physiological limit, where any effect on the specific glucose consumption rate could be expected. "Low" glucose levels (around 2–3 g/L) correlated to high lactic acid production rates, which makes no sense biochemically, but the largest part of the given data did not feature a glucose range in which qGlc was mechanistically affected. Because glucose concentration had the overall lowest weight in the analysis, and because the model was used in an explorative rather than quantitative way, the fit could be, from a statistical point of view, accepted. PLS-R was still very useful to correctly identify the exceptional importance of qGlc and the close relationship between H^+ and LAC. Ranking their importance with regard to the impact on qLac could, furthermore, save valuable time by running the most interesting experiments first, instead of screening all eventualities. Conclusively, MVDA results were helpful to save time, but must be always reviewed carefully by considering whether the used data is really adequate and truly representative for the given problem.

4.3. Hypothesis for the Positive Effect of pH on Productivity

In our opinion, one facet of pH-dependent flux modulation remains overlooked until today and might be used to maximize the production of high-quality therapeutics of tomorrow: reverting the same method which researchers were using for years to reduce metabolic flux [30,76] into lactic acid and redirecting it to the TCA [50,77], we *increase* pH to maximize metabolism in substrate limitations. This is especially important in the light that lactic acid toxicity is actually more pronounced at lower pH than at higher pH [61,78–80] and may be the cause of more cell lysis [81], resulting in the lower productivity of a process. Flux increase is hypothesized to result from the cell's regulation of its internal pH by a plethora of transporters [82–85] in the following manner: many enzymes involved in glycolysis are strongly affected by pH, among them those considered as metabolic bottlenecks, such as HK (hexokinase), PFK (phosphofructokinase) and others [86–89]. Even though the cytosol of the cell is strongly buffered, a certain gradient difference of around 1.5 units cannot be crossed if the cell should stay alive, not even if the cell in question is one of the strongest lactic acid producers [90] we are aware of today. Therefore, it is assumed that the internal pH, however buffered, must be affected by the external pH; thus, metabolism can be directed to different fates. The increase in flux is possibly distributed to all nodes and not just one node of the metabolism, among them the one responsible for product formation [77,91], as symbolized in Figure 9. We believe that the herein presented novel concept may lead to more biomass, which, in turn, may lead to higher titers [19], since more substrate leads to more cells which can produce more product due to an improved utilization of energy and nutrients.

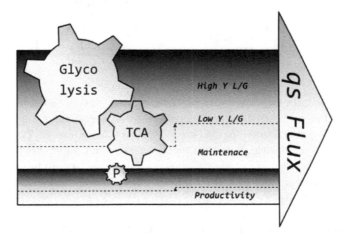

Figure 9. Suggested flux dependency on productivity. A higher overall flux may lead to a higher flux in all metabolic nodes, including lactic acid secretion and the one responsible for product formation. Usually, the basic consumption of cells does not change much because pH is controlled tightly. However, by setting and holding physico-chemical conditions which act on this flux by a high pH, while still operating in a relatively low Y L/G range, the overall higher flux may be cumulatively translated into higher productivity.

4.4. Living with Uncertainty

The biggest risk in the experiments was biomass underestimation, as this might mean starvation and an early stop of the process. In case that biomass estimation comes with a high error, two possibilities can be considered to cope with the uncertainty: changing the target qGlc set-point to a safer operation set-point or giving the target set-point a safe *range*, in which it can move and still satisfy the goals. The maximum error of biomass estimation may depend on the model by which it is assessed and have an asymmetrical distribution, *i.e.*, in an imaginary worst case scenario (Figure 10) of an overestimation of up to +50% and underestimation up to −33% (our biomass estimation in these experiments was in the range of maximum 25% over the whole process time, including the death phase of the culture). We want to point out that such a large uncertainty might not be as problematic as it sounds, as variation is a part of the daily business when dealing with biological systems [92]. In this contribution, we demonstrated how a robust qGlc set-point could be developed for any process using viable cell count estimation to calculate a safe feeding rate: in the historical data set we see that this clone almost always consumed at least 10 pg/ch to maintain metabolic integrity (Figure 7D). We exemplarily calculated a target set-point on the basis of the maximum error from biomass underestimation (Table 4) so, even with a high error on biomass, the culture will be supplied with sufficient nutrients in a glucose-limited fed-batch.

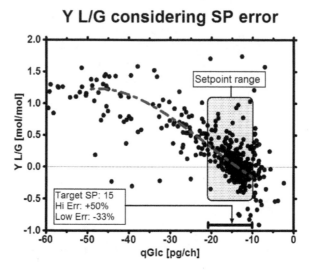

Figure 10. Metabolic control window with a suggested robust qGlc set-point. Data points represent both historical and experimental data. Depending on the task, the set-point range may be selected either conservatively (better overfeeding than starving, higher qGlc) or more courageously (improved metabolic state at the cost of reduced growth, lower qGlc). In general, the error of the online biomass estimation method may serve as a good first estimate for the range. It must be noted that the cells might not truly consume at a high qGlc just because it is desired by the operator and fed in this way, especially in the mid-phase of the culture—a fact which we suggest to solve by applying a high pH.

Table 4. Consequences of a high error for a robust set-point (SP) selection. To prevent starvation of the culture, the lower set-point range must be high enough to compensate for a possible underestimation of biomass.

	SP	*Lower Range SP*	*Upper Range SP*
Error	0%	−33%	+50%
Desired set-point considering error	10	6.7	15
Robust target set-point	**14.9**	**10**	**22.4**

4.5. Limitations and Suggested Improvements

The range for qGlc enabled the introduction of a safety margin, which may be set up according to the error of the biomass estimation if certain conditions are fulfilled (*i.e.*, on-line available metabolic results *or* on-line pH). Especially in the case of a signal as simple as pH, of course not all is well. Especially in a buffered system, pH hardly changes due to cellular activity early in the process, but rather, because of process events, pH probes are known to drift. Therefore, other signals which may replace pH as the set-point switch are, for instance, capacitance, conductivity, pO_2, pCO_2, offgas, mass spectroscopy (MS), turbidity, base, inline microscopy, heat exchange, fluorescence, infrared, metabolic ratios, mass flow controller, online analyzer and many more [35,93]. If expensive high-tech equipment is not available, the controller output from proportional-integrative-derivative (PID) control [94–96] of suitable devices might pose an interesting alternative. A change in metabolic activity might be even better detected by using derivatives [2] or otherwise modified raw signals in combination with a robust signal-processing algorithm which may solve the possible problem with outliers very elegantly [55,97–99].

4.6. Summary and Outlook

The goal of the study was to reduce lactic acid accumulation in fed-batch fermentations in process development to increase overall process performance. We developed an adaptive feeding strategy which was based on real-time signals, capacitance and pH, and found that an on-line analyzer was not necessary for our intended control approach in glucose limitation. We have shown for the first time that the deliberate decrease of pH [26,29,32], which impairs lactic acid production but also cellular proliferation, was not necessary to stop lactic acid accumulation in the bioreactor, if the metabolic state could be controlled with a stoichiometric feeding strategy in which uncertainty is taken into account by design. As high lactic acid levels no longer posed a threat to the culture, a higher glucose consumption rate, favored, *i.e.*, by a high pH set-point, may boost productivity [81]. Experimental demonstration furthermore led to a more consistent pH and glucose profile, which might increase product consistency, as was reported by other authors [32,100,101], or simply process robustness. The uncertainty of biomass estimation was solved by designing a range of uncertainty-based feeding set-points, which may switch the feed rate in real time when combined with the previously suggested on-line-accessible parameters. The basis of our methodology, using specific uptake rates, is independent of scale, location and initial conditions. Therefore, the simplicity of the suggested method allowed a transfer of the desired *metabolic state* between piloting and manufacturing, which fits perfectly in the context of Process Analytical Technology (PAT) and Quality by Design (QbD) with respect to an improvement of robustness [1,102–104].

A process run under such conditions from start to end holds a few yet unmentioned benefits in terms of cost of goods and productivity [105]: Healthy cultures with high viability are likely to create less cell debris because of the lower dead cell count, and the here-proposed feeding strategy is likely to result in a low residual substrate concentration at the end of the fermentation, which may potentially improve harvest efficiency. This naturally has an impact on subsequent unit operations [106], which, for example, include time-intensive separation, washing and purification steps [92,107] (Figure 11). We therefore hypothesize that the proposed strategy may be suitable to increase plant productivity and costs in the unit operations after the actual fermentation [108], which leads to leaner processes with a shorter turnover time and result in an overall improvement in productivity.

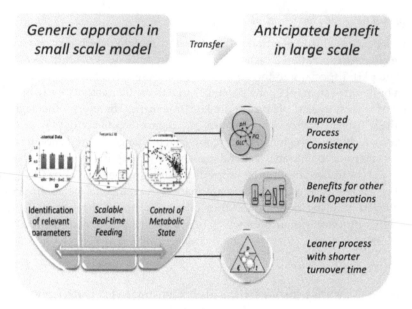

Figure 11. Generic approach for the holistic improvement during mammalian cell culture development.

Acknowledgments: We sincerely acknowledge Civan Yagtu for assistance in the lab, Patrick Wechselberger for providing technical support (Lucullus PIMS) also outside working hours, Christian Dietzsch for configuration of the online analyzer (CubianXC) and Hans Lohninger from the Vienna University of Technology for kindly providing us with a full version of Datalab [54] which was used to generate the model to estimate biomass for our feeding strategy.

Author Contributions: Viktor Konakovsky designed, planned and performed the experiments at the Vienna University of Technology, analyzed the data, performed statistical analysis, prepared tables, Figures, additional files, drafted and wrote the manuscript. Christoph Clemens gave valuable inputs while the manuscript was drafted. Martina Berger, Markus Michael Müller, Christoph Clemens and Jan Bechmann contributed the largest part of the data, helped with critical discussions and gave valuable support also for advanced feeding control. Christoph Herwig passed on personal and professional experience, especially in automation, logic, and structure, helped drafting the manuscript and supported training activities during the project. Christoph Herwig and Stefan Schlatter supervised research and conceived the study. All authors read and approved the final manuscript. Partial financial support from the CATMAT Doctorate College of the Vienna University of Technology is gratefully acknowledged.

Conflicts of Interest: The authors declare no conflict of interest.

Abbreviations

The following abbreviations are used in this manuscript:

CHO: Chinese Hamster Ovary

CHO-DG44: Clone B is derived from this cell line

Clone A: Clone in use for a biomass model based on capacitance

Clone B: Clone in this contribution (historical and experimental data)

CVRMSE: Coefficient of variation of RMSE (Root Mean Square Error)

DoE: Design of Experiments

Feed concentration, $_{Substrate}$: Here: Glucose substrate concentration in the feed [mg/mL]

GLC: Glucose concentration [g/L]

GS-CHO: Glutamine-Synthetase CHO

H^+: Hydrogen ion concentration [mol/L]

Hist: Historical runs from process development

HK: Hexokinase

IVCC: Integrated viable cell concentration [c/mL]

Km: Clone-dependent Monod constant for glucose affinity (0.1–1 g/L) [g/L]

LAC: Lactic acid concentration [g/L]

mAbs: Monoclonal Antibodies

MVDA: Multivariate Data Analysis

N: Number of observations

OD: Optical density [-]

PAT: Process Analytical Technology

pg/ch: Picogram per Cell per hour

PFK: 6-Phosphofructo-1-kinase

PLS: Partial Least Squares

PLS-R: Partial Least Squares Regression

Q^2: Variable prediction coefficient

QbD: Quality by Design

qGlc: Specific glucose uptake rate [pg/ch]

qLac: Specific lactic acid uptake/production rate [pg/ch]

qOUR: Specific oxygen uptake rate [mmol/ch]

qs adapted: New qs set-point after online control action (*i.e.*, pH or online analyzer)

qs: General specific uptake rate notation, identical with qGlc as s stands for glucose [pg/ch]

R^2: Regression Coefficient [-]

R-30: Experiment R-30, broad range for target qGlc set-point

R-31: Experiment R-31, tight range for target qGlc set-point

rlowess: Robust local regression using weighted linear least squares

SP: Set-point

SD: Standard Deviation

$V_{Reactorx}$: Volume of the reactor [mL]

VCC: Viable Cell Concentration [cells/mL]

VIP: Variance importance of the projection, a measure of relevance of the parameter

Y GLC/OX: Yield glucose per oxygen [mol/mol]

Y L/G: Yield lactic acid per glucose [mol/mol]

Y P/G: Yield product per glucose [mg/g]

Y X/GLC: Yield biomass per glucose [cells/g]

Y X/GLN: Yield biomass per glutamine [cells/g]

Appendix A. The Importance of Glucose

A high glucose concentration was not necessarily required to keep the culture well, as was shown in this contribution. Often, cell cultures in manufacturing and process development are operated "between 1 and 6 g/L" without further ado. One reason for this lies in the natural fear to lose a culture if the cells should be exposed to substrate limitations and starve, for instance because a stoichiometric feeding regime did not work as expected. Another reason for running at a high target glucose concentration range may have to do with glycosylation patterns of the final product which may be influenced by GLC levels [109]. However, the intracellular flux and concentrations might surpass the extracellular concentration of substrates in importance for the final structure of a complex protein [110]. Finally, some clones do not grow well below a certain glucose threshold and a limitation strategy is not the right approach for success; however, in this study, it worked very well.

To understand why GLC was often not a crucial parameter for lactic acid metabolism, it is necessary to take a look at the raw data. All historical runs started off with a remarkably high qGlc, which, over the course of a fed-batch fermentation, wore off and approached a certain plateau. In other words, qGlc (Figure 4A) was not solely governed by GLC as often assumed in microbial cultivations and appeared to be somewhat insensitive to a *high* GLC presence (see Figure A1) late in the culture, which means that even concentrated bolus shots would not change qGlc at this time point. GLC levels in cell culture were traditionally almost always operated way above the Monod constant Km, which depends on the genetic traits of the clone and may range between 0.1–1 g/L [111].

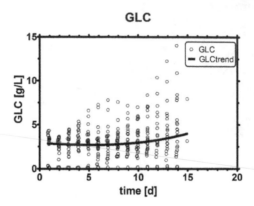

Figure A1. Historical GLC profiles. The GLC level itself was often inadequate to describe metabolic effects, such as the relation to qGlc in fed-batches. A robust third-order polynomial indicated an average trend of all runs, regardless of the individual trend in each cultivation.

Appendix B. Coefficients in Multivariate Data Analysis (MVDA)

The importance of parameters can be obtained by looking at the coefficients of the principal components, which also shows their effect on the target variable. (Here: qLac; positive coefficients indicate a direct correlation, while negative coefficients indicate inverse correlation. All coefficients were negative in this analysis). For their variable importance ranking, the coefficients are sorted regardless of their effect on the target parameter with decreasing importance, and this can be visualized by the VIP plot. In this dataset, the coefficients for glucose concentration (GLC) made apparently no sense. A negative coefficient would imply that a low glucose concentration was related to a high lactic acid production rate. GLC was almost always very high (>>>Km of glucose of the clone) and this clone was capable of consuming lactic acid even when there was plenty of glucose available. As a consequence, this behavior was captured in the statistical analysis and finally led to negative coefficients for glucose in this data set.

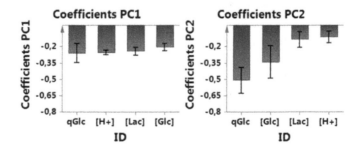

Figure A2. MVDA coefficients of PC1 and PC2 capturing direction of parameter effect on the target variable (qLac). PC1 captures most variance in the following order: qGlc, H+, Lac, GLC, while PC2 captures some remaining variance in qGlc and Glc but not Lac and H+ as they are already well represented by PC1.

Appendix C. Viability

R-31 shows very high viability of the remaining cell population at the end of 14 days, while at the same time the total viable cell count was decreasing. A large portion of cells were dying in a much more different way than in the other process development bioprocesses. Whether this should be traced back to the difference in pH or the metabolic state could not be investigated in more detail in this contribution.

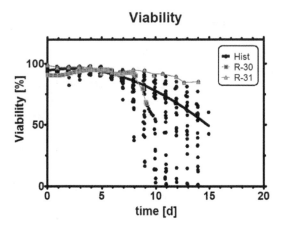

Figure A3. Viability trend in historical data, R-30 and R-31. The latter run had the highest viability at the process end. R-31 was tightly controlled throughout the process time in terms of metabolic state and featured a pH change from 7.4 to 7.2 between days 9 and 10.

Notes: All figures were optimized for printing; however, they are also made available in color in the supplementary materials of this journal. We encourage the reader to take a look at them as tiny differences inside the paper become much clearer when looking at the figures in full scale at a high resolution.

References

1. Rathore, A.S.; Mhatre, R. *Quality by Design for Biopharmaceuticals: Principles and Case Studies*; Auflage: 1. Wiley-Interscience: Hoboken, NJ, USA, 2011.

2. Charaniya, S.P. Systems Analysis of Complex Biological Data for Bioprocess Enhancement. Ph.D. Dissertation, University of Minnesota, Minneapolis, MN, USA, 2008.

3. Croughan, M.S.; Freund, N.W. Strategy to Reduce Lactic Acid Production and Control PH in Animal Cell Culture. U.S. Patent US8470552 B2, 25 June 2013. Available online: http://www.google.com/patents/US8470552 (accessed on 6 January 2016).

4. Le, H.; Kabbur, S.; Pollastrini, L.; Sun, Z.; Mills, K.; Johnson, K.; Karypis, G.; Hu, W.-S. Multivariate analysis of cell culture bioprocess data—Lactate consumption as process indicator. *J. Biotechnol.* **2012**, *162*, 210–223. [CrossRef] [PubMed]

5. Le, H.; Castro-Melchor, M.; Hakemeyer, C.; Jung, C.; Szperalski, B.; Karypis, G.; Hu, W.-S. Discerning key parameters influencing high productivity and quality through recognition of patterns in process data. *BMC Proc.* **2011**, *5*. [CrossRef] [PubMed]

6. Zhou, M.; Crawford, Y.; Ng, D.; Tung, J.; Pynn, A.F.J.; Meier, A.; Yuk, I.H.; Vijayasankaran, N.; Leach, K.; Joly, J.; *et al.* Decreasing lactate level and increasing antibody production in Chinese Hamster Ovary cells (CHO) by reducing the expression of lactate dehydrogenase and pyruvate dehydrogenase kinases. *J. Biotechnol.* **2011**, *153*, 27–34. [CrossRef] [PubMed]

7. Hu, W.-S.; Kantardjieff, A.; Mulukutla, B.C. Cell Lines that Overexpress Lacatate Dehydrogenase c. Patent WO2012075124 A2, 7 June 2012.

8. Brown, N.J.; Higham, S.E.; Perunovic, B.; Arafa, M.; Balasubramanian, S.; Rehman, I. Lactate Dehydrogenase-B Is Silenced by Promoter Methylation in a High Frequency of Human Breast Cancers. *PLoS ONE* **2013**, *8*. [CrossRef] [PubMed]

9. Legmann, R.; Melito, J.; Belzer, I.; Ferrick, D. Analysis of glycolytic flux as a rapid screen to identify low lactate producing CHO cell lines with desirable monoclonal antibody yield and glycan profile. *BMC Proc.* **2011**. [CrossRef] [PubMed]

10. Ma, N.; Ellet, J.; Okediadi, C.; Hermes, P.; McCormick, E.; Casnocha, S. A single nutrient feed supports both chemically defined NS0 and CHO fed-batch processes: Improved productivity and lactate metabolism. *Biotechnol. Prog.* **2009**, *25*, 1353–1363. [CrossRef] [PubMed]

11. Altamirano, C.; Paredes, C.; Illanes, A.; Cairó, J.J.; Gòdia, F. Strategies for fed-batch cultivation of t-PA producing CHO cells: Substitution of glucose and glutamine and rational design of culture medium. *J. Biotechnol.* **2004**, *110*, 171–179. [CrossRef] [PubMed]

12. Wahrheit, J.; Nicolae, A.; Heinzle, E. Dynamics of growth and metabolism controlled by glutamine availability in Chinese hamster ovary cells. *Appl. Microbiol. Biotechnol.* **2014**, *98*, 1771–1783. [CrossRef] [PubMed]

13. Kim, S.H.; Lee, G.M. Down-regulation of lactate dehydrogenase-A by siRNAs for reduced lactic acid formation of Chinese hamster ovary cells producing thrombopoietin. *Appl. Microbiol. Biotechnol.* **2007**, *74*, 152–159. [CrossRef] [PubMed]

14. Müller, D.; Katinger, H.; Grillari, J. MicroRNAs as targets for engineering of CHO cell factories. *Trends Biotechnol.* **2008**, *26*, 359–365. [CrossRef] [PubMed]

15. Luan, Y.-T.; Stanek, T.C.; Drapeau, D. Controlling Lactic Acid Production in Fed-Batch Cell Cultures via Variation in Glucose Concentration; Bioreactors and Heterologous Gene Expression. U.S. Patent US7429491 B2, 30 September 2008. Available online: http://www.google.com/patents/US7429491 (accessed on 6 January 2016).

16. Drapeau, D.; Luan, Y.-T.; Stanek, T.C. Restricted Glucose Feed for Animal Cell Culture. Patent WO2004104186 A1, 2 December 2004. Available online: http://www.google.com/patents/WO2004104186A1 (accessed on 6 January 2016).

17. Basch, J.O.; Gangloff, S.; Joosten, C.E.; Kothari, D.; Lee, S.S.; Leister, K.; Matlock, L.; Sakhamuri, S.; Schilling, B.M.; Zegarelli, S.G. Product Quality Enhancement in Mammalian Cell Culture Processes for Protein Production. Patent WO2004058944 A2, 15 July 2004. Available online: http://www.google.com/patents/WO2004058944A2 (accessed on 6 January 2016).

18. Sauer, P.W.; Burky, J.E.; Wesson, M.C.; Sternard, H.D.; Qu, L. A high-yielding, generic fed-batch cell culture process for production of recombinant antibodies. *Biotechnol. Bioeng.* **2000**, *67*, 585–597. [CrossRef]

19. Templeton, N.; Dean, J.; Reddy, P.; Young, J.D. Peak antibody production is associated with increased oxidative metabolism in an industrially relevant fed-batch CHO cell culture. *Biotechnol. Bioeng.* **2013**, *110*, 2013–2024. [CrossRef] [PubMed]

20. Mulukutla, B.C.; Khan, S.; Lange, A.; Hu, W.-S. Glucose metabolism in mammalian cell culture: New insights for tweaking vintage pathways. *Trends Biotechnol.* **2010**, *28*, 476–484. [CrossRef] [PubMed]

21. Yongky, A.; Lee, J.; Le, T.; Mulukutla, B.C.; Daoutidis, P.; Hu, W.-S. Mechanism for multiplicity of steady states with distinct cell concentration in continuous culture of mammalian cells. *Biotechnol. Bioeng.* **2015**, *112*, 1437–1445. [CrossRef] [PubMed]

22. Mulukutla, B.C.; Yongky, A.; Daoutidis, P.; Hu, W.-S. Bistability in Glycolysis Pathway as a Physiological Switch in Energy Metabolism. *PLoS ONE* **2014**, *9*. [CrossRef] [PubMed]

23. Wahrheit, J.; Niklas, J.; Heinzle, E. Metabolic control at the cytosol-mitochondria interface in different growth phases of CHO cells. *Metab. Eng.* **2014**, *23*, 9–21. [CrossRef] [PubMed]

24. Miller, W.M.; Blanch, H.W.; Wilke, C.R. A kinetic analysis of hybridoma growth and metabolism in batch and continuous suspension culture: Effect of nutrient concentration, dilution rate, and pH. *Biotechnol. Bioeng.* **1988**, *32*, 947–965. [CrossRef] [PubMed]

25. Ivarsson, M.; Noh, H.; Morbidelli, M.; Soos, M. Insights into pH-induced metabolic switch by flux balance analysis. *Biotechnol. Prog.* **2015**, *31*, 347–357. [CrossRef] [PubMed]

26. Trummer, E.; Fauland, K.; Seidinger, S.; Schriebl, K.; Lattenmayer, C.; Kunert, R.; Vorauer-Uhl, K.; Weik, R.; Borth, N.; Katinger, H.; *et al.* Process parameter shifting: Part I. Effect of DOT, pH, and temperature on the performance of Epo-Fc expressing CHO cells cultivated in controlled batch bioreactors. *Biotechnol. Bioeng.* **2006**, *94*, 1033–1044. [CrossRef] [PubMed]

27. Osman, J.J.; Birch, J.; Varley, J. The response of GS-NS0 myeloma cells to pH shifts and pH perturbations. *Biotechnol. Bioeng.* **2001**, *75*, 63–73. [CrossRef] [PubMed]

28. Mulukutla, B.C.; Gramer, M.; Hu, W.-S. On metabolic shift to lactate consumption in fed-batch culture of mammalian cells. *Metab. Eng.* **2012**, *14*, 138–149. [CrossRef] [PubMed]

29. Joosten, C.E.; Leist, C.; Schmidt, J. Cell Cultivation Process. U.S. Patent US8765413 B2, 1 July 2014. Available online: http://www.google.com/patents/US8765413 (accessed on 6 January 2016).

30. Ozturk, S.S.; Palsson, B.O. Growth, metabolic, and antibody production kinetics of hybridoma cell culture: 2. Effects of serum concentration, dissolved oxygen concentration, and medium pH in a batch reactor. *Biotechnol. Prog.* **1991**, *7*, 481–494. [CrossRef] [PubMed]

31. Patel, S.D.; Papoutsakis, E.T.; Winter, J.N.; Miller, W.M. The lactate issue revisited: Novel feeding protocols to examine inhibition of cell proliferation and glucose metabolism in hematopoietic cell cultures. *Biotechnol. Prog.* **2000**, *16*, 885–892. [CrossRef] [PubMed]

32. Ivarsson, M.; Villiger, T.K.; Morbidelli, M.; Soos, M. Evaluating the impact of cell culture process parameters on monoclonal antibody N-glycosylation. *J. Biotechnol.* **2014**, *188*, 88–96. [CrossRef] [PubMed]

33. L'Allemain, G.; Paris, S.; Pouysségur, J. Growth factor action and intracellular pH regulation in fibroblasts. Evidence for a major role of the Na^+/H^+ antiport. *J. Biol. Chem.* **1984**, *259*, 5809–5815. [PubMed]

34. Rathore, A.S.; Bhambure, R.; Ghare, V. Process analytical technology (PAT) for biopharmaceutical products. *Anal. Bioanal. Chem.* **2010**, *398*, 137–154. [CrossRef] [PubMed]

35. Luttmann, R.; Bracewell, D.G.; Cornelissen, G.; Gernaey, K.V.; Glassey, J.; Hass, V.C.; Kaiser, C.; Preusse, C.; Striedner, G.; Mandenius, C.-F. Soft sensors in bioprocessing: A status report and recommendations. *Biotechnol. J.* **2012**, *7*, 1040–1048. [CrossRef] [PubMed]

36. Wechselberger, P.; Sagmeister, P.; Herwig, C. Real-time estimation of biomass and specific growth rate in physiologically variable recombinant fed-batch processes. *Bioprocess Biosyst. Eng.* **2013**, *36*, 1205–1218. [CrossRef] [PubMed]

37. Lu, F.; Toh, P.C.; Burnett, I.; Li, F.; Hudson, T.; Amanullah, A.; Li, J. Automated dynamic fed-batch process and media optimization for high productivity cell culture process development. *Biotechnol. Bioeng.* **2013**, *110*, 191–205. [CrossRef] [PubMed]

38. Ozturk, S.S.; Thrift, J.C.; Blackie, J.D.; Naveh, D. Real-time monitoring and control of glucose and lactate concentrations in a mammalian cell perfusion reactor. *Biotechnol. Bioeng.* **1997**, *53*, 372–378. [CrossRef]

39. Franze, R.; Link, T.; Takuma, S.; Takagi, Y.; Hirashima, C.; Tsuda, Y. Method for the Production of a Glycosylated Immunoglobulin. U.S. Patent US20110117087 A1, 19 May 2011. Available online: https://www.google.com/patents/US20110117087 (accessed on 6 January 2016).

40. Lenas, P.; Kitade, T.; Watanabe, H.; Honda, H.; Kobayashi, T. Adaptive fuzzy control of nutrients concentration in fed-batch culture of mammalian cells. *Cytotechnology* **1997**, *25*, 9–15. [CrossRef] [PubMed]

41. Luan, Y.; Stanek, T.C.; Drapeau, D. Restricted Glucose Feed for Animal Cell Culture. U.S. Patent US7429491. Available online: http://www.sumobrain.com/patents/us/Restricted-glucose-feed-animal-cell/US7429491.html (accessed on 6 January 2016).

42. Gagnon, M.; Hiller, G.; Luan, Y.-T.; Kittredge, A.; DeFelice, J.; Drapeau, D. High-End pH-controlled delivery of glucose effectively suppresses lactate accumulation in CHO Fed-batch cultures. *Biotechnol. Bioeng.* **2011**, *108*, 1328–1337. [CrossRef] [PubMed]

43. KWlaschin, F.; Hu, W.-S. Fedbatch culture and dynamic nutrient feeding. *Adv. Biochem. Eng. Biotechnol.* **2006**, *101*, 43–74.

44. Aehle, M.; Schaepe, S.; Kuprijanov, A.; Simutis, R.; Lübbert, A. Simple and efficient control of CHO cell cultures. *J. Biotechnol.* **2011**, *153*, 56–61. [CrossRef] [PubMed]

45. Lin, H.; Bezaire, J. Pre-Programmed Non-Feedback Controlled Continuous Feeding of Cell Cultures. Patent WO2013040444 A1, 21 March 2013.

46. Zhou, W.; Hu, W.-S. On-line characterization of a hybridoma cell culture process. *Biotechnol. Bioeng.* **1994**, *44*, 170–177. [CrossRef] [PubMed]

47. Noll, T.; Biselli, M. Dielectric spectroscopy in the cultivation of suspended and immobilized hybridoma cells. *J. Biotechnol.* **1998**, *63*, 187–198. [CrossRef]

48. JDowd, E.; Jubb, A.; Kwok, K.E.; Piret, J.M. Optimization and control of perfusion cultures using a viable cell probe and cell specific perfusion rates. *Cytotechnology* **2003**, *42*, 35–45.

49. Zhou, W.; Rehm, J.; Hu, W.-S. High viable cell concentration fed-batch cultures of hybridoma cells through on-line nutrient feeding. *Biotechnol. Bioeng.* **1995**, *46*, 579–587. [CrossRef] [PubMed]

50. Europa, A.F.; Gambhir, A.; Fu, P.-C.; Hu, W.-S. Multiple steady states with distinct cellular metabolism in continuous culture of mammalian cells. *Biotechnol. Bioeng.* **2000**, *67*, 25–34. [CrossRef]

51. Mi, L.; Feng, Q.; Li, L.; Wang, X.H. Method for Parameter Control of the Process for Culturing Serum-Suspension Free Animal Cell. Patent CN1557948 A, 29 December 2004. Available online: http://www.google.com/patents/CN1557948A (accessed on 6 January 2016).

52. Konakovsky, V.; Yagtu, A.C.; Clemens, C.; Müller, M.M.; Berger, M.; Schlatter, S.; Herwig, C. Universal Capacitance Model for Real-Time Biomass in Cell Culture. *Sensors* **2015**, *15*, 22128–22150. [CrossRef] [PubMed]

53. Dietzsch, C.; Spadiut, O.; Herwig, C. On-line multiple component analysis for efficient quantitative bioprocess development. *J. Biotechnol.* **2013**, *163*, 362–370. [CrossRef] [PubMed]

54. Lohninger, H. Datalab 3.5, A Programme for Statistical Analysis.   2000.   Available online: http://datalab.epina.at.

55. MATLAB, Inc. Robust Local Regression Using Weighted Linear Least Squares in Matlab (RLOWESS). Available online: http://de.mathworks.com/help/curvefit/smoothing-data.html#bq_6ys3-3 (accessed on 6 January 2016).

56. Jobé, A.M.; Herwig, C.; Surzyn, M.; Walker, B.; Marison, I.; von Stockar, U. Generally applicable fed-batch culture concept based on the detection of metabolic state by on-line balancing. *Biotechnol. Bioeng.* **2003**, *82*, 627–639. [CrossRef] [PubMed]

57. Wold, S.; Sjöström, M.; Eriksson, L. PLS-regression: A basic tool of chemometrics. *Chemom. Intell. Lab. Syst.* **2001**, *58*, 109–130. [CrossRef]

58. Glassey, J.; Gernaey, K.V.; Clemens, C.; Schulz, T.W.; Oliveira, R.; Striedner, G.; Mandenius, C.-F. Process analytical technology (PAT) for biopharmaceuticals. *Biotechnol. J.* **2011**, *6*, 369–377. [CrossRef] [PubMed]

59. Haenlein, M.; Kaplan, A.M. A Beginner's Guide to Partial Least Squares Analysis. *Underst. Stat.* **2004**, *3*, 283–297. [CrossRef]

60. Halestrap, A.P.; Price, N.T. The proton-linked monocarboxylate transporter (MCT) family: structure, function and regulation. *Biochem. J.* **1999**, *343*, 281–299. [CrossRef] [PubMed]

61. Smerilli, M.; Neureiter, M.; Wurz, S.; Haas, C.; Frühauf, S.; Fuchs, W. Direct fermentation of potato starch and potato residues to lactic acid by Geobacillus stearothermophilus under non-sterile conditions. *J. Chem. Technol. Biotechnol.* **2015**, *90*, 648–657. [CrossRef] [PubMed]

62. Kurano, N.; Leist, C.; Messi, F.; Kurano, S.; Fiechter, A. Growth behavior of Chinese hamster ovary cells in a compact loop bioreactor: 1. Effects of physical and chemical environments. *J. Biotechnol.* **1990**, *15*, 101–111. [CrossRef]

63. Ansorge, S.; Esteban, G.; Ghommidh, C.; Schmid, G. Monitoring Nutrient Limitations by Online Capacitance Measurements in Batch & Fed-batch CHO Fermentations. In *Cell Technology for Cell Products*; Smith, R., Ed.; Springer: Houten, Netherlands, 2007; pp. 723–726.

64. Yongky, A. Analysis of Central Metabolic Pathways in Cultured Mammalian Cells. Ph.D. Dissertation, University of Minnesota, Minneapolis, MN, USA, 2014.

65. Ündey, C.; Ertunç, S.; Mistretta, T.; Looze, B. Applied advanced process analytics in biopharmaceutical manufacturing: Challenges and prospects in real-time monitoring and control. *J. Process Control* **2010**, *20*, 1009–1018. [CrossRef]

66. Spadiut, O.; Rittmann, S.; Dietzsch, C.; Herwig, C. Dynamic process conditions in bioprocess development. *Eng. Life Sci.* **2013**, *13*, 88–101. [CrossRef]

67. Kantardjieff, A. Transcriptome Analysis in Mammalian Cell Culture: Applications in Process Development and Characterization. Ph.D. Dissertation, University of Minnesota, Minneapolis, MN, USA, 2009.

68. Fadok, V.A.; Bratton, D.L.; Guthrie, L.; Henson, P.M. Differential effects of apoptotic *versus* lysed cells on macrophage production of cytokines: role of proteases. *J. Immunol. (Baltim. Md.: 1950)* **2001**, *166*, 6847–6854. [CrossRef]

69. Rathmell, J.C.; Fox, C.J.; Plas, D.R.; Hammerman, P.S.; Cinalli, R.M.; Thompson, C.B. Akt-directed glucose metabolism can prevent Bax conformation change and promote growth factor-independent survival. *Mol. Cell. Biol.* **2003**, *23*, 7315–7328. [CrossRef] [PubMed]

70. Zheng, L.; Dengler, T.J.; Kluger, M.S.; Madge, L.A.; Schechner, J.S.; Maher, S.E.; Pober, J.S.; Bothwell, A.L. Cytoprotection of human umbilical vein endothelial cells against apoptosis and CTL-mediated lysis provided by caspase-resistant Bcl-2 without alterations in growth or activation responses. *J. Immunol. (Baltim. Md.: 1950)* **2000**, *164*, 4665–4671. [CrossRef]

71. Ansorge, S.; Esteban, G.; Schmid, G. On-line monitoring of responses to nutrient feed additions by multi-frequency permittivity measurements in fed-batch cultivations of CHO cells. *Cytotechnology* **2010**, *62*, 121–132. [CrossRef] [PubMed]

72. Ansorge, S.; Esteban, G.; Schmid, G. On-line monitoring of infected Sf-9 insect cell cultures by scanning permittivity measurements and comparison with off-line biovolume measurements. *Cytotechnology* **2007**, *55*, 115–124. [CrossRef] [PubMed]

73. Dabros, M.; Dennewald, D.; Currie, D.J.; Lee, M.H.; Todd, R.W.; Marison, I.W.; Stockar, U. Cole–Cole, linear and multivariate modeling of capacitance data for on-line monitoring of biomass. *Bioprocess Biosyst. Eng.* **2008**, *32*, 161–173. [CrossRef] [PubMed]

74. Cannizzaro, C.; Gügerli, R.; Marison, I.; von Stockar, U. On-line biomass monitoring of CHO perfusion culture with scanning dielectric spectroscopy. *Biotechnol. Bioeng.* **2003**, *84*, 597–610. [CrossRef] [PubMed]

75. Carvell, J.P.; Dowd, J.E. On-line Measurements and Control of Viable Cell Density in Cell Culture Manufacturing Processes using Radio-frequency Impedance. *Cytotechnology* **2006**, *50*, 35–48. [CrossRef] [PubMed]

76. De Jesus, M.J.; Bourgeois, M.; Baumgartner, G.; Tromba, P.; Jordan, D.M.; Amstutz, M.H.; Wurm, P.F.M. The Influence of pH on Cell Growth and Specific Productivity of Two CHO Cell Lines Producing Human Anti Rh D IgG. In *Animal Cell Technology: From Target to Market*; Lindner-Olsson, D.E., Chatzissavidou, M.N., Lüllau, D.E., Eds.; Springer: Houten, Netherlands, 2001; pp. 197–199.

77. Gambhir, A.; Korke, R.; Lee, J.; Fu, P.-C.; Europa, A.; Hu, W.-S. Analysis of cellular metabolism of hybridoma cells at distinct physiological states. *J. Biosci. Bioeng.* **2003**, *95*, 317–327. [CrossRef]

78. Eyal, A.M.; Starr, J.N.; Fisher, R.; Hazan, B.; Canari, R.; Witzke, D.R.; Gruber, P.R.; Kolstad, J.J. Lactic Acid Processing; Methods; Arrangements; and, Product. U.S. Patent US6320077 B1, 20 November 2001. Available online: http://www.google.com/patents/US6320077 (accessed on 6 January 2016).

79. Abdel-Rahman, M.A.; Tashiro, Y.; Sonomoto, K. Recent advances in lactic acid production by microbial fermentation processes. *Biotechnol. Adv.* **2013**, *31*, 877–902. [CrossRef] [PubMed]

80. Wu, C.; Huang, J.; Zhou, R. Progress in engineering acid stress resistance of lactic acid bacteria. *Appl. Microbiol. Biotechnol.* **2014**, *98*, 1055–1063. [CrossRef] [PubMed]

81. Klein, T.; Heinzel, N.; Kroll, P.; Brunner, M.; Herwig, C.; Neutsch, L. Quantification of cell lysis during CHO bioprocesses: Impact on cell count, growth kinetics and productivity. *J. Biotechnol.* **2015**, *207*, 67–76. [CrossRef] [PubMed]

82. Boron, W.F. Regulation of intracellular pH. *Adv. Physiol. Educ.* **2004**, *28*, 160–179. [CrossRef] [PubMed]

83. Dechant, R.; Binda, M.; Lee, S.S.; Pelet, S.; Winderickx, J.; Peter, M. Cytosolic pH is a second messenger for glucose and regulates the PKA pathway through V-ATPase. *EMBO J.* **2010**, *29*, 2515–2526. [CrossRef] [PubMed]

84. Boyer, M.J.; Tannock, I.F. Regulation of intracellular pH in tumor cell lines: Influence of microenvironmental conditions. *Cancer Res.* **1992**, *52*, 4441–4447. [PubMed]

85. Olsnes, S.; Tønnessen, T.I.; Sandvig, K. pH-regulated anion antiport in nucleated mammalian cells. *J. Cell Biol.* **1986**, *102*, 967–971. [CrossRef] [PubMed]

86. Pilkis, S.J.; Claus, T.H. Hepatic Gluconeogenesis/Glycolysis: Regulation and Structure/Function Relationships of Substrate Cycle Enzymes. *Annu. Rev. Nutr.* **1991**, *11*, 465–515. [CrossRef] [PubMed]

87. Okar, D.A.; Wu, C.; Lange, A.J. Regulation of the regulatory enzyme, 6-phosphofructo-2-kinase/fructose-2,6-bisphosphatase. *Adv. Enzym. Regul.* **2004**, *44*, 123–154. [CrossRef] [PubMed]

88. Somberg, E.W.; Mehlman, M.A. Regulation of gluconeogenesis and lipogenesis. The regulation of mitochondrial pyruvate metabolism in guinea-pig liver synthesizing precursors for gluconeogenesis. *Biochem. J.* **1969**, *112*, 435–447. [CrossRef] [PubMed]

89. Mulquiney, P.J.; Kuchel, P.W. Model of 2,3-bisphosphoglycerate metabolism in the human erythrocyte based on detailed enzyme kinetic equations: Equations and parameter refinement. *Biochem. J.* **1999**, *342*, 581–596. [CrossRef] [PubMed]

90. Hutkins, R.W.; Nannen, N.L. pH Homeostasis in Lactic Acid Bacteria. *J. Dairy Sci.* **1993**, *76*, 2354–2365. [CrossRef]

91. Nolan, R.P.; Lee, K. Dynamic model of CHO cell metabolism. *Metab. Eng.* **2011**, *13*, 108–124. [CrossRef] [PubMed]

92. Shivhare, M.; McCreath, G. Practical Considerations for DoE Implementation in Quality by Design. *BioProcess Int.* **2010**, 22–30. Available online: http://www.bioprocessintl.com/wp-content/uploads/bpi-content/BPI_A_100806AR03_O_98037a.pdf (accessed on 6 January 2016).

93. Carrondo, M.J.T.; Alves, P.M.; Carinhas, N.; Glassey, J.; Hesse, F.; Merten, O.-W.; Micheletti, M.; Noll, T.; Oliveira, R.; Reichl, U.; *et al.* How can measurement, monitoring, modeling and control advance cell culture in industrial biotechnology? *Biotechnol. J.* **2012**, *7*, 1522–1529. [CrossRef] [PubMed]

94. Oeggerli, A.; Eyer, K.; Heinzle, E. On-line gas analysis in animal cell cultivation: I. Control of dissolved oxygen and pH. *Biotechnol. Bioeng.* **1995**, *45*, 42–53. [CrossRef] [PubMed]

95. Åström, K.J.; Murray, R.M. Feedback Systems Web Site. 5 April 2008. Available online: http://www.cds.caltech.edu/~murray/amwiki. (accessed on 29 May 2015).

96. Dumont, G. EECE Courses Prof. Guy Dumont: Lecture Notes. Available online: http://www.phoneo ximeter.org/ece-courses/eece-460/lecture-notes/ (accessed on 29 May 2015).

97. Press, W.H.; Teukolsky, S.A.; Vetterling, W.T.; Flannery, B.P. *Numerical Recipes 3rd Edition: The Art of Scientific Computing*, 3rd ed.; Cambridge University Press: Cambridge, UK; New York, NY, USA, 2007.

98. Motulsky, H. *Intuitive Biostatistics: A Nonmathematical Guide to Statistical Thinking*; Oxford University Press: New York, NY, USA, 2013.

99. Motulsky, H. *Fitting Models to Biological Data Using Linear and Nonlinear Regression: A Practical Guide to Curve Fitting*; 1. Aufl.; Oxford University Press: New York, NY, USA, 2004.

100. Aghamohseni, H.; Ohadi, K.; Spearman, M.; Krahn, N.; Moo-Young, M.; Scharer, J.M.; Butler, M.; Budman, H.M. Effects of nutrient levels and average culture pH on the glycosylation pattern of camelid-humanized monoclonal antibody. *J. Biotechnol.* **2014**, *186*, 98–109. [CrossRef] [PubMed]

101. Fan, Y.; Del Val Jimenez, I.; Müller, C.; Wagtberg Sen, J.; Rasmussen, S.K.; Kontoravdi, C.; Weilguny, D.; Andersen, M.R. Amino acid and glucose metabolism in fed-batch CHO cell culture affects antibody production and glycosylation. *Biotechnol. Bioeng.* **2015**, *112*, 521–535. [CrossRef] [PubMed]

102. Wechselberger, P.; Herwig, C. Model-based analysis on the relationship of signal quality to real-time extraction of information in bioprocesses. *Biotechnol. Prog.* **2012**, *28*, 265–275. [CrossRef] [PubMed]

103. Zalai, D.; Dietzsch, C.; Herwig, C. Risk-based Process Development of Biosimilars as Part of the Quality by Design Paradigm. *PDA J. Pharm. Sci. Technol. PDA* **2013**, *67*, 569–580. [CrossRef] [PubMed]

104. Gnoth, S.; Jenzsch, M.; Simutis, R.; Lübbert, A. Process Analytical Technology (PAT): Batch-to-batch reproducibility of fermentation processes by robust process operational design and control. *J. Biotechnol.* **2007**, *132*, 180–186. [CrossRef] [PubMed]

105. Li, F.; Vijayasankaran, N.; Shen, A.; Kiss, R.; Amanullah, A. Cell culture processes for monoclonal antibody production. *mAbs* **2010**, *2*, 466–477. [CrossRef] [PubMed]

106. Abu-Absi, S.F.; Yang, L.; Thompson, P.; Jiang, C.; Kandula, S.; Schilling, B.; Shukla, A.A. Defining process design space for monoclonal antibody cell culture. *Biotechnol. Bioeng.* **2010**, *106*, 894–905. [CrossRef] [PubMed]

107. Gronemeyer, P.; Ditz, R.; Strube, J. Trends in Upstream and Downstream Process Development for Antibody Manufacturing. *Bioengineering* **2014**, *1*, 188–212. [CrossRef]

108. Gao, Y.; Kipling, K.; Glassey, J.; Willis, M.; Montague, G.; Zhou, Y.; Titchener-Hooker, N.J. Application of agent-based system for bioprocess description and process improvement. *Biotechnol. Prog.* **2010**, *26*, 706–716. [CrossRef] [PubMed]

109. Green, A.; Glassey, J. Multivariate analysis of the effect of operating conditions on hybridoma cell metabolism and glycosylation of produced antibody: Effect of operating conditions on mAb glycosylation. *J. Chem. Technol. Biotechnol.* **2015**, *90*, 303–313. [CrossRef]

110. Hossler, P.; Mulukutla, B.C.; Hu, W.-S. Systems Analysis of N-Glycan Processing in Mammalian Cells. *PLoS ONE* **2007**, *2*, e713. [CrossRef] [PubMed]

111. Le, H.T.N. Mining High-Dimensional Bioprocess and Gene Expression Data for Enhanced Process Performance. Ph.D. Dissertation, University of Minnesota, Minneapolis, MN, USA, 2012.

12

Optimization and Characterization of Chitosan Enzymolysis by Pepsin

Bi Foua Claude Alain Gohi, Hong-Yan Zeng * and A Dan Pan

Biotechnology Institute, College of Chemical Engineering, Xiangtan University, Xiangtan 411105, Hunan, China; claudefouabi@hotmail.fr (B.F.C.A.G.); jessciapan@hotmail.com (A.D.P.)
* Correspondence: hongyanzeng99@hotmail.com

Academic Editor: Christoph Herwig

Abstract: Pepsin was used to effectively degrade chitosan in order to make it more useful in biotechnological applications. The optimal conditions of enzymolysis were investigated on the basis of the response surface methodology (RSM). The structure of the degraded product was characterized by degree of depolymerization (DD), viscosity, molecular weight, FTIR, UV-VIS, SEM and polydispersity index analyses. The mechanism of chitosan degradation was correlated with cleavage of the glycosidic bond, whereby the chain of chitosan macromolecules was broken into smaller units, resulting in decreasing viscosity. The enzymolysis by pepsin was therefore a potentially applicable technique for the production of low molecular chitosan. Additionally, the substrate degradation kinetics of chitosan were also studied over a range of initial chitosan concentrations (3.0~18.0 g/L) in order to study the characteristics of chitosan degradation. The dependence of the rate of chitosan degradation on the concentration of the chitosan can be described by Haldane's model. In this model, the initial chitosan concentration above which the pepsin undergoes inhibition is inferred theoretically to be about 10.5 g/L.

Keywords: chitosan; enzymolysis; pepsin; response surface methodology; inhibition; kinetic

1. Introduction

Chitosan is a natural polysaccharide, which is widely distributed among living organisms in nature and has been studied extensively in the last few decades. The promising utilization of chitosan in various fields, including medicine, pharmacology and the food industry, is due to the combination of its excellent biological properties, its biocompatibility, biodegradability and low toxicity. However, its limited application in medicine and the food industry is attributed to its high molecular weight, giving it low solubility in aqueous media. These limitations can be overcome by the hydrolysis of chitosan leading to the production of low molecular weight (LMW) chitosan (oligosaccharides) [1–3]. Hydrolysis of chitosan involves physical, chemical and enzymatic dissociation. Compared to the physical and chemical methods, enzymatic degradation has some advantages of specificity, mildness, easy control, no wastewater and easy separation of reactants. Enzymolysis can improve the functional properties of chitosan without affecting either its glucose ring, or its biological activity, while producing high quantities of chitooligosaccharides [1,4–6]. The degree of hydrolysis (DH) is an important parameter for determining the functional properties of LMW chitosan. The LMW chitosan with a free-amine group possesses high solubility in acid-free water and low viscosity. Water-solubility enables efficient modification for medical and agricultural applications [7]. LMW chitosan with a free amine was prepared by enzymolysis. It had an average molecular weight of 18,579 Da and a degree of depolymerization (DD) of 93% [8]. With free amine groups, the LMW chitosan has great solubility, which makes it a good candidate for DNA and drug delivery systems [9,10]. It also shows potential germicidal activities against pathogenic bacteria, yeast and filamentous fungus [11]. Many enzymes with different original specificities, such as cellulase, pectinase, pepsin, papain and lipase, have been reported to have the ability to hydrolyze chitosan [1]. Among these enzymes, pepsin has attracted the most attention from researchers on account of its capability to produce the highest yield of LMW chitosan in enzymolysis [12]. In spite of the merits of the proposed approach, the enzymolysis process of degrading chitosan by pepsin is not fully described in detail in the literature. The enzymolysis process of chitosan degradation by papain however has been exhaustively described in our previous study (Pan et al., 2016) [13].

The aim of this study was to understand and improve the enzymolysis process of degrading chitosan into the LMW chitosan. In order to investigate the relationships between the reaction variables (reaction temperature, initial pH, chitosan and pepsin concentrations) and the response (the degree of hydrolysis (DH)), the optimum enzymolysis condition was achieved using the statistical experimental design called the response surface methodology (RSM) analysis. Furthermore, in order to understand the reaction process and to choose the most suitable technology for chitosan enzymolysis, it was necessary to investigate the enzymolysis kinetics. Based on the results of this investigation, a comparative study will be established between our previous study [13] and this work. It would be useful to provide a scientific approach to a theoretic basis for chitosan enzymolysis with high performance and low consumption.

2. Materials and Methods

2.1. Materials

Chitosan from shrimp shells ($\geq$91% deacetylated) and pepsin (EC 3.4.23.1, 3000–3500 units/mg protein) from porcine gastric mucosa (Amresco type A) were purchased from Sinopharm Chemical Reagent Co., Ltd. (Shanghai, China). All other reagents used were of analytical grade and they were used without further purification. All solutions were made with redistilled and ion-free water.

2.2. Chitosan Hydrolysis

Stock solution of pepsin (1.0 g/L) was obtained by dissolving the enzyme in Tris-HCl buffer (0.1 mol/L, pH 7.0). Stock solution of chitosan (10.0 g/L) was prepared in 0.2 mol/L acetic acid/sodium acetate (1:1, v/v) buffer, pH 4, and diluted into different concentrations for the assays of chitosan hydrolysis. In the preliminary experiment, the standard assay contained 1% w/v. Chitosan ($\geq$91% deacetylated) was dissolved in 100 mM sodium acetate buffer at pH 4, and pepsin was dissolved in an enzyme/substrate ratio of 1:100 (w/w) at 50 °C. It was found that the reaction attained equilibrium after 70 min, so the reaction time for the hydrolysis was set to 70 min. The pepsin solutions were initially added to the buffers at different chitosan concentrations, respectively. The mixed solution was then maintained in a thermostatic water bath at a specific temperature and pH while stirring at 500 rpm for 70 min and heated to 95 °C for 10 min to terminate the reaction. After the reaction, the mixture was cooled to room temperature then centrifuged at 800 rpm for 5 min to remove the enzyme. The supernatant was stored to determine reducing sugars (SRSs) using a total organic carbon analyzer (TOC-5000A, Shimadzu, Japan). The tests were made in triplicate, and the results were recorded as an average. The SRS yield was calculated as follows:

$$\text{SRSs yield (\%)} = (\text{carbon mass of SRSs})/(\text{carbon mass of chitosan}) \times 100\%$$

The response surface methodology (RSM) represents a statistical method that uses quantitative data from an appropriate experimental design to determine and simultaneously solve multivariate equations. The main advantage of RSM is the reduced number of experimental trials needed to evaluate multiple parameters and their interactions.

After approximation of the best conditions by the "one-factor-at-a-time" method in our preliminary experiments, RSM was used to test the effect of initial pH, reaction temperature, pepsin concentration and chitosan concentration on the SRS yield in the chitosan enzymolysis. the

Box–Behnken design (BBD) was chosen for the experiment with four independent variables of initial pH (P), reaction temperature (T, °C), pepsin concentration (E, g/L) and chitosan concentration (C, g/L), while optimizing one response variable, SRS yield (Y), from the enzymolysis process. Each independent variable was coded at three levels between −1 and +1, while the variables p, T, E and C were changed in the ranges shown in Table 1. A set of 29 experiments was augmented with three replications at the design center to evaluate the pure error. The experiments were carried out in a randomized order as required in many design procedures. After reaction, the response Y was measured, and the statistical software package Design Expert (Version 8.0.6) was used for regression analysis of the experimental data and to plot the response surface. Conducting an experiment on the given optimal setting validated the model generated during RSM implementation. The second-order polynomial model was applied to predict the response variable (Y) as shown below,

$$Y = \beta_0 + \sum_{j=1}^{4} \beta_i X_i + \sum_{ij=1}^{4} \beta_{ij} X_i^2 + \sum_{i}^{2} \sum_{j=i+1}^{4} \beta_{ij} X_i X_j \tag{1}$$

where Y is the response value (relative activity) and β_0, β_i, β_{ii} and β_{ij} are the regression coefficients for the interception, linear, quadratic and interaction terms, respectively. X_i and X_j were the independent variables.

Table 1. Experimental range and levels of the independent variables.

Independent Variables	Symbols	Units	Code Levels		
			−1	**0**	**1**
pH	P		2	4	6
Temperature	T	°C	30	50	70
Enzyme concentration	E	mg/L	50	100	150
Chitosan concentration	C	g/L	5.0	10.0	15.0

2.3. Characterization of Chitosan

Under the optimal conditions, chitosan was hydrolyzed by pepsin, and the mixed solution after reaction was concentrated to about one-twentieth with a rotary evaporator under reduced pressure. The mixture was neutralized to pH 9.0 and precipitated by adding ethanol. The precipitate was collected after drying over phosphorus pentoxide in a vacuum to get sample LMW chitosan (LMWC) for structural characterization. IR spectral studies were performed in a Perkin Elmer spectrum 2000 spectrometer (CT, Livonia, MI, USA) under dry air at room temperature using KBr pellets. For chitosan and LMWC (2-mg samples in 100 mg of KBr), 20 mg of the mixture were palletized and subject to IR spectroscopy. Scanning electron micrograph (SEM) was obtained with a JEOL JSM-6700F instrument (Tokyo, Japan). For SEM, 0.5-mL aliquots from the above assay tubes were centrifuged in micro-centrifuge tubes. The pellets obtained were treated with phosphate buffer (pH 7.0, 0.3 M), fixed with glutaraldehyde (1%) for 1 h at 4 °C, further treated with 10%–96% alcohol in a sequential manner then dried [3]. The samples (chitosan and LMWCs solids) were spread on a double-sided conducting adhesive tape pasted to a metallic stub, subjected to gold (100 µg) covering and observed at an accelerating voltage of 20 kV. Room temperature UV-VIS spectra of the chitosan and LMWCs solids were recorded on a SHIMADZU UV-2550 spectrophotometer. UV-VIS spectroscopy was performed on the solution of chitosan perchlorate. A solution of chitosan (10^{-2} g/L) was prepared by adding a stoichiometric amount of 10^{-1} M perchloric acid to a calculated dry weight of chitosan in the solid state. Taking into account the water content and degree of deacetylation of the chitosan, it was then stirred to complete dissolution. Weight-average molecular weight (M_W) was measured by GPC. The GPC equipment consisted of connected columns (TSK G5000-PW and TSK G 3000-PW), a TSP P100 pump and an RI 150 refractive index indicator detector. The eluent was 0.2 M CH_3COOH/0.1 M

CH_3COONa. The eluent and the chitosan sample solution were filtered through 0.45-μm Millipore filters. The flow rate was maintained at 0.1 mL/min. The sample concentration was 0.4 mg/mL, and the standards used to calibrate the column were TOSOH pulman. All data provided by the GPC system were collected and analyzed using the Jiangshen workstation software package. A decrease in the pepsin-catalyzed viscosity of the highly viscous chitosan solution during the two-hour reaction was measured continuously in a Cannon-Fensk (Schott Geraete, Model GMBH—D65719, Mainz, Germany) capillary viscosimeter. The solutions were filtered before determining the viscosity, which was carried out at the lowest shear velocities permitted within the experimental error and the Newtonian plateau. The determination of the depolymerization degree (DD) of chitosan samples was carried out by the linear potentiometric method. This analysis was carried out by dissolving 0.25 g of chitosan in 20 mL of HCl solution, 0.1 N, then filling it up to 100 mL with distilled water. Titration was performed until the chitosan solution reached a pH of approximately 6.5 (range of chitosan non-protonation). Concerning the polydispersity index study, the aqueous solution of sodium alginate (0.1% w/v) was sprayed into the chitosan solution obtained after 1 h and 2 h of hydrolysis (0.1% w/v) containing Pluronic F-68 (0.5% w/v) under continuous magnetic stirring at 1000 rpm for 30 min. Nanoparticles were formed as a result of the interaction between the negative groups of sodium alginate and the positively-charged amino groups of chitosan (ionic gelation). Nanoparticles were collected by centrifugation (REMI high speed, cooling centrifuge, REMI Corp., Mumbai, India) at 18,000 rpm for 30 min at 4 °C. For the particle size and size distribution study, these nanoparticles were redispersed in 5 mL of HPLC grade water. The sample volume used for the analysis was kept constant, i.e., 5 mL to nullify the effect of stray radiations from sample to sample. Studies were carried out in triplicate ($n = 3$), and the standard deviation (SD) was recorded.

3. Results and Discussion

3.1. Enzymolysis of Chitosan by RSM

3.1.1. Box–Behnken Design Analysis

The hydrolysis yield experiment was conducted using the Box–Behnken design, and the results are presented in Table 2. Statistical analysis of variance (ANOVA) was performed in order to investigate not only the fitness and significance of the model, but also the effects of the individual variables and interaction effects on the response. It was also noticed in this study that the SSR yield (Y) was higher within the first 60 min of hydrolysis. This was attributed to the accessibility of the pepsin active sites to the glycosidic bonds of chitosan. According to the ANOVA results (Table 3), the model is highly significant, so it has an important effect on the SSR yield (Y) with a p-value of less than 0.0001 to predict the response values. In terms of the significant coefficients, the independent variables p (initial pH), T (reaction temperature) and E (pepsin concentration) were highly significant terms ($p \leqslant 0.0001$); therefore, pH P affects the solubility of chitosan before all other reactants, while pepsin through its concentration E under control of temperature T proceeds to the hydrolysis of chitosan. In terms of interaction, the terms of PE, TE and EC ($p < 0.05$) were significant terms influencing SSR yield (Y). It is quite remarkable that all of the significant interaction terms contain the independent variables E. E interacts with all other reaction components, implying that the process of hydrolysis does in fact depend on E. All of the quadratic terms of the P^2, T^2, E^2 and C^2 were highly significant terms ($p < 0.0001$). The elimination of the insignificant terms could improve the regression model, and the quadratic model was given as:

$$Y = 91.23 + 6.15\,P - 7.30\,T + 10.84\,E - 4.97\,PE + 6.22\,TE + 7.19\,EC - 32.14\,P^2$$
$$- 28.96\,T^2 - 28.60\,E^2 - 32.24\,C^2 \tag{2}$$

Equation (2) is in terms of the coded factors.

Equation (2) confirms that those linear and interaction terms were significant in affecting the SSR yield. The positive coefficients (+6.15 and +10.84) of P (initial pH) and E (pepsin concentration) in Equation (2) signify a linear effect of increasing from P 2–4.2 and E 50–103.76 mg/L on the SSR yield (Y) and then reaching equilibrium when the P to E further increased. The negative coefficient (-7.30) of T (reaction temperature) in Equation (2) indicates a linear effect of decreasing when T goes over 50 °C on the SSR yield. Moreover, the interaction coefficient of PE in Equation (2) shows a negative effect of decreasing the SSR yield (Y), whereas the interaction coefficients of TE and EC for the equation have a positive effect that increases the yield (Y).The determination coefficient (R^2) of the regression model equation was evaluated by the F-test for analysis of variance (ANOVA), and the ANOVA statistics for the response Y are shown in Table 3. The value of the determination coefficient (R^2 0.9891) indicated that the quadratic model was statistically significant, advocating for a high correlation between observed and predicted values. The predicted R^2 is a measure of how good the model predicts the values for the response, and the adjusted R^2 verifies the experimental data and model precision. The predicted R^2 and adjusted R^2 were 0.9406 and 0.9782 (both close to one), respectively, which indicates the adequacy of the model and showing that the 0.9406% variability of the response Y is capable of explaining the model. The "lack of fit tests" compare residual error to "pure error" from replicated experimental design points. Its p-values were greater than 0.05; this response indicated that the lack of fit for the model was insignificant; that is to say, the quadratic model was valid for the present study. Adequate precision measures the signal to noise ratio, and a ratio greater than four is desirable. The adequate precision for Y was 30.039, demonstrating an adequate signal. This model can be used to navigate the design space. On the other hand, a relatively lower value of the coefficient of variation (CV 9.39%) indicated the good precision and reliability of the experiments [14]. ANOVA results of these quadratic models indicated that the model can be used to predict the process of chitosan hydrolysis.

Table 2. Experimental Box–Behnken design matrix and its response and predicted value.

Run	Experimental Variables				Response Y (%)	
	P	T (°C)	E (mg/L)	C (g/L)	Expt.	Predicted
1	6.00	30.00	100.00	10.00	39.76	40.092
2	4.00	50.00	100.00	10.00	89.58	91.232
3	6.00	50.00	100.00	15.00	30.06	27.680
4	2.00	70.00	100.00	10.00	13.52	13.202
5	4.00	70.00	150.00	10.00	43.74	43.440
6	4.00	50.00	50.00	5.00	32.58	28.142
7	2.00	50.00	50.00	10.00	4.950	8.538
8	4.00	50.00	100.00	10.00	92.28	91.232
9	6.00	50.00	100.00	5.00	40.23	38.320
10	6.00	50.00	150.00	10.00	44.78	42.515
11	4.00	30.00	150.00	10.00	47.58	45.598
12	2.00	50.00	150.00	10.00	45.02	40.158
13	4.00	50.00	100.00	10.00	89.94	91.232
14	4.00	50.00	150.00	5.00	30.48	35.437
15	4.00	50.00	50.00	15.00	15.91	10.967
16	6.00	50.00	50.00	10.00	24.59	30.775
17	4.00	50.00	100.00	10.00	90.14	91.232
18	4.00	70.00	50.00	10.00	8.680	9.325
19	6.00	70.00	100.00	10.00	32.44	32.478
20	4.00	50.00	150.00	15.00	42.58	47.032
21	4.00	70.00	100.00	5.00	25.49	24.970
22	4.00	30.00	100.00	15.00	34.93	36.773
23	2.00	50.00	100.00	5.00	17.13	18.173
24	4.00	70.00	100.00	15.00	20.05	20.505
25	4.00	30.00	50.00	10.00	37.39	36.353
26	2.00	30.00	100.00	10.00	34.80	34.775
27	4.00	50.00	100.00	10.00	94.22	91.232
28	2.00	50.00	100.00	15.00	22.66	23.233
29	4.00	30.00	100.00	5.00	37.02	37.888

Table 3. ANOVA analysis for the response surface quadratic model ($\alpha = 0.05$)

Source	Sum of Squares	DF	Mean Square	F	p-Value
Model	18,573.62	14	1326.69	90.55	<0.0001
P	453.62	1	453.62	30.96	<0.0001
T	638.90	1	638.90	43.61	<0.0001
E	1410.07	1	1410.07	96.24	<0.0001
C	23.35	1	23.35	1.59	0.2274
PT	48.72	1	48.72	3.33	0.0896
PE	98.80	1	98.80	6.74	0.0211
PC	61.62	1	61.62	4.21	0.0595
TE	154.63	1	154.63	10.55	0.0058
TC	2.81	1	2.81	0.19	0.6683
EC	206.93	1	206.93	14.12	0.0021
P2	6699.96	1	6699.96	457.30	<0.0001
T2	5438.75	1	5438.75	371.22	<0.0001
E2	5304.36	1	5304.36	362.05	<0.0001
C2	6742.76	1	6742.76	460.22	<0.0001
Residual	205.12	14	14.65		
Lack of Fit	189.50	10	18.95	4.85	0.0708
Pure Error	15.62	4	3.90		
Cor Total	18,778.73	28			
R^2	0.9891				
Adjusted R^2	0.9782				
Predicted R^2	0.9406				
Adeq precision	30.039				
CV	9.39				

3.1.2. Interactions between the Variables

Three-dimensional response surfaces were plotted on the basis of the graphical representations of the regression equation in order to investigate the interaction between the variables, as well as to determine the optimum condition of each factor for maximum enzymolysis for the production of low molecular weight (LMW) chitosan. The model suggested the presence of significant interaction principally between E (pepsin concentration) and the three other terms P (initial pH), T (reaction temperature) and C (chitosan concentration). We further characterize the interaction in the range of the process variables. Figure 1A represents the combined effect of P and E on SRS yield (Y), while the other two variables were held at zero. The elliptical nature of the contour plot between P and E indicates that significant interaction between these two variables had an effect on SRS yield (Y). Pepsin concentration E demonstrated a quadratic effect on the response, where the SRS yield increased at lower concentrations (<103.76 mg/L), followed by a slight decline with an increase in pepsin concentration. The trend also observed that the SRS yield increased to a maximum with the increase in pH and then gradually decreased to a higher pH (<4.2). It is clear from Figure 1B that the combined effect of reaction temperature T and pepsin concentration E was significant with the contour curve of an oval shape. The enzyme concentration E had almost no direct influence on SRS yield (Y); E relies rather on the temperature to influence Y. However, the temperature demonstrated a quadratic effect on the response, where the SRSs yield (Y), increased to the maximum (89.58%) with the increase in temperature and then decreased gradually at a higher temperature (>50 °C). The combined effect of the concentrations of chitosan and pepsin, C and E, respectively, on the SRS yield (Y) is shown in Figure 1C. The contour line with an elliptical shape demonstrates that the combined effect of the chitosan and pepsin concentrations C and E on the SRS yield (Y) is significant. Both concentrations have effects on the SRS yield (Y) when the chitosan concentration C and pepsin concentration E were under 10 g/L and 110 mg/L, respectively, and then achieve a balance with increasing chitosan concentration C and pepsin concentration E. Pepsin concentration E plays a pivotal role in this reaction process. Although the actual situation might be more complicated than what we reported, an attempt for the optimization of enzymolysis was made by RSM.

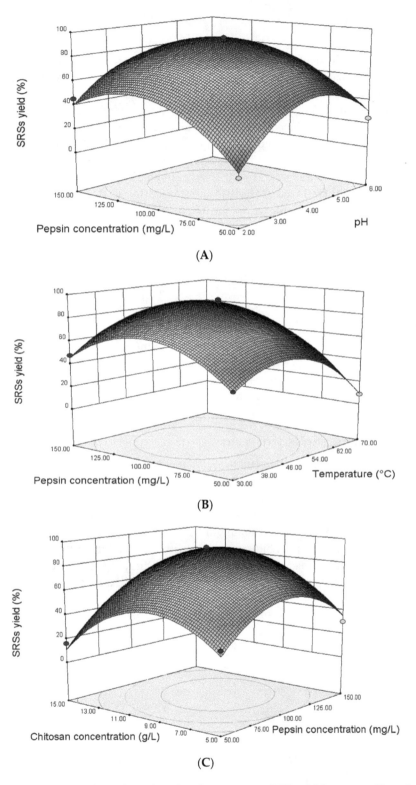

Figure 1. Response surface of predicted reducing sugar (SRS) yield versus pH and pepsin concentration (**A**); reaction temperature and pepsin concentration (**B**); chitosan concentration and pepsin concentration (**C**).

3.1.3. Optimization Analysis

Canonical analysis is one of the multivariate linear statistical analyses used to locate the stationary point of the response surface and to determine whether it represents a maximum, minimum or saddle point. It is also used to characterize the nature of the stationary points [15]. RSM was used to optimize the desired response of the system, which was the SRS yield (Y), and to keep all of the variables in range of the experimental values. The optimal conditions for the production of SRSs by pepsin-catalyzed chitosan enzymolysis by the model equation were as follows: pH 4.16, reaction temperature 47.95 °C, pepsin concentration 108.56 mg/L and chitosan concentration 9.97 g/L. This result differs from the one proposed by Kumar et al., 2007b (pH 5.0 and 45 °C) [6], as well as the one proposed by Tomas. R et al., 2007 (pH 4.5 and 40 °C) [12]. This could be due to the difference in pepsin origin. The theoretical SRS yield predicted under the above conditions was 92.77%. In order to verify the optimization results, experiments with three independent replicates were performed under the predicted optimal conditions. The SRS yield was 91.1% ± 0.4% in optimized conditions of pH 4.0, temperature 50 °C, pepsin concentration 110 mg/L and chitosan concentration 10.0 g/L, which was close to the predicted response and confirmed the efficacy of the predicted model. The results of previous work [13] (Pan et al., 2016) and this one confirm the efficiency of the RSM in the process of optimizing chitosan enzymolysis.

3.2. Enzymolysis Kinetic

The hydrolysis kinetics is essential in supplying the basic information for the design and operation of the enzymolysis with the aim of gaining good soluble chitosan products. In order to investigate the enzymolysis kinetics, pepsin was used to degrade chitosan in a solution (pH 4.0) containing initial substrate concentrations ranging from 2.0 to 18.0 g/L with 110 mg/L pepsin at 50 °C for 120 min. The SRS yield in the reaction solution was monitored at regular intervals.

3.2.1. Effect of Chitosan Substrate Concentration

The effect of substrate concentration on the catalytic performance of pepsin is shown in Figure 2, which indicates that substrate concentration did not have a significant effect on the equilibrium time. At lower substrate concentrations (2.0~10.0 g/L), the SRS yield increased with substrate concentration. At higher substrate concentrations (>10.0 g/L), the SRS yield decreased with the increase of substrate concentrations. This might be because the viscosity of the reaction system increased with the concentration of chitosan, which further slowed the diffusion of chitosan to the active center of the enzyme molecule resulting in the reduction of enzymatic activity. The results suggest that the chitosan concentration greater than about 10.0 g/L inhibits the activity of pepsin.

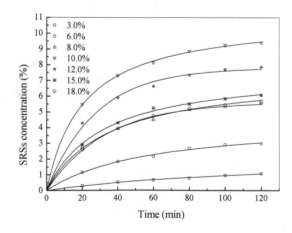

Figure 2. Enzymolysis of chitosan at different substrate concentrations.

3.2.2. Kinetic Constants

For the data obtained in the present study (Figure 2), the kinetic parameters were calculated on the basis of first-order and second-order models [16,17], and the results are shown in Table 4. The results show that all of the correlation coefficients R^2 in the two models are above 0.95, demonstrating their applicability. Comparing the two models, the second-order model (R^2 0.984~0.989) was more suitable for describing the process of chitosan enzymolysis based on a higher R^2. For the second-order model, the theoretical values (Q_e) were in good agreement with the corresponding experiment values ($Q_{e,exp}$). The hydrolysate SRS concentration increased with substrate concentration, and the maximum value Q_e reached was 10.03 g/L, after which there was a decrease with further increasing substrate concentration. A similar trend is also found in the hydrolysis rate constant k_2, demonstrating that chitosan had an inhibitory effect on pepsin activity at high concentrations (> about 10.0 g/L). In the second-order model, the constant k_2 was the specific hydrolysis rate (v) and was used to calculate the initial hydrolysis rate h, at $t \rightarrow 0$ [18], as follows.

$$h = k_2 Q_e^2 \tag{3}$$

where Q_e is the SRS concentration in the reaction solution at equilibrium.

Table 4. Kinetic parameters for the chitosan enzymolysis by pepsin at 50 °C.

Chitosan Concentration (g/L)	First-Order $Q_t = Q_e(1 - \exp(-k_1 t))$ *		Second-Order $t/Q_t = 1/(k_2 Q_e^2) + t/Q_e$ *		
	k_1 (1/min)	R^2	k_2 (L/(g·min))	H (g/(L·min))	R^2
2.0	0.00946	0.99543	0.00251	2.48935	0.99632
6.0	0.01966	0.99522	0.00353	4.62264	0.99618
8.0	0.0301	0.99593	0.00426	7.15633	0.99904
10.0	0.04086	0.99465	0.00441	11.07718	0.99954
12.0	0.03626	0.99513	0.00425	9.54351	0.99935
15.0	0.03109	0.99857	0.00406	7.76309	0.99813
18.0	0.02874	0.99956	0.00374	7.48846	0.99882

* Q_t and Q_e are the SRS concentration at t and equilibrium, respectively; k_1 and k_2 are the rate constants of the first-order and second-order models, respectively.

The initial hydrolysis rate h increased with substrate concentration from 3.0 to 10.0 g/L and then decreased due to the inhibitory effect of the chitosan as its concentration was further increased. The enzymolysis process in the presence of the inhibition of a substrate to the enzyme could be described by the Haldane model, and the Haldane equation is presented by Equation (4) [19,20],

$$v = \frac{V_{max}S}{K_m + S + \frac{S^2}{K_{ss}}} \tag{4}$$

where v is the specific rate of hydrolysis, which is equal to the hydrolysis rate constant k_2. K_{ss} is the inhibition constant. V_{max} is the maximum rate of hydrolysis. K_m is the Michaelis rate hydrolysis constant. S is the substrate concentration.

The relationship between the specific hydrolysis rate (v) and chitosan concentrations is shown in Figure 3, where the experimental v sharply increased when chitosan concentrations were lower than 10.0 g/L, whereas the inhibition effect of chitosan gradually became prominent at above 10.0 g/L. The value of R^2 is 0.9463, which suggests that the Haldane model had excellent fits to the experimental data. Using a nonlinear least squares regression analysis of Origin 9.0, the kinetic parameters of the experimental data in Figure 2 were determined as follows: V_{max} = 0.0043 g/(L·min), K_m = 46.6244 g/L and K_{ss} = 2.3395 g/L. It can be noted that the substrate had been inhibitory, making it impossible to observe an actual V_{max}. Therefore, K_m takes on a hypothetical meaning. When d_v/d_S equals zero,

Equation (5) will go through a maximum value at substrate concentration S^* and specific hydrolysis rate v^*. The values can be calculated as follows [21],

$$S^* = \sqrt{K_m K_{ss}} \tag{5}$$

$$v^* = \frac{V_{max}}{1 + 2\sqrt{K_m/K_{ss}}} \tag{6}$$

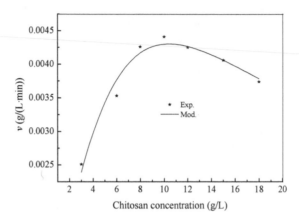

Figure 3. Maximum specific hydrolysis rate for chitosan substrate concentrations from 3.0 to 18.0 g/L.

Equation (6) reflects that the hydrolysis potential of pepsin was determined by the K_m/K_{ss} ratio, not just by K_{ss} alone. The larger the K_m/K_{ss} was, the lower the hydrolysis potential was. According to these values (V_{max}, K_m and K_{ss}), S^* and v^* were computed as 10.4440 g/L and 0.0002 g/(L·min), and the minimum chitosan inhibitive concentration was 10.44 g/L. Below this concentration, hydrolysis seemed to be increasingly suboptimal, and above this concentration, hydrolysis was inhibited increasingly due to substrate inhibition. At high chitosan concentrations, the enhancement of the viscosity in the reaction system increased the hindrance to diffusion of chitosan molecules. It also restrained the diffusion between the product and activity sites of pepsin, leading to decreased catalytic activity. The calculated results suggested that in order to avoid substrate inhibition, the process of enzymolysis should be operated at a chitosan concentration below 10.5 g/L, which is close to the experimental value of 10.5 g/L. These results could be significant towards understanding the capacities of pepsin for chitosan degradation. Considering the results from two different works on the subject (Pan et al., 2007 [13], and this one), the process of chitosan enzymolysis seems to generally follow the pseudo-second-order and Haldane models.

3.3. The Structural Properties of the LMW Chitosan

3.3.1. FTIR Analysis

Figure 4 shows the FTIR spectra of the initial chitosan and degraded chitosan LMWC samples where the initial chitosan and LMWCs show basically similar FTIR spectra. The peaks at around 3462, 2960, 1638, 1387, 1081 and 892 cm^{-1} represent the presence of the –OH group, the –CH$_2$– group (aliphatic group), the –C=O group, the C–O stretching of the primary alcoholic group (–CH$_2$-OH is considered to be a potential site for cross-linking), the –OH group and the β (1→4) glucoside bond in chitosan, respectively [5,22]. These results demonstrated that the structures of the main chain of the initial chitosan and LMWCs were the same. The NH$_2$ amino groups had a characteristic peak near 3462 cm^{-1}, which was overlapping by the peak due to the –OH group [23]. The occurrence of absorption peaks at around 2900 cm^{-1} was assigned to the asymmetric stretching vibration of the

–CH$_2$–, and the rapid reduction in the intensity for the LMWCs was probably attributed to degradation of chitosan after hydrolysis. In contrast with the initial chitosan, a significant new peak at 1571 cm^{-1} in the LMWCs, the N–H bending mode of –NH$_2$, was noticeably split at around 1638 cm^{-1}, which suggests that the –C=O groups had more opportunity to form stronger hydrogen bonds, and the scission of polymer chains led to the decrease of the chitosan molecular weight [8]. The results indicated that there was no significant difference between the main structures of the two samples before and after the enzymatic hydrolysis, but the molecular weight of the main hydrolysis products decreased, which was in good agreement with Kumar et al., 2007b [6].

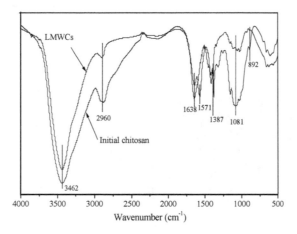

Figure 4. FTIR spectra of the initial chitosan and low molecular weight chitosan (LMWC) samples.

3.3.2. UV-VIS Analysis

Figure 5 shows the optical transmittance spectra for the initial chitosan and LMWCs in the range of 300–800 cm^{-1}. The two samples showed very good transmittance in the visible region. For the initial chitosan, low transmission intensity can be observed below the 317-cm^{-1} wavelength; therefore, the transmission intensity started to increase at 317 cm^{-1}, until it reached 70.16% at 463 nm. The transmittance of the LMWCs immediately started to increase at 300 cm^{-1}, up to 70.80% at 380 cm^{-1}. A new absorption peak not appearing in the initial chitosan sample was observed at around 344 cm^{-1}, which suggested the existence of a chemical reaction in the LMWCs. The peak could be assigned to the n→π* transition for the carboxyl group in the LMWCs formed after the main chain scission of chitosan [23]. The result of FTIR analysis further confirmed that carbon-oxygen double bonds formed after the degradation of chitosan occurred by the ring opening, turning chitosan into one with low molecular weight.

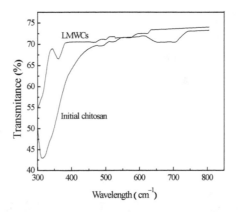

Figure 5. UV-VIS transmittance spectra of initial chitosan and LMWC samples.

3.3.3. SEM Analysis

The surface and internal structure of the LMWCs were evaluated using SEM, and the results are shown in Figure 6. The initial chitosan presented a heterogeneous structure, which consisted of random-sized, loose particles with irregular edges. For the LMWCs, the loose granular nature of chitosan was observed after enzymolysis, which seemed to have taken place homogeneously in bulk. The enzymolysis made the surface of the hydrolyzed chitosan denser and more porous, leading to a change in the surface morphology, thus exhibiting a thick, dense, but porous structure with small cavities distributed on the entire surface of the LMWCs. The discrepancies in the observed microstructure between the initial chitosan and LMWCs might also be attributed to their different inter- or intra-particle hydrogen-bonding systems. It was further confirmed by SEM analysis that chitosan was degraded into small molecular weight chitosan (LMWC) when the chain of chitosan macromolecules was broken into smaller unit.

Figure 6. SEM images of the initial chitosan (**A**) and LMWC (**B**) samples, ×50.

3.3.4. Depolymerization Degree, Viscosity, Molecular Weight and Polydispersity Index Analysis

The DD, M_W, average viscosity and polydispersity index of chitosan and degraded chitosans (LMWCs and chitosan oligosaccharides (COS)) are listed in Table 5. The DD of chitosan increased with the decrease in molecular weight and viscosity. The average viscosity and molecular weight value of chitosan decreased by approximately 86% in relation to the molecular weight value during the first 60 min of reaction [6]. In the presence of the optimal conditions of hydrolysis, the polysaccharide chains are submitted to degradation due to the efficiency of the different parameter and the prolonged times necessary for obtaining advanced depolymerization. The polydispersity index (PDI = M_W/M_n) was studied, and after 2 h of hydrolysis, there was no trace of monomers; contrary to our previous study with papain [13], where trace levels of monomers were detected. There were only chitosan oligosaccharides (COS) and low molecular weight chitosan (LMWC) in different proportions according to the time of hydrolysis (Table 5). These could be interpreted as confirming that chitosan was degraded into smaller molecular weight units.

Table 5. Properties of degraded chitosan.

Source	M_w (×10³)	DD (%)	Viscosity Decrease (%)	Yield (%)
Native	300	-	-	-
CH₁	195.4	74.60	86	-
CH₂	65.9	92.00	93	-
Monomers [2]	-	-	-	n.t.
LMWC [1]	25–20	-	-	28.26
LMWC [2]	13–9	-	-	35.04
COS [1]	90–85	-	-	71.74
COS [2]	65–50	-	-	64.94

[1]: after 1 h; [2]: after 2 h; CH₁: chitosan hydrolyzed after 1 h; CH₂: chitosan hydrolyzed after 2 h; monomers: sum of GlcN and GlcNAc; COS: chitosan oligosaccharides; LMWC: low molecular weight chitosan; n.t.: no trace.

4. Conclusions

The high molecular weight chitosan shows poor solubility in aqueous solutions, and the high viscosity of its solution limits its applications. To improve its solubility, as well as biological, chemical and physical properties, enzymolysis by pepsin was employed to prepare low molecular chitosan. RSM was launched to investigate the influence of process variables on the DH followed by a BBD approach. The optimized conditions were a 10.0-g/L chitosan concentration of pH 4.0 at 50 °C, a 110-mg/L pepsin concentration and enzymolysis time of 70 min, where the predicted value of the DH was 91.1%.

The enzymolysis process of chitosan follows the pseudo-second order and Haldane models, wherein for more than a 10.5-g/L concentration chitosan, pepsin undergoes severe inhibition due to the viscosity increase, causing an increase in the barrier released in the reaction system. Based on the characteristic analyses by FTIR, UV-VIS and SEM, hydrolyzed product LMWCs almost retained the backbone of the chitosan macromolecular structure. The breaking of the C-O-C glycosidic bond led to chain scission and the formation of carbonyl groups. Therefore, the degradation method was feasible, convenient and potentially applicable.

Acknowledgments: This work was supported by the National Natural Science Foundation of China (21105085, 31270988) and the Key Project of Hunan Provincial Natural Science Foundation of China (12JJ2008).

Author Contributions: Bi Foua Claude Alain Gohi and Hong-Yan Zeng conceived and designed the experiments; Bi Foua Claude Alain Gohi performed the experiments; Bi Foua Claude Alain Gohi, Hong-Yan Zeng and A Dan Pan analyzed the data; Hong-Yan Zeng contributed reagents/materials/analysis tools; Bi Foua Claude Alain Gohi wrote the paper." Authorship must be limited to those who have contributed substantially to the work reported.

Conflicts of Interest: The authors declare no conflict of interest.

References

1. Xia, W.; Liu, P.; Liu, J. Advance in chitosan hydrolysis by non-specific cellulases. *Bioresour. Technol.* **2008**, *99*, 6751–6762. [CrossRef] [PubMed]
2. Wang, B.; He, C.; Tang, C.; Yin, C. Effects of hydrophobic and hydrophilic modifications on gene delivery of amphiphilic chitosan based nanocarriers. *Biomaterials* **2011**, *32*, 4630–4638. [CrossRef] [PubMed]
3. Kumar, A.V.; Varadaraj, M.C.; Gowda, L.R.; Tharanathan, R.N. Low molecular weight chitosan-preparation with the aid of pronase, characterization and their bactericidal activity towards *Bacillus cereus* and *Escherichia coli*. *Biochim. Biophys. Acta* **2007**, *1771*, 495–505. [CrossRef] [PubMed]
4. Su, P.; Wang, S.; Shi, Y.; Yang, Y. Application of cellulase-polyamidoamine dendrimer-modified silica for microwave-assisted chitosan enzymolysis. *Process. Biochem.* **2013**, *48*, 614–619. [CrossRef]
5. Wu, S. Preparation of water soluble chitosan by hydrolysis with commercial α-amylase containing chitosanase activity. *Food Chem.* **2011**, *128*, 769–772. [CrossRef]
6. Kumar, A.V.; Varadaraj, M.C.; Gowda, L.R.; Tharanathan, R.N. Low molecular weight chitosan-preparation with aid of pepsin, characterization, and its bactericidal activity. *Biomacromolecules* **2007**, *8*, 566–572. [CrossRef] [PubMed]
7. Ilyina, A.V.; Tikhonov, V.E.; Albulov, A.I.; Varlamov, V.P. Enzymic preparation of acid-free-water-soluble chitosan. *Process. Biochem.* **2000**, *35*, 563–568. [CrossRef]
8. Nah, J.W.; Jang, M.K. Spectroscopic characterization and preparation of low molecular, water-soluble chitosan with free-amine group by novel method. *J. Polym. Sci. Part A Polym. Chem.* **2002**, *40*, 3796–3803. [CrossRef]
9. Kim, T.H.; Park, I.K.; Nah, J.W.; Choi, Y.J.; Cho, C.S. Galactosylated chitosan/DNA nanoparticles prepared using water-soluble chitosan as a gene carrier. *Biomaterials* **2004**, *25*, 3783–3792. [CrossRef] [PubMed]
10. Jang, M.K.; Jeong, Y.I.; Nah, J.W. Characterization and preparation of core-shell type nanoparticle for encapsulation of anticancer drug. *Colloids Surf. B Biointerfaces* **2010**, *81*, 530–536. [CrossRef] [PubMed]
11. Tikhonov, V.E.; Stepnova, E.A.; Babak, V.G.; Yamskov, I.A.; Palma-Guerrero, J.; Jansson, H.B.; Lopez-Llorca, L.V.; Salinas, J.; Gerasimenkoc, D.V.; Avdienko, I.D.; et al. Bactericidal and antifungal activities of a low molecular weight chitosan and its N-/2(3)-(dodec-2-enyl) succinoyl/-derivatives. *Carbohydr. Polym.* **2006**, *64*, 66–72. [CrossRef]

12. Roncal, T.; Oviedo, A.; de Armentia, I.L.; Fernández, L.; Villarán, M.C. High yield production of monomer-free chitosan oligosaccharides by pepsin catalyzed hydrolysis of a high deacetylation degree chitosan. *Carbohydr. Res.* **2007**, *342*, 2750–2756. [CrossRef] [PubMed]

13. Pan, A.D.; Zeng, H.Y.; Gohi, B.F.C.A.; Li, Y.Q. Enzymolysis of Chitosan by Papain and Its Kinetics. *Carbohydr. Polym.* **2016**, *135*, 199–206. [CrossRef] [PubMed]

14. Tanyildizi, M.S.; Özer, D.; Elibol, M. Optimization of α-amylase production by *Bacillus* sp. using response surface methodology. *Process. Biochem.* **2005**, *40*, 2291–2296. [CrossRef]

15. Alim, M.A.; Lee, J.H.; Akoh, C.C.; Choi, M.S.; Shin, J.A.J.; Lee, K.T. Enzymatic transesterification of fractionated rice bran oil with conjugated linoleic acid: Optimization by response surface methodology. *LWT Food Sci. Technol.* **2008**, *41*, 764–770. [CrossRef]

16. Vavilin, V.A.; Fernandez, B.; Palatsi, J.; Flotats, X. Hydrolysis kinetics in anaerobic degradation of particulate organic material: An overview. *Waste Manag.* **2008**, *28*, 939–951. [CrossRef] [PubMed]

17. Kusvuran, E.; Irmak, S.; Yavuz, H.I.; Samil, A.; Erbatur, O. Comparison of the treatment methods efficiency for decolorization and mineralization of Reactive Black 5 azo dye. *J. Hazard. Mater.* **2005**, *119*, 109–116. [CrossRef] [PubMed]

18. Vadivelan, V.; Kumar, K.V. Equilibrium, kinetics, mechanism, and process design for the sorption of methylene blue onto rice husk. *J. Colloid Interface Sci.* **2005**, *286*, 90–100. [CrossRef] [PubMed]

19. Dona, A.C.; Pages, G.; Gilbert, R.G.; Kuchel, P.W. Digestion of starch: In vivo and in vitro kinetic models used to characterise oligosaccharide or glucose release. *Carbohydr. Polym.* **2010**, *80*, 599–617. [CrossRef]

20. Ayhan, F.; Ayhan, H.; Pişkin, E.; Tanyolac, A. Optimization of urease immobilization onto non-porous HEMA incorporated poly (EGDMA) microbeads and estimation of kinetic parameters. *Bioresour. Technol.* **2002**, *81*, 131–140. [CrossRef]

21. Nuhoglum, A.; Yalcin, B. Modeling of phenol removal in a batch reactor. *Process. Biochem.* **2005**, *40*, 1233–1239. [CrossRef]

22. Brugnerotto, J.; Lizardi, J.; Goycoolea, F.M.; Argüelles-Monal, W.; Desbrières, J. An infrared investigation in relation with chitin and chitosan characterization. *Polymer* **2001**, *42*, 3569–3580. [CrossRef]

23. Ma, Z.; Wang, W.; He, Y.W.Y.; Wu, T. Oxidative degradation of chitosan to the low molecular water-soluble chitosan over peroxotungstate as chemical scissors. *PLoS ONE* **2014**, *9*, e100743. [CrossRef] [PubMed]

Permissions

All chapters in this book were first published in Bioengineering, by MDPI; hereby published with permission under the Creative Commons Attribution License or equivalent. Every chapter published in this book has been scrutinized by our experts. Their significance has been extensively debated. The topics covered herein carry significant findings which will fuel the growth of the discipline. They may even be implemented as practical applications or may be referred to as a beginning point for another development.

The contributors of this book come from diverse backgrounds, making this book a truly international effort. This book will bring forth new frontiers with its revolutionizing research information and detailed analysis of the nascent developments around the world.

We would like to thank all the contributing authors for lending their expertise to make the book truly unique. They have played a crucial role in the development of this book. Without their invaluable contributions this book wouldn't have been possible. They have made vital efforts to compile up to date information on the varied aspects of this subject to make this book a valuable addition to the collection of many professionals and students.

This book was conceptualized with the vision of imparting up-to-date information and advanced data in this field. To ensure the same, a matchless editorial board was set up. Every individual on the board went through rigorous rounds of assessment to prove their worth. After which they invested a large part of their time researching and compiling the most relevant data for our readers.

The editorial board has been involved in producing this book since its inception. They have spent rigorous hours researching and exploring the diverse topics which have resulted in the successful publishing of this book. They have passed on their knowledge of decades through this book. To expedite this challenging task, the publisher supported the team at every step. A small team of assistant editors was also appointed to further simplify the editing procedure and attain best results for the readers.

Apart from the editorial board, the designing team has also invested a significant amount of their time in understanding the subject and creating the most relevant covers. They scrutinized every image to scout for the most suitable representation of the subject and create an appropriate cover for the book.

The publishing team has been an ardent support to the editorial, designing and production team. Their endless efforts to recruit the best for this project, has resulted in the accomplishment of this book. They are a veteran in the field of academics and their pool of knowledge is as vast as their experience in printing. Their expertise and guidance has proved useful at every step. Their uncompromising quality standards have made this book an exceptional effort. Their encouragement from time to time has been an inspiration for everyone.

The publisher and the editorial board hope that this book will prove to be a valuable piece of knowledge for researchers, students, practitioners and scholars across the globe.

List of Contributors

Shraddha Patel
Center for Biomedical Engineering and Rehabilitation Science, Louisiana Tech University, Ruston, LA 71272, USA
Wayne State University, St. John Hospital & Medical Center, 22101 Moross Rd, Detroit, MI 48236, USA

Uday Jammalamadaka, Lin Sun and Karthik Tappa
Center for Biomedical Engineering and Rehabilitation Science, Louisiana Tech University, Ruston, LA 71272, USA

David K. Mills
Center for Biomedical Engineering and Rehabilitation Science, Louisiana Tech University, Ruston, LA 71272, USA
School of Biological Sciences, Louisiana Tech University, Ruston, LA 1272, USA

Israr Bin Muhammad Ibrahim and Parya Aghasafari
Graduate Student, College of Engineering, University of Georgia, 597 DW Brooks Drive, Athens, GA 30602, USA

Ramana M. Pidaparti
College of Engineering, University of Georgia, 132A Paul D. Coverdell Center, Athens, GA 30602, USA

Michael Lebuhn and Bernhard Munk
Bavarian State Research Center for Agriculture, Department for Quality Assurance and Analytics, Lange Point 6, 85354 Freising, Germany

Jaqueline Derenkó, Antje Rademacher and Michael Klocke
Leibniz Institute for Agricultural Engineering Potsdam-Bornim, Department Bioengineering, Max-Eyth-Allee 100, 14469 Potsdam, Germany

Susanne Helbig and Steffen Prowe
Beuth University of Applied Sciences, Department of Life Sciences and Technology, Luxemburger Strasse 10, 13353 Berlin, Germany

Alexander Pechtl, Wolfgang H. Schwarz and Wolfgang Liebl
Department of Microbiology, Technische Universität München, Emil-Ramann-Str. 4, D-85354 Freising-Weihenstephan, Germany

Yvonne Stolze and Andreas Schlüter
Institute for Genome Research and Systems Biology, CeBiTec, Bielefeld University, Bielefeld, Germany

Chao Ma, Jianfa Ou, Ningning Xu and Xiaoguang (Margaret) Liu
Department of Chemical and Biological Engineering, The University of Alabama, 245 7th Avenue, Tuscaloosa, AL 35401, USA

Janna L. Fierst
Department of Biological Science, The University of Alabama, 300 Hackberry Lane, Tuscaloosa, AL 35487, USA

Shang-Tian Yang
Department of Chemical and Biomolecular Engineering, The Ohio State University, 151West Woodruff Avenue, Columbus, OH 43210, USA

Ahmed I. El-Batal
Drug Radiation Research Department, National Center for Radiation Research and Technology (NCRRT), Atomic Energy Authority, Cairo 11371, Egypt

Ayman A. Farrag
Botany and Microbiology Department, Faculty of Science, Al-Azhar University, Cairo 11371, Egypt

Mohamed A. Elsayed and Ahmed M. El-Khawaga
Chemical Engineering Department, Military Technical College, Cairo 11371, Egypt

Joy Edward Larvie, Mohammad Gorji Sefidmazgi, Abdollah Homaifar and Ali Karimoddini
Department of Electrical and Computer Engineering, North Carolina A&T State University, 1601 E. Market Street, Greensboro, NC 27411, USA

Scott H. Harrison
Department of Biology, North Carolina A&T State University, 1601 E. Market Street, Greensboro, NC 27411, USA

Anthony Guiseppi-Elie
Department of Biomedical Engineering, Texas A&M University, 5045 ETB, College Station, TX 77843, USA

Paolo Di Sia
Department of Philosophy, Education and Psychology, University of Verona, Lungadige Porta Vittoria 17, Verona 37129, Italy
ISEM, Institute for Scientific Methodology, Palermo 90146, Italy

Ignazio Licata
ISEM, Institute for Scientific Methodology, Palermo 90146, Italy

Jonathan L. Robinson and Mark P. Brynildsen
Department of Chemical and Biological Engineering, Princeton University, Princeton, NJ 08544, USA

Jaya Shankar Tumuluru
Idaho National Laboratory, 750 University Blvd., Energy Systems Laboratory, PO: Box: 1625, Idaho Falls, ID 83415, USA

Rong Zeng
College of Chemistry and Chemical Engineering, Hubei University, Wuhan 430062, China

Xiao-Yan Yin, Tao Ruan, Qiao Hu, Ya-Li Hou, Zhen-Yu Zuo, Hao Huang and Zhong-Hua Yang
College of Chemical Engineering and Technology, Wuhan University of Science and Technology, Wuhan 430081, China

Viktor Konakovsky and Christoph Herwig
Institute of Chemical Engineering, Division of Biochemical Engineering, Vienna University of Technology, Gumpendorfer Strasse 1A 166-4, 1060 Vienna, Austria

Christoph Clemens, Markus Michael Müller, Jan Bechmann, Martina Berger and Stefan Schlatter
Boehringer Ingelheim Pharma GmbH & Co. KG Dep. Bioprocess Development, Biberach, Germany

Bi Foua Claude Alain Gohi, Hong-Yan Zeng and A Dan Pan
Biotechnology Institute, College of Chemical Engineering, Xiangtan University, Xiangtan 411105, Hunan, China

Index

Printed in the USA
CPSIA information can be obtained
at www.ICGtesting.com
JSHW051323221024
72173JS00006B/1284